Lecture Notes in Physics

Edited by H. Araki, Kyoto, J. Ehlers, München, K. Hepp, Zürich
R. Kippenhahn, München, H. A. Weidenmüller, Heidelberg
and J. Zittartz, Köln

Managing Editor: W. Beiglböck

252

Local and Global Methods of Nonlinear Dynamics

Proceedings of a Workshop
Held at the Naval Surface Weapons Center
Silver Spring, MD, July 23–26, 1984

Edited by A. W. Sáenz, W. W. Zachary and R. Cawley

Springer-Verlag
Berlin Heidelberg GmbH

Editors

A. W. Sáenz
Code 4603 S, Naval Research Laboratory
Washington, D. C. 20375, USA
and
Physics Department, Catholic University of America
Washington, D. C. 20064, USA

W. W. Zachary
Code 4603 S, Naval Research Laboratory
Washington, D.C. 20375, USA

R. Cawley
Naval Surface Weapons Center, White Oak Laboratory,
Silver Spring, MD 20910, USA

ISBN 978-3-540-16485-2

Library of Congress Cataloging-in-Publication Data. Local and global methods of nonlinear dynamics. (Lecture notes in physics; 252) 1. Dynamics–Congresses. 2. Nonlinear theories–Congresses. 3. Hamiltonian systems–Congresses. 4. Global analysis (Mathematics)–Congresses. 5. Mathematical physics–Congresses. I. Sáenz, Albert William, 1923-. II. Zachary, W. W., 1935-. III. Gawley, R. (Robert), 1936-. IV. Naval Surface Weapons Center. V. Series.
QC133.L63 1986 531'.11 86-11890
ISBN 978-3-540-16485-2 ISBN 978-3-540-39824-0 (eBook)
DOI 10.1007/978-3-540-39824-0

Originally published by Springer-Verlag Berlin Heidelberg New York Tokyo in 1986
Softcover reprint of the hardcover 1st edition 1986

2153/3140-543210

Preface

The papers contained in this volume were presented at the workshop on "Local and Global Methods of Nonlinear Dynamics", 23-26 July, 1984, at the Naval Surface Weapons Center in Silver Spring, Maryland. This was the first workshop sponsored by the Navy Dynamics Institute Program, a joint effort of NSWC, the Naval Research Laboratory, and the Office of Naval Research.

Initially, nonlinear dynamics in its modern form was almost an exclusive province of mathematicians, but in recent years an explosive interest in the subject has developed among physicists and other physical and biological scientists. The roots of nonlinear dynamics are manifold, and may be regarded as going back to Newton. Increasing sophistication in the methods of mechanics gave rise by the nineteenth century to considerably improved techniques of perturbation theory, which, however, remained essentially local in nature because they did not generically lead to solutions valid for all time. The subject underwent a significant modernizing change with the work of Poincaré and Birkhoff, in which qualitative rather than quantitative methods were emphasized. The result of this shift, namely global dynamics, is the principal perspective of modern dynamical systems theory. In recent years, dynamical systems has become an area of increasing interest also among mathematicians. While local theory continues to yield useful and valuable physical applications, the real historical progress has surely been towards global dynamics, with the first important modern result being KAM theory begun in 1954 by Kolmogorov. For an account of this, see the lecture by R. Barrar.

Our concern in this volume is primarily, and almost (but not quite) exclusively, with conservative systems. In this we have adhered to the "Doctrine of the Unities," at the expense of a significant body of recent research in dissipative systems, to be sure. In addition, we have addressed our efforts to contінous rather than discrete dynamical systems. But, we hope that this concentration will accomplish our main purpose, which has been to focus attention upon the dichotomy of local and global methods and results, to build bridges between the two, and to make at least some of the latter more accessible to physicists. It is too often not realized by physicists that issues of convergence are vital, or that abstract methods, such as reduction of symplectic manifolds with symmetry, have something to say that is physically interesting.

Among the many distinguished contributions of Poincaré to the understanding of dynamics was his work on periodic orbits, a subject that mathematicians are even today pursuing intensively. M. S. Berger's lecture considers global aspects of such solutions for nonlinear dynamical systems.

Resonant Hamiltonian systems, an active area of mathematical and physical research, are discussed in M. Kummer's first lecture in this volume. His second lecture deals with an important subject in contemporary nonlinear dynamics, namely, realizations of the reduced space of a Hamiltonian system with symmetry, in which ideas of Marsden and Weinstein play a central role.

This is one of several contributions to this volume which show the pervasive influence of Sophus Lie on nonlinear dynamics. During the last twenty years, Lie's ideas have inspired Deprit and others to develop efficient perturbation methods—Lie-transform methods—which have been successfully applied in celestical mechanics, in particle-accelerator, electron-optics, and geometrical-optics design, and in plasma physics. A. A. Kamel's lecture relates the history of Lie-transform methods in celestial mechanics. A discussion of Lie-transform perturbative methods in a rather general setting and of their application to selected problems in Hamiltonian mechanics, will be found in J. M. Finn's lecture. A. N. Kaufman's lecture considers the application of Lie transforms in studying the behavior of a plasma of charged particles in a self-consistent electromagnetic wave.

A global perspective on symplectic dynamical systems with continous symmetry emerged in the early seventics. A pedagogical presentation of perturbation theory in this modern geometrical setting is contained in S. Omohundro's contribution to this volume.

The original, classical approach of Lie for solving systems of differential equations has been applied in hydrodynamics, general relativity, plasma physics, and other subjects with considerable success. In this volume, B. Abraham-Shrauner explains how this approach can be exploited to obtain exact solutions of the nonlinear Vlasov-Maxwell equations.

Explicit solutions of the Hamilton-Jacobi equations for conservative systems are known in very few cases. F. H. Molzahn and T. A. Osborn have recently derived such a solution rigorously for an N-body system by variational and combinatorial (tree-graph) methods, as reported in their contribution to this volume.

R. Cawley's lecture concerns his generalization of Hamiltonian theory to dissipative dynamical systems, which is based on local methods first used by Dirac and on a novel Hamilton-Jacobi equation. The difficulties involved in constructing a global version of this generalized theory are also considered in his lecture.

Channeling motions of particles in crystals give rise to interesting and difficult problems of Hamiltonian dynamics. The contribution of H. S. Dumas and J. A. Ellison discusses rigorously the application of averaging theory to channeling. That of A. W. Sáenz outlines a rigorous proof of orbital stability of certain channeling motions using canonical maps and the Moser twist theorem.

Two contributions to this volume deal with invariants of Hamiltonian systems. L. S. Hall discusses approximate invariants for such systems in the autonomous case, while J. Goedert and H. R. Lewis consider the existence of exact invaniants for certain nonautonomous Hamiltonian systems.

Most of the research reported in this volume is new or recent, but several of the lectures are at least partially pedagogical. Our purpose in organizing the workshop and in publishing the present volume was to inform physical and mathematical scientists of developments in an old field that is experiencing an extraordinary revival, and simultaneously undergoing a remarkable transformation into a very diverse and bright modern subject.

We wish to thank the NSWC Training Office, whose very generous support made the workshop possible, the secretarial staff at NRL, whose time and effort have made this volume come together, and Dr. M. Shlesinger of ONR for supporting the Navy Dynamics Institute Program. Special thanks are due to Dr. William Condell of ONR for his imaginative foresight in a number of matters, and for his advice and help in our larger efforts. We are also grateful to Dr. J. Frazier and Dr. T. Coffey, the respective Directors of Research of NSWC and NRL for giving welcoming addresses at the workshop.

A.W. Sáenz and W.W. Zachary, Naval Research Laboratory, Washington, D.C.
R. Cawley, Naval Surface Weapons Center, Silver Spring, Md.

TABLE OF CONTENTS

Preface

M.S. Berger	Global Aspects of Periodic Solutions of Nonlinear Conservative Systems	1
M. Kummer	On Resonant Hamiltonian Systems with Finitely Many Degrees of Freedom	19
M. Kummer	Realizations of the Reduced Phase Space of a Hamiltonian System with Symmetry	32
R. Barrar	KAM Today	40
A.A. Kamel	Note on the Evolution of the Lie-Deprit Transform	49
J.M. Finn	Lie Transforms: A Perspective	63
A.N. Kaufman	The Covariant Lie-Transformed Plasma Action Principle	87
S. Omohundro	Geometric Hamiltonian Structures and Perturbation Theory	91
B. Abraham-Shrauner	Lie Point Transformation Group Solutions of the Nonlinear Vlasov-Maxwell Equations	121
F.H. Molzahn and T.A. Osborn	A Constructive Solution to the Hamilton-Jacobi Equation	146
R. Cawley	Local and Global Aspects of a Generalized Hamiltonian Theory	169
H.S. Dumas and J.A. Ellison	Particle Channeling in Crystals and the Method of Averaging	200
A.W. Sáenz	Rigorous Stability Results on Crystal Channeling via Canonical Maps	231
L.S. Hall	Some Considerations for a Theory of Approximate Invariants	237
J. Goedert and H.R. Lewis	Exact Invariants in the Form of Momentum Resonances for Particle Motion in One-Dimensional, Time-Dependent Potentials	253

GLOBAL ASPECTS OF PERIODIC SOLUTIONS OF NONLINEAR CONSERVATIVE SYSTEM

M. S. Berger *
Dept. of Mathematics
University of Massachusetts
Amherst, Massachusetts 01003

1. INTRODUCTION

The study of periodic solutions of nonlinear Hamiltonian systems has a fascinating and long history in both physical science and mathematics. The modern nonlinear dynamics once again focuses attention on periodic motions and their perturbations. The importance of periodic motions of dynamical systems is clear to anyone who studies celestial mechanics, and has been emphasized by many researchers, notably Poincaré, who emphasized the density of periodic motions in many nonlinear dynamical systems.

The mathematical study of periodic motions of Hamiltonian systems had for many years been focused on fixed point theorems, notions of analytic continuation, numerous and ingenious power series expansions, problems of small divisors, and various perturbative methods. Some of these methods were constructive and computable and led to numerous significant achievements. Nonetheless, beginning about 16 years ago, in 1968, I began publishing research on using a global Calculus of Variations approach to study periodic solutions of conservative systems. The present article is a review of my ideas in this field and their subsequent development by numerous researchers. Indeed, many researchers have taken up this global Calculus of Variations idea in this decade, and have achieved real successes. I hope to relate some of these advances in the sequel.

Of course, Poincaré attempted to use Calculus of Variations arguments in his distinguished work in celestial mechanics. However, in his day, the Calculus of Variations was not sufficiently far advanced to enable him to carry out the mathematical aspects of his ideas (see Section 3). Moreover, it is interesting that new global families of periodic solutions for Hamiltonian systems and constructive methods for finding them await discovery despite centuries of study.

2. THE DIFFICULTY OF UTILIZING THE GLOBAL CALCULUS OF VARIATIONS IN NONLINEAR HAMILTONIAN SYSTEMS AND A NEW IDEA IN OVERCOMING THE PROBLEM

---

* Research partially supported by the NSF and AFOSR.

Of course, the key difficulty, known to many researchers, in utilizing the global ideas I have in mind, is that interesting periodic motions are not absolute minimizers of the Lagrangian of nonlinear Hamiltonian systems, but rather appear as saddle points in either the Lagrangian or Hamiltonian formulations. The mathematical study of periodic motions by the global Calculus of Variations ideas is then forced to consider mathematical theories concerned with the so-called critical points of functionals, an area of mathematics that is distinguished by the combined use of all the resources of modern research, namely a combination of analysis, algebra, and topology. There are basically three theories of critical points of functionals: Morse-theory, Ljusternik-Schnirelmann theory, and Natural Constraints. In this article I shall try to discuss contemporary research in the computable utilization of each of these theories to discover new periodic motions. I wish to begin with an example of the last method of natural constraints from a historical point of view dating back to Poincaré's research ideas of 1905 [0].

Problem (Poincaré [0]) : Find the shortest closed (nonconstant) geodesic on a smooth ovaloid. Determine its properties.

Fig. 1

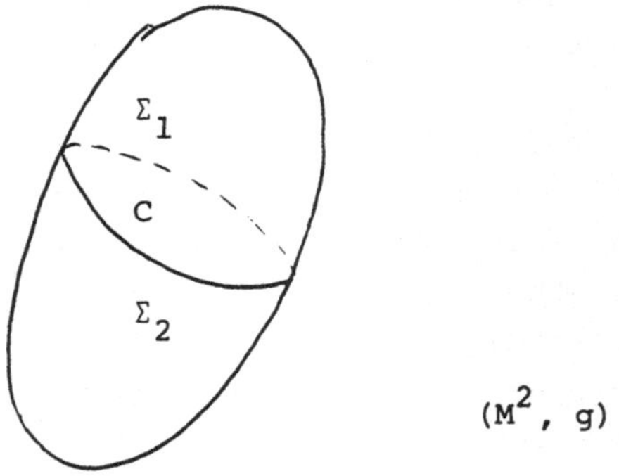

This problem is connected with periodic motions of conservative systems because we are searching for nonconstant closed geodesics. The manifold (M^2, g), a two-dimensional manifold with Riemannian metric g in our case, is chosen to be an ovaloid that is the boundary of a convex bounded two-dimensional set, and one can see that the natural problem of minimizing the arc length functional on closed curves yields zero as minimum, and zero corresponds to the noninteresting closed geodesics, the solution in which the curve C reduces to a point. The interesting periodic solution is thus not an absolute minimum of the are length functional. In fact, it is a saddle point of this functional, and to determine this geodesic in a constructive way has proved to be, in the nearly 80 years since Poincaré's approach, a difficult problem. The natural global topological invariant for the problem, namely the homotopy class of closed curves on M^2, is zero because M^2 is an ovaloid, and thus since $\pi_1(M^2) = 0$ a substitute must be found for homotopy. Moreover, this substitute must be sufficiently strong to reveal the properties of the shortest nonconstant closed geodesic C. Poincaré's idea was to use differential geometry and to replace homotopy theory by differential geometry, namely the Gauss-Bonnet theorem. The basic idea was that this special geodesic would have no self-intersections and would possess (in restricted cases) special stability properties as a geometric object. To begin with a mathematical treatment, we assume a simple closed curve C exists and apply the Gauss-Bonnet theorem to it (see Fig. 1). We obtain

$$\int_{\Sigma_1} K + \oint K_g = 2\pi \ . \tag{1}$$

Here K is the Gauss curvature and K_g is the geodesic curvature. Now we assume that C is a geodesic so that $K_g = 0$ and we obtain from (1)

$$\int_{\Sigma_1} K(x)\,dV = 2\pi \ . \tag{2}$$

This equation is the natural constraint we are searching for because we can prove that minimizing the arc length functional relative to the constraint (2) yields a simple closed geodesic, and on the other hand, as we have shown above, every simple closed geodesic on (M^2, g) satisfies (2). This was first carried out a few years ago in my paper [1] with E. Bombieri for manifolds with curvature restrictions, and has subsequently been shown to be valid without such restrictions by Allard and Pitts. For example, here is the proof that shows that a smooth minimum of the arc length functional constrained by (2) is a closed geodesic. The Euler-Lagrange equation for a smooth critical point of the arc length functional subject to (2) is

$$K_g = \lambda K, \tag{3}$$

where λ is a Lagrange multiplier. Integrating this equation over the smooth curve C we find $\oint K_g = \lambda \oint_C K$.

Utilizing the Gauss-Bonnet theorem (1) and the constraint (2) we find that $\oint_C K_g = 0$. Thus by (3), $\lambda\oint K = 0$. Since (M^2, g) is an ovaloid $K > 0$ so this implies that $\lambda = 0$ and the Euler-Lagrange equation (3) reduces to $K_g = 0$, which implies that C is a geodesic. This means that the constraint (2) does not affect the Euler-Lagrange equation and it is therefore called a natural constraint. Thus, the shortest nonconstant closed geodesic can be characterized as a minimum of the arc length functional over a general class of closed curves subject to this natural constraint. This idea of using isoperimetric variational principles has a number of virtues. First, the minimizing curves in question can be shown to be smooth and have no self-intersections because of their minimizing properties. Secondly, such extremal curves can be computed by constrained optimization methods. Thirdly, the stability of such geodesics can be studied carefully by perturbing the metric g. Fourth, this idea can be extended to nonlinear dynamical systems and periodic solutions. In fact, we now turn to extending this idea to other nonlinear Hamiltonian systems.

3. EXTENSION OF NATURAL CONSTRAINTS TO OTHER DYNAMICAL SYSTEMS

Problem: To find the analogue of the "first normal mode" for

$$\frac{d^2\underline{x}}{dt^2} + \nabla V(\underline{x}(t)) = 0 \tag{4}$$

where $x(t) \in \mathbb{R}^N$ and $V(x)$ is a convex function (solution: Berger [3]).

To find the natural constraint for this problem, we set $t = \lambda s$ and seek 2π-periodic solutions in s. Integrating the transformed equation over a period, we find that a constraint for the problem is

$$\int_0^{2\pi} \nabla V(\underline{x}(s))\,ds = 0 . \tag{5}$$

The natural Lagrangian for the transformed problem is

$$L(\underline{x}(s)) = \frac{1}{2}\int_0^{2\pi} |\underline{x}(s)|^2 - \lambda^2\int_0^{2\pi} V(\underline{x}(s)) \tag{6}$$

and is unbounded above and below when studied over the class of 2π-

periodic vector functions $\underline{x}(s)$, so that all maxima or minima of L are infinite. However, the method of natural constraints suggests the isoperimetric variational principle of minimizing the "kinetic energy" over the constraint Σ:

$$K(x(s)) = \int_0^{2\pi} \left|\frac{d\underline{x}}{ds}\right|^2 , \tag{7}$$

$$\text{where} \quad \Sigma = \{\int_0^{2\pi} V(\underline{x}(s))ds = R, \quad \int_0^{2\pi} \nabla V(\underline{x}(s))ds = 0\} .$$

In this variational principle, the period λ appears as a Lagrange multiplier and the solutions of the isoperimetric variational principle automatically yield a nonconstant periodic solution when the vector functions $\underline{x}(t)$ in Σ are defined relative to the Sobolev space $W_{1,2}((0, 2\pi), \mathbb{R}^N)$.

What can be said about the solution of this isoperimetric variational principle? First, the minimum of the problem exists and is attained at a periodic solution $\underline{x}_R(t)$. This periodic solution is smooth, and as R varies over all real numbers, one-parameter families of periodic solutions are obtained. This family of solutions is computable by the methods of constrained optimization. Moreover, the parameter λ that occurs can be shown to be the minimal period of the periodic motion. Finally, assume $\nabla V(\underline{x}) = \Sigma_i \lambda_i \underline{x}_i + 0(|\underline{x}|^2)$. Then, as $R \to 0$, the solution and its minimal period $(\underline{x}_R(S), \lambda) \to (0, 2\pi/\lambda_N)$. This last fact signifies that the periodic solution of the isoperimetric variational problem (7) is the analogue of the first normal mode that we are seeking since its period is close to the period of the first normal mode of the associated linearized system at $\underline{x} = 0$. The important thing to notice in this case is that, contrary to all power series developments, there is no small divisor restriction on the existence of the one parameter family of periodic solutions emanating from the first normal mode of the linearized problem. This leads to one of the great achievements of the contemporary Calculus of Variations ideas, an extension of the famous Liapunov's theorem for periodic solutions near a singular point. We shall take up this circle of ideas in the next section, but first here is a verification of some of the results we stated in the above paragraphs.

1) The constraint (5) given by $\int_0^{2\pi} \nabla V(\underline{x}(s))ds = 0$ is a natural one for the first normal mode of (4) in case the potential $V(X)$ is strictly convex

Proof: The key idea here is showing that the smooth critical points $\{\underline{x}(s)\}$ of the Lagrangian (6) are unaffected by the constraint (5). Notice that the constraint (5) is a system of N nonlinear equations and so is associated with N Lagrange multipliers $\beta = (\beta_1, \beta_2, \ldots, \beta_N)$. The associated Euler-Lagrange equations for the extremals of (7) are

$$x'' + \lambda_i \nabla V + \beta \cdot \nabla(\nabla V) = 0 \ . \qquad (7^+)$$

Thus, integrating this last equation over a period $(0,2\pi)$, we find the N equations

$$\beta \cdot \int_0^{2\pi} \nabla(\nabla V) = 0 \ .$$

The strict convexity of V then forces the constant vector $\beta = 0$.

Moreover, the strict convexity of V also forces the Lagrange multiplier λ_1 to be positive so that $\lambda_1 = \lambda^2$ for some real nonzero number λ. To see this, set $\beta = 0$ in (7+) and take the inner product with $\underline{x}(s)$ and then integrate over the period $(0,2\pi)$ to find

$$\int_0^{2\pi} |x|^2 ds = \lambda_1 \int_0^{2\pi} \nabla V(\underline{x}) \cdot \underline{x}(s) ds \ .$$

2) <u>The variational problem (7) has a smooth periodic solution when it is confined to the Sobolev space</u> $W_{1,2}((0, 2\pi), \mathbb{R}^N)$.

Proof: This result is attained by analyzing the isoperimetric problem (7) by a combination of Sobolev space ideas and the utilization of Hilbert space techniques of weak convergence. The key idea is that the natural constraint (5) uniquely determines the constant vector part of any admissable vector function $\underline{x}(s)$ in the constraint set Σ in terms of its component with mean value zero. In other words, N directions, causing a saddle point to appear are excised by the natural constraint. Consequently, normal methods (see my book [4], Chap. 6) lead directly to the attainment of a smooth extremal.

The case of convex potentials (not necessarily strictly convex) is handled by an approximation argument.

3) <u>The Lagrange multiplier λ that occurs in the variational problem (7) is directly related to the minimal period of the periodic solution found for (4).</u>

<u>Proof:</u> This result follows by supposing the extremal $\underline{x}(t)$ does not have minimal period $2\pi\lambda$. Then in terms of the independent variable s, for integer n, $\underline{x}(s/n) \in \Sigma$ would be a periodic extremal of the constrained variational problem (7). However, this contradicts the minimality of the $\underline{x}(s)$ as a global minimum of (7) with the constraint Σ.

4) <u>Assuming</u> $V(x) = \frac{1}{2}\Sigma\lambda_i{}^2 x_i{}^2 + 0(|x|^3)$, <u>with</u> $\lambda_i \to 0$ $(i = 1,\ldots, n)$ <u>as $R \to 0$, the periodic solution</u> $x_R(s)$ <u>and the minimal period $2\pi\lambda$ have the following properties,</u>

$(x_R(s),\ 2\pi\lambda) \to (0,\ 2\pi/\lambda_N)$ where λ_N is the largest of the eigenvalues λ_i, i.e., the minimal period of $x_R(s)$ tends to the minimal period of the "first" normal mode of the linearized problem

$$\frac{d^2x}{dt^2} + V''(0)x = 0 .$$

<u>Proof:</u> To prove this result we merely let $R \to 0$ in (7) and notice that the quadratic terms in (7) are then dominant. In particular, as $R \to 0$, $|x_R(s)| \to 0$ and the lowest eigenvalue $2\pi/\lambda_N$ of the quadratic variational problem

$$\inf\int_0^{2\pi} |\dot{x}(s)|^2 \quad \text{over} \quad \tilde{\Sigma} = \{\int_0^{2\pi} \sum_{i=1}^{N} \lambda_1 x_1{}^2 = 1,\ \int_0^{2\pi} x = 0\}$$

is the correct approximation for the minimal period $\lambda(R)$ as $R \to 0$, in (7).

<u>Significance of the Results Found</u>

Let us consider the importance of these results as stated briefly in 1)-4). They show that there is a global constructive method for determining the analogue of the first normal mode of linear Hamiltonian systems of the form (4). This first normal mode is a direct extension of the problem associated with the linear Hamiltonian system

$$\frac{d^2x}{dt^2} + Ax = 0 \ . \tag{8}$$

It shows also that the method of determining this nonlinear normal mode is a global Calculus of Variations one since the conventional approaches, as we shall see in the next section, are perturbative and utilize power series expansion techniques. Moreover, the idea does not require that we are near the stationary point $x = 0$, but rather requires only a global geometric property of the Hamiltonian potential $V(x)$.

4. MODERN EXTENSIONS OF LIAPUNOV'S THEOREM VIA THE GLOBAL CALCULUS OF VARIATIONS.

We begin by reformulating Liapunov's theorem in terms of our point of view. We consider two Hamiltonian systems of second order equations. First, the linearized system

$$x'' + Ax = 0 \ , \tag{9}$$

where A is a nonsingular $n \times n$ self-adjoint matrix with eigenvalues $\lambda_1^2 \leq \lambda_2^2 \leq \lambda_3^2 \ldots \leq \lambda_n^2$, and then the nonlinear system

$$x'' + Ax + \nabla U(x) = 0 \ , \tag{10}$$

where $\nabla U(x) = 0(|x|)$ at $x = 0$. Now, the standard Liapunov theorem discusses whether the jth normal mode of the linearized system is preserved near the equilibrium point $x = 0$ under the nonlinear perturbation $\nabla U(x)$. The standard Liapunov result is as follows.

Standard Liapunov Theorem: The jth normal mode is preserved (near equilibrium) provided $\lambda_i/\lambda_j \neq$ integer where $i = 1, 2, \ldots, \hat{j}, \ldots, N$, with $\hat{j}$ meaning that the jth index is deleted from the comparison. (11)

The work in Section 2 shows that when $j = N$ the method of natural constraints demonstrates that the first normal mode, that is, the normal mode for the smallest period, is always preserved. This leads us to suspect that for systems like (10) the restrictions of the standard Liapunov theorem might be removed alltogether.

The key idea that we have just mentioned can be extended to first order Hamiltonian systems of a finite number of degrees of freedom where the Hamiltonian H is positive definite. This is exactly the case treated by Weinstein [5] and Moser [6]. In this case the Hamiltonian system can be written

$$z' = JH'(z) , \tag{12}$$

where J is defined below (15), $z = (p,q)$, and the linearized system at $z = 0$ is

$$y' = JH''(0)y . \tag{13}$$

The Liapunov theorem in this case compares (12) and (13) relative to periodic solutions, and the idea is that the normal modes of the periodic solutions for the linear system (13) in some sense are close approximations to certain periodic solutions for the solutions of (12), at least in the case of positive definite Hamiltonian systems. This theorem is stated more precisely below, but the key point here is to notice that it completely avoids the ideas of power series expansions associated with the usual proofs of the Liapunov theorem. For example, see the book of Siegel and Moser [7]. The idea being that the Calculus of Variations methods we use are <u>nonperturbative</u> and do not depend on good approximations for the solution at $z = 0$, but rather are characterized intrinsically by variational principles as in Section 3. Here is the basic outline of Moser's proof, the extension of Liapunov's theorem.

5. AN OUTLINE OF THE PROOF OF THE EXTENDED LIAPUNOV THEOREM

The theorem discusses the behavior near an equilibrium point $z = (p,q) = 0$ of the periodic solutions and compares the linearized system

$$\dot{z} = JH_{zz}(0)z \tag{14}$$

with the nonlinear system

$$\dot{z} = JH_z . \tag{15}$$

Here the matrix J has the form $J = \begin{pmatrix} 0 & I \\ -I & 0 \end{pmatrix}$, $H(0) = 0$, and $H_z(0) = 0$. The theorem due to Weinstein and Moser is as follows.

Theorem: Suppose H is C^2 near $z = 0$ with Hessian $H_{zz}(0)$ positive definite. Then for $\varepsilon > 0$ small, $H(z) = \varepsilon$ contains N periodic orbits of (14) with periods close to (16) those of (15).

Here is a sketch of Moser's proof [6], which is an extension of the ideas that we are discussing. We begin with the functional

$$\mathcal{J}(u) = \int_0^T [\dot{\underline{z}}\cdot J\underline{z} - h(\underline{z})]\, dt. \tag{17}$$

All periodic solutions of the linearized equation are assumed to have the same period. This avoids having to deal with periodic solutions which may be multiple coverings of a more fundamental periodic solution. In addition, we use invariants of the linear equation under time translation. Indeed, the system is autonomous.

Putting these two ideas together, we begin searching for critical points of the $\mathcal{J}(u)$. However, since we are working near the stationary point $z = 0$, we can reduce this global problem to a local one of bifurcation theory as we have described above. In particular, we use the Liapunov-Schmidt reduction technique to reduce the infinite dimensional problem to a finite dimensional one. The key point here is that the Calculus of Variations structure of the problem is preserved by this transformation. The general idea was first established in 1970 in a Ph.D. thesis of one of my students and used by me in a study of periodic solutions a few years thereafter. The final step is to use the invariants under time translation by using the group property associated with it. Indeed, the reduced functional is invariant under the S^1 action of the group, and so the Ljusternik-Schnirelmann theory can be applied to a small ball of radius ε about the singular point $z = 0$. The approximation for the periods of the linearized problems comes directly by comparing the reduced variational problem with the quadratic one associated with the linearized system.

6. IDEAS ASSOCIATED WITH LIAPUNOV'S THEOREM APPLICABLE TO OTHER PROBLEMS

Our idea here is to use the theory of nonlinear eigenvalue problems and the Calculus of Variations in the large to study the relationships between the following equations defined on a Hilbert space H,

$$u = \lambda L u, \tag{18a}$$

$$u = \lambda(\underline{L}u + \underline{N}u), \tag{18b}$$

at an eigenvalue λ_0 of the compact self-adjoint operator L. In particular, the standard (conventional) Liapunov theorem deals with the case when λ_0 is a simple eigenvalue of L, and the modern extension of Liapunov's theorem deals with the case when λ_0 is a multiple eigenvalue of L.

The basic idea here is that the parameter λ is directly connected with the periods of the periodic solutions to the Hamiltonian system (12). When $||u|| \to 0$, the periodic solutions of the nonlinear problem must tend to the solutions of the linear problem. Thus the question becomes,

"Is the multiplicity of the solutions for the linear problem preserved for the nonlinear problem?" (19)

To analyze this question, one must use advanced methods of studying critical point theory and utilize, in the first place, standard methods of studying critical points generalizing the minimax characterization of the eigenvalues of the linear problem (18a). The method most easily used is the Ljusternik-Schnirelmann technique. This technique is indeed a direct extension of the minimax technique for the computation of the eigenvalues for the equation (18a). In fact, the analogue of these eigenvalues is determined exactly by the critical points C_n of the functional $A(u)$ restricted to the sphere $||u||^2 = R$ and is given by the formula

$$C_n = \min_{[C]} \max_{[C]_n} A(u) . \tag{20}$$

In this case, $A(u)$ is derived by the Hilbert space inner product

$$A(u) = \frac{1}{2}(Lu,u) + \int_0^1 (N(su),u)ds , \tag{21}$$

and the classes $[C]$ and $[C]_n$ are generalizations of subspaces of dimension n based on a topological invariant called Ljusternik-Schnirelmann category. In order for this topological invariant to be large, so that the classes are well defined, one usually assumes that the variational problem is invariant under a group of transformations. In the late 1960s, for example, in my papers [8] and [9] I assumed the discrete group was Z_2. This required me to put a condition of oddness on the nonlinear operator $N(x)$, and consequently to restrict the potential $U(x)$ in (10) to be even. Subsequent work by Weinstein [5], Moser [6], and Rabinowitz [10] showed that it is more appropriate to use the circle group S^1 when discussing periodic solutions. Thus they were able to prove the Liapunov theorem without an assumption of symmetry on $U(x)$, or more generally, on the Hamiltonian H .

The fundamental problem in extending Liapunov's results globally can now be clearly understood on the basis of the above discussion. One hopes that the critical point theory of Ljusternik-Schnirelmann via minimax principles guarantees that the nonlinear normal modes determined near the stationary point as above can be extended globally, independently of the norm of the solutions and depending only on the convexity of the Hamiltonian functional. I tried this approach in the late 1960s [8], but came across the problem that mathematically, periodic solutions produced by this method might not be distinct. The difficulty of multiple coverings has subsequently been resolved but only at the cost of putting severe limitations on the Hamiltonian systems involved. In particular, in the article of Ekeland-Lasry [11] convex Hamiltonian systems of the form (12) are considered in which the origin $z = 0$ is in isolated equilibrium. It is also assumed that a certain level surface $H(\underline{z})$ = constant stays between the spheres of radius r and $\sqrt{2}r$ for small r. This condition guarantees that the normal modes found by the minimax principle are not multiple coverings and so give rise to distinct periodic solutions. Unfortunately, the assumption on the Hamiltonian functions involved is very strict and consequently not too useful in applications. In particular, suppose the Hamiltonian system is written in the form

$$\dot{\underline{z}} = J\{H_{zz}(0)\underline{z} + R(\underline{z})\} \quad \text{with } R(\underline{z}) = O(|\underline{z}|^2). \tag{22}$$

For example, suppose $H_{zz}(0)$ is positive definite with eigenvalues ω_i, $i = 1,2,3,\dots,n$. Then the assumption of Ekeland-Lasry implies that $\omega_i < 2\omega_1$ for all i. (Clearly, a very strong assumption.) Moreover, the characterization of the periodic solutions found and their computation in specific cases is extremely awkward because a Legendre argument is used (in addition to the basic existence proofs of the topological minimax method).

7. NONAUTONOMOUS HAMILTONIAN SYSTEMS

The periodic solutions for nonautonomous Hamiltonian systems of arbitrary finite dimension can also be found by global variational principles. The simplest such result appeared in a paper of mine and Schechter in 1977, although we had proven the result long before in 1973 (see my book [4], p. 304, Theorem 6.1.8). In this result we discuss periodic solutions for the system

$$\ddot{\underline{x}} = \nabla U(\underline{x},t). \tag{23}$$

Here, $\underline{x}(t)$ is an N-vector and the function $U(\underline{x},t)$ is a C^1 real-valued function of x and is T-periodic in t. We seek T-periodic solutions of this equation and we find that, provided

$$U(\underline{x},t) \longrightarrow \infty \quad \text{as} \quad |x| \rightarrow \infty , \tag{24}$$

such solutions exist and can be determined explicity as the minimal of the functional

$$\mathcal{J}(\underline{x}) = \int_0^T \{\tfrac{1}{2}\dot{\underline{x}}^2 + U(\underline{x},t)\}\, dt. \tag{25}$$

Once again, very general topological arguments can be carried out to prove the existence of multiple solutions at the cost of some explicitness. A particularly interesting example recently appeared and is due to Bahri-Berestiсki [12]. They discuss the equation

$$\ddot{\underline{x}} + \nabla U(\underline{x}) = \underline{f}(t) \tag{26}$$

where the vector function $\underline{f}(t)$ is T-periodic in t. They seek distinct T-periodic solutions. The system is clearly conservative with the Lagrangian

$$L(x(t)) = \int_0^T [\tfrac{1}{2} \; |x(t)|^2 - U(x) + f(t)\cdot x(t)]\, dt. \tag{27}$$

They prove that provided $U(0) = 0$ and $\nabla U(0) = 0$ and

$$U(x) \geq C|x|^{2+\varepsilon} + C \quad \text{for some fixed } \varepsilon > 0 , \tag{28}$$

then the equation has an infinite number of distinct periodic solutions. Once again the periodic solutions of this equation are saddle points of the functional (27). They use topological methods of minimax type and Morse theory for their determination as well as Hilbert space context such as we have discussed earlier. The methods are nonconstructive but determine saddle points of the functional (27) that were difficult to find previously.

Perhaps the main result or rather consequence about nonautonomous Hamiltonian systems for large finite numbers of degrees of freedom is due to Birkhoff and Lewis and states that on a fixed energy surface in a neighborhood of a periodic solution of a broad class of nonlinear Hamiltonian systems there are an infinite number of distinct periodic solutions. This conjecture was established in two dimensions by Birkhoff and Poincaré and for higher dimensional situations recent progress was made on this result by Conley and Zehnder [13]. They established for nonlinear Hamiltonian systems of the form

$$\dot{\underline{z}} = JH_z(t,\underline{z}), \tag{29}$$

where H is T-periodic in t, the existence of 2n+1 periodic solutions of fixed period T. By using the associated variational formulation with the functional

$$f(\underline{z}) = \int_0^1 \{<\dot{\underline{z}}, j\underline{z}> - h(t,\underline{z}(t))\}\, dt, \tag{30}$$

they make use of a Sobolev space formulation to express the functional in terms of functional analysis. They then work near a given periodic solution and so, using Liapunov-Schmidt formulation for bifurcation, reduced the problem to a finite-dimensional one. Then they used Conley's generalized Morse theory and Morse inequalities to produce the given number of periodic solutions.

Closely related to these results is the major problem in this field, namely that of extending the famous KAM Theorem globally. This very important extension is a major research problem and I hope that the methods discussed in this article may lead to the desired global extension. Indeed, the implicit function theorem on which the KAM Theorem is based does not make use of the global infinite dimensional Calculus of Variations.

8. PERIODIC SOLUTIONS FOR INFINITE DIMENSIONAL HAMILTONIAN SYSTEMS

The global methods we have discussed already are extremely useful in infinite dimensional Hamiltonian systems as well as those involving a finite number of degrees of freedom. As a particular case, we consider the non-linear Schrödinger equation

$$i\frac{\partial u}{\partial t} = \Delta u + |u|^{\sigma} u. \tag{31}$$

Here $i^2 = -1$ and Δ is the Laplace operator in n dimensions. We seek periodic solutions in t that have the form ic solutions in t that have the form

$$U(\underline{x},t) = e^{i\lambda t} V(\underline{x}) \quad \text{where } V(\underline{x}) \to 0 \quad \text{as } x \to 0. \tag{32}$$

We call these solutions "stationary states" in analogy with Quantum Mechanics, although they are sometimes called solitary waves or "solitons". It turns out that the equation (31) for a fixed positive λ (all positive values allowed) has an infinite number of such smooth periodic solutions provided the number of space variables involved is larger than one. Once again the method of studying this problem is by using the Calculus of Variations. In particular, the function V(x) in (32) satisfies the reduced nonlinear elliptic equation

$$\Delta V - \lambda y + V^{\sigma+1} = 0 \tag{33}$$

and nontrivial solutions of this equation can be formulated in terms of the Isoperimetric Variational Principle which can be formulated as

$$I(v) = \int_{\mathbb{R}^n} (|\nabla v|^2 + \lambda v^2)\, dv \quad \text{subject to the constraint}$$
$$\int_{\mathbb{R}^n} v^{\sigma+2} = \text{constant.} \tag{34}$$

I analyzed this problem in 1971 in my paper [14] based on the topological minimax principle described in Section 5. The utilization of this topological principle is slightly complicated because of the noncompactness of the Euclidean space $\mathbb{R}^n$. This can be overcome by the symmetry and the Ljusternik-Schnirelmann topological minimax principle can be applied in this case. In fact, we can prove

<u>Theorem</u>: The equation (33) has an infinite number of distinct solutions characterized by explicit isoperimetric variational principles for each fixed positive λ. Moreover, these solutions are smooth and characterized by the number of times each vanished on the x_1-interval $(0,\infty)$. The variational characterization is by means of the formulas (20) and (21).

<u>Proof</u>: The proof is discussed in the papers [14] and [15].

BIBLIOGRAPHIC NOTES

Here is a survey of some of the background articles involving the global Calculus of Variations arguments of periodic solutions that we are discussing here. My contributions started in 1968 with the publication of my book [16] <u>Perspectives in Nonlinearity</u>. I then continued this with my articles [2], [8], and [9]. This work was also taken up by W. B. Gordon [17] using a somewhat different approach. In [18], he studied the Kepler problem by these methods and this problem was also studied by J. Moser in [19]. The perturbation of this problem is also mentioned in my book [4].

My former student Westreich took up the problem of periodic orbits between the years of 1970 and 1975. A list of his papers will be found in the bibliography of the book by Chow and Hale [20], which contains many other references and proofs in the direction discussed here.

Rabinowitz has taken up the study of variational methods for periodic orbits of Hamiltonian systems and greatly added to it. His contributions are surveyed in [21]. Van Groesen is currently extending the Method of Natural Constraints. His contributions are discussed in [22]. Numerous other authors have extended the results surveyed here. See the bibliography in Zehnder [23], where over one hundred papers on the theme of the present survey are listed.

OUTSTANDING PROBLEMS

It might be thought that with so many research papers on this subject the subject might be depleted. Quite the contrary is true. First, there is the mammoth research problem consisting of tying up this work with the problems of celestial mechanics. See for example the beautiful article of Deprit and Henrard [24] and Deprit's other papers. There is an immense amount of numerical work in this direction and many new fundamental ideas to discover. Typical is the "Principle of Natural Termination" of the Scandinavian astronomer Strömgren.

Then there is the important extension of our idea to infinite dimensional Hamiltonian systems representing periodic motions in space variables of higher dimensions. The entire realm of nonlinear normal modes for a vibrating plate and higher dimensional objects such as a shell is completely open.

The recent flurry of attention to solitons in the KdV equation has shielded the outstanding problem of "breathers" for such one dimensional evolutionary equations; a "breather" being a periodic solution in t and decaying appropriately in x. An outstanding problem in this direction is for nonintegrable evolutionary equations such as the ϕ^4 equation. Huge sums of money have been spent on numerical evidence for the existence of such a periodic solution, but to date very few theoretical results are available concerning it.

Generalizing the discussion in Section 8 is the problem of periodic solutions for the Yang-Mills equations of gauge theory. Immense amount of attention has been placed on the theoretical study of "instantons" for such equations. On the other hand, very scant attention has been devoted to the global study of the associated time-dependent periodic solutions.

REFERENCES

[1] H. Poincaré, "Sur les lignes géodésiques des surfaces convexes," Trans. Amer. Math. Soc. 6, 237-274 (1905).

[2] M. S. Berger and E. Bombieri, "On Poincaré's isoperimetric variational problem for closed simple geodesics", J. Functional Anal. 42, 274-298 (1981).

[3] M. S. Berger, "Periodic solutions of second order dynamical systems and isoperimetric variational problems", Am. J. Math. 93, 1-10 (1971).

[4] M. S. Berger, Nonlinearity and Functional Analysis (Academic, New York, 1977).

[5] A. Weinstein, "Lagrangian submanifolds and Hamiltonian systems", Ann. Math. 98, pp. 377-410 (1973).

[6] J. Moser, "Periodic orbits near an equilibrium and a theorem by Alan Weinstein", Comm. Pure Appl. Math. 29 (1976).

[7] C. L. Siegel and J. K. Moser, Lectures on Celestial Mechanics (Springer, New York, 1971).

[8] M. S. Berger, "On periodic solutions of second order Hamiltonian systems (I)", J. Math. Anal. and Appl. 29, 512-522 (1970).

[9] M. S. Berger, "A bifurcation theory for real solutions of nonlinear elliptic partial differential equations", in Bifurcation Theory and Nonlinear Eigenvalues, J. Keller and S. Antman, Eds., (Benjamin, New York, 1969), pp. 113-216.

[10] P. H. Rabinowitz, "Periodic solutions of Hamiltonian systems", Comm. Pure Appl. Math. 31, pp. 157-184 (1978).

[11] I. Ekeland and J. Lasry, "On the number of periodic trajectories for a Hamiltonian flow on a convex energy surface", Ann. Math. 112, pp. 283-319 (1980).

[12] A. Bahri and H. Berestick i, "Existence of forced oscillations for some nonlinear differential equations", Comm. Pure Appl. Math. 37, 403-442 (1984).

[13] C. C. Conley and E. Zehnder, "The Birkhoff-Lewis fixed point theorem and a conjecture of V. Arnold", Invent. Math., 33,49 (1983).

[14] M. S. Berger, "On the existence and structure of stationary states for a nonlinear Klein-Gordon equation", J. Functional Anal. 9, 249-261 (1972).

[15] C. Jones, Preprint.

[16] M. S. Berger and Marion Berger, Perspectives in Nonlinearity, (Benjamin, New York, 1968).

[17] W. B. Gordon, "A Theorem on the existence of periodic solutions to Hamiltonian systems with convex potentials", J. Diff. Eq., 10, 324-335 (1971).

[18] W. B. Gordon, "A minimizing property of Kepler orbits", Am. J. Math 99, 961-971 (1977).

[19] J. Moser, "Regularization of Kepler's problem and the averaging method on a manifold", Comm. Pure Appl. Math. 23, 609-636 (1970).

[20] S. N. Chow and J. K. Hale, Methods of Bifurcation Theory (Springer, New York, 1982).

[21] P. H. Rabinowitz, "Periodic solutions of Hamiltonian systems: a survey", SIAM J. Math. Anal. 13, 343-352 (1982).

[22] E. W. C. Groesen, Applications of Natural Constraints to Natural Hamiltonian Systems, 1983.

[23] E. Zehnder, "Periodic solutions of Hamiltonian equations", Lecture Notes in Mathematics 1031 (Springer, New York, 1981).

[24] A. Deprit and J. Henrard, "A Manifold of periodic orbits", Adv. Astron. Astrophys. 6, 1-124 (1968).

Lecture 1: ON RESONANT HAMILTONIAN SYSTEMS WITH FINITELY MANY DEGREES OF FREEDOM.

Martin Kummer
Department of Mathematics
University of Toledo
Toledo, Ohio 43606

0.   INTRODUCTION

A typical example of a Hamiltonian system, with which we shall concern ourselves is afforded by the elastic pendulum, depicted in Fig. 1. The corresponding Hamilitonian is the following function over its phase space $\mathbb{R}^4$ (position variables q_1, q_2; corresponding momenta p_1, p_2):

$$H = \frac{1}{2m} (p_1^2 + p_2^2) + mgq_2 + \sigma(L). \tag{1}$$

Here the first term represents the kinetic energy, the second term the gravitational, and the third term the elastic energy. It is no restriction to assume $\sigma(L_0) = 0$, and in order to guarantee that the origin 0 of $\mathbb{R}^4$ is an equilibrium point we must require $\sigma'(L_0) = mg$. We also introduce the abbreviations $c = \sigma''(L_0)$, $d = \sigma'''(L_0)$, $f = \sigma^{1V}(L_0)$, $\omega_1 = (\frac{g}{L_0})^{1/2}$, $\omega_2 = (\frac{c}{m})^{1/2}$, $\Delta = \frac{m}{L_0} (\omega_2^2 - \omega_1^2)$. Here, ω_1, ω_2 are the frequencies of the horizontal and vertical motions in the limit of small deviations from the equilibrium. The Hamiltonian in these variables becomes $H = H_2 + H_3 + H_4 + O_5$, where O_5 is a convergent power series in the phase variables, beginning with a term

$$L = [q_1^2 + (L_0 - q_2)^2]^{1/2}$$

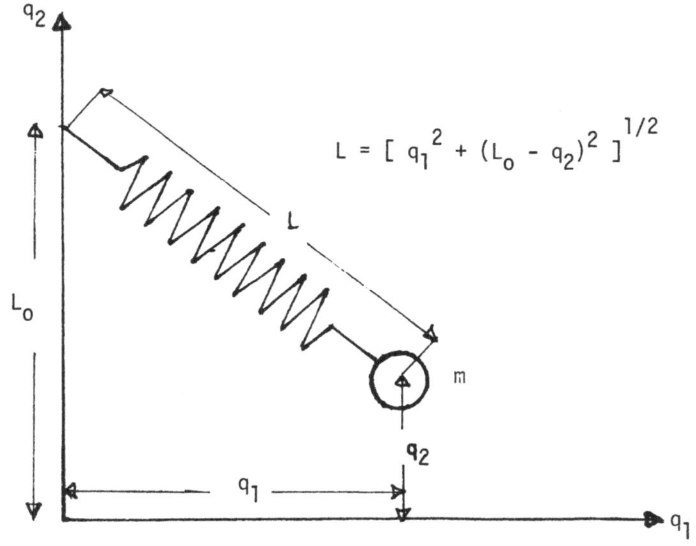

Fig. 1. The elastic pendulum.

of order 5, and H_k (k = 2, 3, 4) are the following homogeneous polynomials of degree k:

$$H_2 = \frac{1}{2m} p_1^2 + \frac{m}{2} \omega_1^2 q_1^2 + \frac{1}{2m} p_2^2 + \frac{m}{2} \omega_2^2 q_2^2 ,$$

$$H_3 = - \frac{\Delta}{2} q_2 q_1^2 - \frac{d}{6} q_2^3 ,$$

$$H_4 = L_o^{-1} \cdot \frac{\Delta}{8} \cdot q_1^4 + L_o^{-1} [\frac{d}{4} - \frac{\Delta}{2}] q_1^2 q_2^2 + \frac{f}{24} q_2^4 .$$

Our goal is to study the flow induced by H near the equilibrium point 0 in $\mathbb{R}^4$ in the two resonant cases

(i) $\omega_2 = 2\omega_1$, (ii) $\omega_1 = \omega_2$.

If we introduce the complex variables

$$z_\nu = (\frac{\omega_\nu m}{2})^{\frac{1}{2}} q_\nu - i(2m\omega_\nu)^{-\frac{1}{2}} p_\nu \quad (\nu = 1,2) \tag{2}$$

together with the c.c. (= conjugate complex) variables $\overline{z_\nu}$, then in the new variables H_2 takes the simple form

$$H_2 = \omega_1 N_1 + \omega_2 N_2, \quad N_\nu = |z_\nu|^2, \quad \omega_\nu > 0, \tag{3}$$

and the higher order terms become homogeneous polynomials in z_ν, $\overline{z_\nu}$. The associated differential equations are

$$\dot{z}_\nu = i \frac{\partial H}{\partial \overline{z}_\nu} \quad \text{and c.c. .} \tag{4}$$

At this point it is natural to consider the more general Hamiltonian $H = H_2 + \mathcal{O}_3$, where H_2 was defined in (3) and $\mathcal{O}_3$ represents a convergent power series in the variables z_ν, $\overline{z}_\nu$ (ν = 1,2) beginning with a term of order 3. We speak of the compact $k:\ell$ -resonance or of the nonresonant case depending on whether $\omega_1/\omega_2 = k/\ell$, where k,ℓ are positive integers so that the surfaces H_2 = const. are compact, or ω_1/ω_2 is irrational. In the first case we shall assume w. l. o. g. that either k and ℓ are mutually prime and $\ell > k \geq 1$ or $k = \ell = 1$. Actually, as long as $k + \ell \geq 5$ and the (Birkhoff) normal form of our Hamiltonian contains nonzero terms of order $< k + \ell$, the flow pattern in the compact $k:\ell$ - resonance does not differ essentially from the one of the nonresonant case. Indeed, if in addition some nondegeneracy condition (involving the coefficients of the normal form) is satisfied, KAM - theory asserts that, given $\varepsilon > 0$, there exists a neighborhood of the origin 0 of $\mathbb{R}^4$ which is filled with invariant tori, except for a set of relative measure $< \varepsilon$ (see, e. g., [1]). The latter set contains two 2-manifolds through 0 which are filled with families of periodic solutions with frequencies near ω_1 and ω_2 , respectively.

For this reason the three resonances: 1:1, 1:2, 1:3 are the most interesting ones. Due to the special form of the Hamiltonian the 1:3 resonance cannot be realized in the elastic pendulum. This explains why in the case of the elastic pendulum the resonances (i) and (ii) above are the only ones that are worth our attention.

At this point some comments on the history of the subject are in order. The 1:2 resonance of the elastic pendulum was first studied by Duistermaat [2]. A unified treatment of the $k:\ell$ resonance with k,ℓ mutually prime integers was presented by Schmidt and Sweet in [3]. The 1:1 resonance (with compact energy surface) was first analyzed by Braun [4]. However, since Braun's normal form is not the most general one, his study is incomplete. My own study [5] of the 1:1-resonance completes Braun's work and extends the method to the noncompact case [6]. It was then realized by Churchill, Rod, and myself [7] that the same method could also be applied to the other resonances. For a related approach to the $k:\ell$ - resonance ($k\neq\ell$) see also [15].

The outline of this lecture is as follows:

1.) Our method is explained in the compact $k:\ell$ resonance and then illustrated with help of the resonances (i), (ii) of the elastic pendulum. It seems that the results pertaining to the 1:1-resonance of the elastic pendulum are new.

2.) As in [8] and [9] we show that certain perturbation problems of the Kepler problem lead to Hamiltonians describing a 1:1 resonance.

3.) Our methods can also be applied to the 1:2:2 resonance (and more generally to the 1:2:2:...:2 resonance). This resonance was first studied by Verhulst and van der Aa [10]. We achieve a considerable simplification of their proof of the near integrability of this resonance and present a general analysis of possible periodic orbits near the origin of the phase space $\mathbb{R}^6$.

1.) THE $k:\ell$ RESONANCE.

By rescaling of the time variable, it is no restriction to assume $\omega_1 = k$, $\omega_2 = \ell$ so that $H_2 = kN_1 + \ell N_2$. The surface $k|z_1|^2 + \ell|z_2|^2 = I$ (> 0) can graphically be represented as a solid double cone of height π over a circle of radius $(I/k)^{1/2}$ centered at the origin of the z_1-plane (compare [7] and [11]). If $\alpha_2 = \arg z_2$ ($-\pi \leq \alpha_2 \leq \pi$), the level surfaces $\alpha_2 =$ const. are represented by cones of height α_2 over the given circle. Correspondingly, the cone $\alpha_2 = \pi$ and $\alpha_2 = -\pi$ have to be identified (see Fig. 2). The solutions corresponding to H_2,

$$z_1 = z_{10}e^{ikt} , \qquad z_2 = z_{20}e^{i\ell t} , \tag{5}$$

lie on a torus (represented by a cylinder in Fig. 2) on which they wind k-times along one cycle and ℓ-times along the other cycle before they close up. All solutions are periodic with primitive period 2π, except possibly the solution $z_1=0$ ($z_2=0$) which in the case $\ell > 1$ ($k > 1$) have shorter period $2\pi/\ell$ ($2\pi/k$). These exceptional periodic solutions are precisely the ones which, according to a classical result of Liapunov, continue to exist after the non-linear terms have been "turned on".

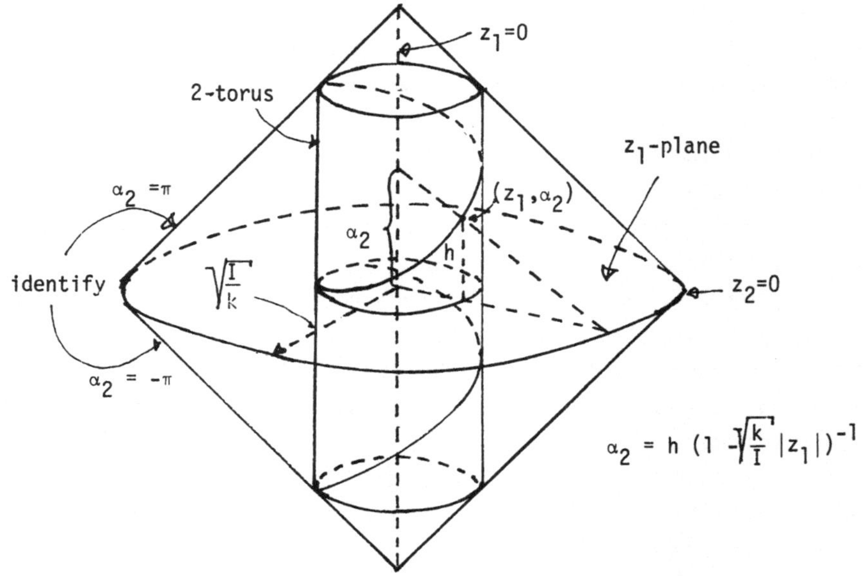

Fig. 2. The surface $k|z_1|^2 + \ell|z_2|^2 = I$ with 2-torus on which a periodic orbit winds around k-times along one cycle and ℓ-times along the other one.

In order to describe the influence of the perturbation, we bring it into normal form K (as to how this can be done see [7], [12], [13], or the contribution of Dr. Finn and the references given by him) up to the lowest term that does not vanish and drop the remaining terms. Since K by its very definition has zero Poisson bracket w. r. to H_2, K defines a flow on the manifold which is obtained by viewing each orbit of H_2 lying on a surface $H_2 = I$ (> 0) as a point (= orbit manifold). A quite explicit realization of this orbit manifold which is independent of I (as long as $I > 0$) is obtained as the image of the map π: $\mathbb{C}^2 \setminus \{0\} \to \mathbb{R}^3 \setminus \{0\}$ that is defined by means of the formulas:

$$\vec{x} := (x_1 + ix_2, x_3) = \pi(z) = (M_+ \; J^{-(\frac{k+\ell}{2})}, \; M_3 \, J^{-1}) . \tag{6}$$

Here $M_+ = z_1^{\ell} \bar{z}_2^{k}$, $M_3 = \frac{1}{2} \left(\frac{N_1}{\ell} - \frac{N_2}{k}\right)$ and $J = \frac{1}{2} \left(\frac{N_1}{\ell} + \frac{N_2}{k}\right) = \frac{1}{2k\ell} I$.

It is immediate that π collapses each orbit (5) to a point, but also the converse, namely that $\pi^{-1}(x)$ for $\vec{x} \in$ Image π consists of an orbit (5), is not difficult to show (see [7]). The image of π can most conveniently be described as the 0-set of the function

$$g(x_1, x_2, x_3) = \frac{1}{2} \left(x_1^2 + x_2^2 - f(x_3)\right), \tag{7}$$

Fig. 3. Orbit manifolds for different values of k,ℓ . Singular points $x_3 = \pm 1$ represent periodic solutions of period $\frac{2\pi}{k}$ and $\frac{2\pi}{\ell}$, respectively.

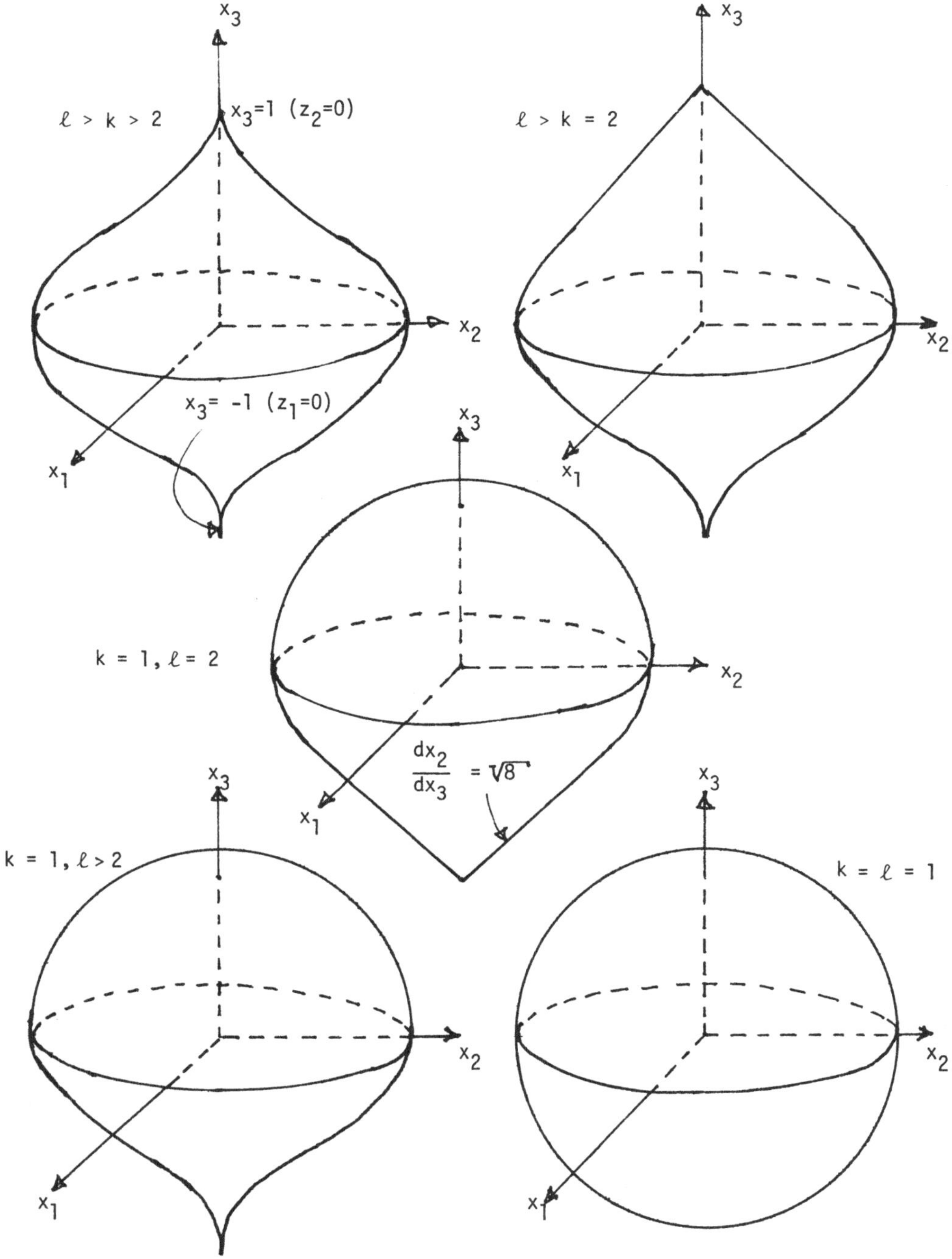

where

$$f(x_3) = [k(1-x_3)]^k \, [\ell(1+x_3)]^\ell . \tag{8}$$

Correspondingly, the orbit manifolds look as in Fig. 3, i.e. they have the shape of an onion ($\ell > k \geq 2$) or balloon ($\ell > k = 1$) or sphere ($k = \ell = 1$). Note that the singular points represent precisely the "Liapunov-modes" (= periodic with period less than 2π) that were mentioned earlier. Besides the Liapunov orbits there may exist orbits of period close to 2π. They correspond to equilibrium points of K on the orbit manifold and therefore are characterized by the condition that ∇K, ∇g are parallel vectors. In fact, the flow induced by K on the orbit manifold, the so called quotient flow, can be shown to be governed by the Euler-like equations

$$\dot{\vec{x}} = J^{-1} (\nabla K \times \nabla g). \tag{9}$$

A quick qualitative picture of the quotient flow lines (i.e., the flow lines of K on $g = 0$) is obtained by cutting the "onion", "balloon", or sphere with surfaces K=const.

For instance in the case of the 1:2-resonance it can be shown that the lowest order normal form term is always of the form $A\,J^{3/2}\,x_1$. (Actually, $A = -\frac{3}{4} L_0^{-1} (\frac{\omega_1}{m})^{1/2}$ in the case of the elastic pendulum.) Accordingly, the flow lines on our "balloon" are obtained by cutting it with planes x_1 = const. (see Fig. 4). The figure shows that the pure horizontal mode is nonexistent and pure vertical motion (a Liapunov orbit) is unstable. The pendulum settles in a stable mode which combines both motions. There are two periodic solutions with period close to 2π: $\vec{f}^{\pm}$. (Their frequencies are $1 \pm A\sqrt{\frac{2}{3}}\, J^{1/2}$.) Detuning can also be studied using this graphical representation: The plane x_1 = const. is replaced by the plane $\nu x_3 - x_1$ = const. where ν is proportional to the detuning parameter, i.e., the plane is tilted away from its vertical position. If $\nu > 2\sqrt{2}$ (= slope of "balloon strings" at $x_3 = -1$ w. r. to the vertical direction) the resonance disappears and the vertical oscillation becomes stable again. The situation should be contrasted with the one prevailing in the 1:3 resonance in which the surfaces K=const. generically are parabolic cylinders, rather than planes. Intersecting them with our "balloon" leads for instance to a quotient flow as depicted in Fig. 5. The Liapunov mode with period $\frac{2\pi}{3}$ remains stable. Another stable periodic orbit of period 2π is always present due to the fact that at least one parabolic cylinder touches the "balloon" externally. There may or may not be a second parabolic cylinder that does that, in which case there are two stable periodic orbits of period near 2π. For instance this is the case in Fig. 5 which depicts the most complex possibility: A third parabolic cylinder touches the balloon internally, giving rise to an unstable periodic orbit together with its stable and unstable submanifolds.

We now turn to a discussion of the 1:1-resonance of the elastic pendulum, i.e. we assume $\omega_1 = \omega_2$, i.e. $mg = cL_0$. The quantity that naturally enters as dimension-

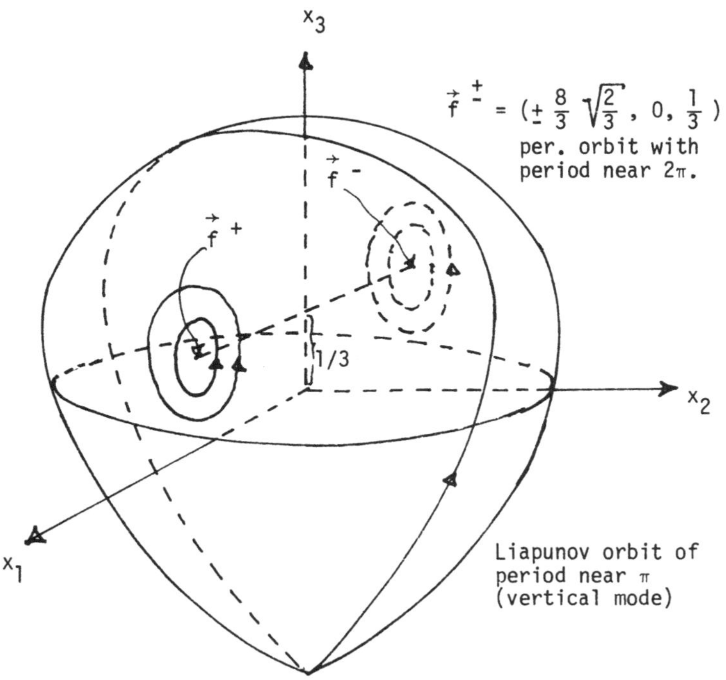

Fig. 4. The quotient flow in the 1:2 resonance.

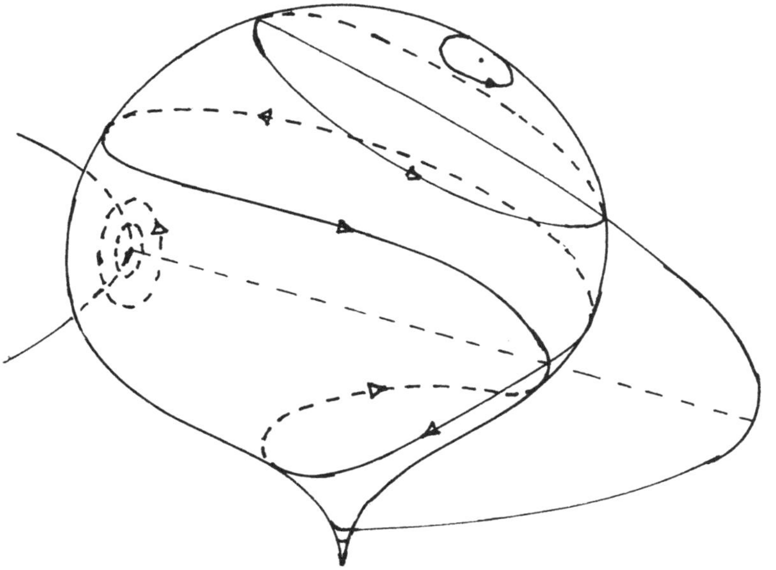

Fig. 5. Typical quotient flow in the 1:3 resonance with 2 stable and one unstable periodic orbit with period near 2π.

Figs. 6, 7. The quotient flow in the 1:1 resonance of the elastic pendulum for
1 < r < 3 (top) and 0 < r < 1 (bottom).

$$\vec{f_1}^{\pm} = (6-r)^{-1}(\pm\ 6(1 - \tfrac{r}{3})^{1/2},\ 0,\ -r)$$

$$x_2 \pm \left(\frac{3-r}{r-1}\right)^{1/2} x_1 = 0$$

$$\vec{f_2}^{\pm} = (2-r)^{-1}\ (0,\ \pm\ 2(1-r)^{1/2},\ -r)$$

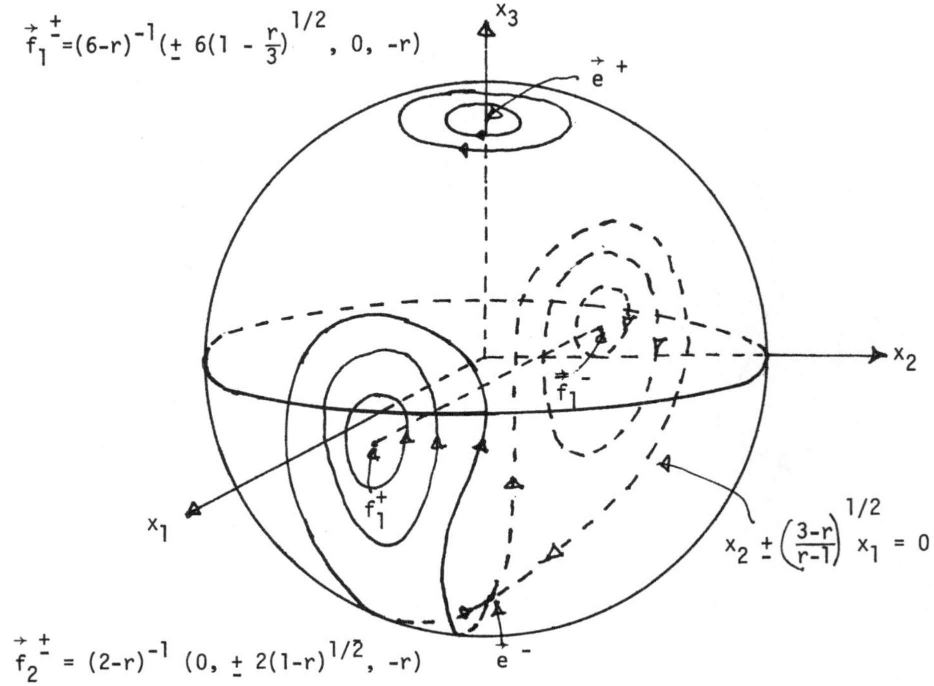

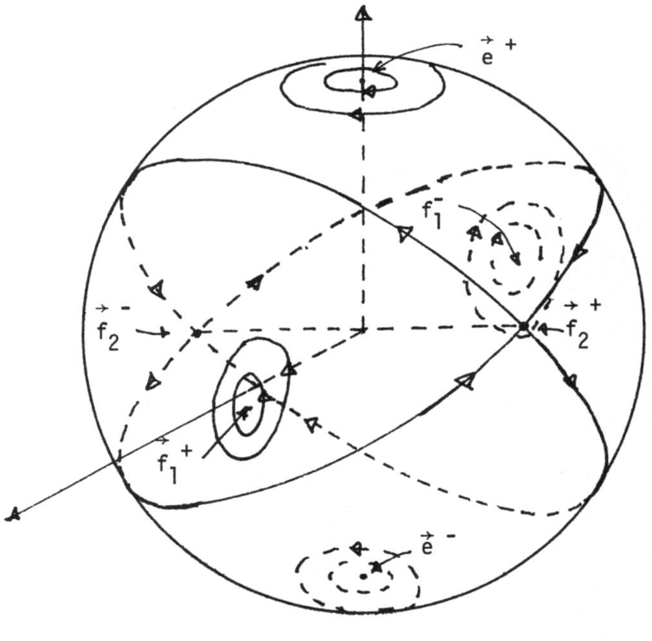

less bifurcation parameter is $r := L_0(\frac{f}{d} - \frac{5}{3}\frac{d}{c})$, where we recall the definitions of c,d,f, given in the introduction. Pure horizontal motion represented by the north pole $\vec{e}^{\,+}$) is always stable. The same is true for pure vertical motion, except if $1 < r < 3$, in which case the quotient flow corresponds to Fig. 6. As in the 1:2 resonance the pure vertical motion ($\vec{e}^{\,-}$) becomes unstable and is replaced by two modes $\vec{f}_1^{\,\pm}$ which combine vertical and horizontal motion. In the case $0 < r < 1$ the quotient flow on the sphere resembles the flow induced by the Hamiltonian of an asymmetric top on the angular momentum sphere, i.e. there are two pairs of stable $\vec{e}^{\,\pm}$, $\vec{f}_1^{\,\pm}$ and one pair of unstable periodic orbits: $\vec{f}_2^{\,\pm}$ (Fig. 7). For all other values of r only the pure modes (horizontal and vertical) are present and they are stable as in the non-resonant case.

2. THE PERTURBED KEPLER PROBLEM IN 2 DIMENSIONS.

We consider the Hamiltonian

$$H = H_0 + \varepsilon V(q,p,\varepsilon),\ H_0 = \frac{1}{2}|p|^2 - \frac{1}{r},$$

($r = |q|$) where $q \neq 0$, p are complex variables and H_0 is the Kepler Hamiltonian. The differential equations associated with H are

$$\frac{dq}{dt} = 2\frac{\partial H}{\partial \bar{p}},\quad \frac{dp}{dt} = -2\frac{\partial H}{\partial \bar{q}}$$

and c.c. . It is well known and easily checked that if the "fictitious time" s is introduced via the recipe rds=dt the flow induced by H on the energy surface $H = -\frac{1}{2}$ is the same as the one induced by the Hamiltonian

$$K = \frac{1}{2} r\,(|p|^2 + 1) + \varepsilon r V(q,p,\varepsilon).$$

Observe also that $H = -\frac{1}{2}$ corresponds to $K = 1$. The Levi-Civita transformation $q = v^2$, $p = iw\bar{v}^{\,-1}$, followed by the linear transformation $z_1 = 2^{-1/2}(v+w)$, $\bar{z}_2 = 2^{-1/2}(v-w)$ produces a Hamiltonian describing a 1:1 resonance, namely

$$H = J + \varepsilon\tilde{V}(z,\bar{z},\varepsilon),\ J = \frac{1}{2}(N_1 + N_2). \tag{10}$$

If we assume that $V(q,p,\varepsilon)$ is linear in p, i.e. has the form

$$V(q,p,\varepsilon) = V_0(q,\varepsilon) + \mathrm{Re}(\bar{p}\,V_1(q,\varepsilon)),$$

where $V_0(q,\varepsilon)$, $V_1(q,\varepsilon)$ are real and complex valued analytic functions of q,ε, then $\tilde{V}(z,\bar{z},\varepsilon)$ is analytic in z_ν, $\bar{z}_\nu$, $\varepsilon(\nu = 1,2)$ and for $\bar{z}_\nu$ c.c. to z_ν and ε-real, takes real values. The differential equations associated with Hamiltonian (10) are again of type (4) so that the Kepler flow is represented by the flow $z_\nu \to z_\nu e^{i\frac{s}{2}}$ on the "energy surface" $S^3 = \{z: |z_1|^2 + |z_2|^2 = 2\}$. Notice however, that since the Levi-Civita transformation has doubled the states (it doubly covers the phase space), antipodal points of S^3 have to be identified, that is to say, the compactified energy surface $H = -\frac{1}{2}$ has topological character P^3 (= projective 3-space).

The introduced quantities are closely related to the elements of a Kepler ellipse. Indeed, setting $z_\nu = (N_\nu)^{1/2} e^{i\alpha_\nu}$ ($\nu = 1,2$) we obtain

$$q = \frac{1}{2}(z_1 + \bar{z}_2)^2 = \frac{1}{2} N_1 e^{2i\alpha_1} + \frac{1}{2} N_2 e^{2i\alpha_2} + (N_1 N_2)^{1/2} e^{ig}$$

$$= e^{ig} [J \cos u + iM_3 \sin u + \sqrt{J^2 - M_3^2}\,],$$

where $g := \alpha_1 - \alpha_2$ is the aphelion advance and $u := \alpha_1 + \alpha_2$ the eccentric anomaly. J (=1 for $\varepsilon = 0$) is the major axis and $M_3 = \frac{1}{2}(N_1 - N_2)$ is the angular momentum. Finally, $e := \sqrt{1 - \frac{M_3^2}{J^2}}$ is the eccentricity. The orbit manifold (i.e., the manifold in which each oriented Kepler ellipse is represented by a point) is again S^2. In fact, it again can be obtained as the image of the map π , which was introduced in Section 1 (where $k = \ell = 1$). It now takes the form $\vec{x} = \pi(z) = J^{-1}(M_+, M_3)$, where $M_+ = z_1\bar{z}_2$ is the Laplace vector and M_3 is the angular momentum. Since $x_3 = J^{-1}M_3$ $= \operatorname{sgn} M_3 \cdot \sqrt{1 - e^2}$ we see that the poles of S^2 (i.e.: $x_1 = x_2 = 0$, $x_3 = \pm 1$) represent direct and rectrograde circular orbits, whereas the equator represents the collision orbits. The quotient flow determined by the normal form Hamiltonian $K(\vec{x}, I)$, associated with our perturbation V, is again governed by the equations (9) which for $k = \ell = 1$ assume the simple form

$$\dot{\vec{x}} = J^{-1} (\nabla K \times \vec{x}). \tag{11}$$

A treatment of the planar restricted three body problem along these lines is presented in [14] and the foregoing considerations together with the application to the restricted 3-body problem have been generalized in [15] .

3. THE 1:2:2: RESONANCE

Van der Aa and Verhulst [10] prove that if a Hamiltonian describing a 1:2:2 resonance is truncated beyond the third order term of its normal form it becomes completely integrable. Our methods lead to a simple proof of this fact. It is immediate that the normal form up to third order must be of the general form:

$$K = I + \operatorname{Re} [z_1^2 (A\bar{z}_2 + B\bar{z}_3)], \quad I := N_1 + 2N_2 + 2N_3.$$

Here A,B are in general complex constants. However by a change of the phases of z_2, z_3, i.e., a symplectic transformation, they can always be made real. A further symplectic change of coordinates

$$\begin{pmatrix} z_2' \\ z_3' \end{pmatrix} = \begin{pmatrix} \cos\phi & \sin\phi \\ -\sin\phi & \cos\phi \end{pmatrix} \begin{pmatrix} z_2 \\ z_3 \end{pmatrix}$$

with $\phi = \arg(A + iB)$ will bring K into the form $K = I + \sqrt{A^2 + B^2}\ \mathrm{Re}\,(z_1{}^2\bar{z}_2)$ for which it is obvious that N_3 is a third integral. (We have dropped the primes in the variables z_2, z_3 again.) The flow induced by K in $\mathbb{C}^3$ is now determined by its projection onto the z_1, z_2-space which in turn is the same as in the 1:2 resonance studied in Section 1. Therefore it can be reconstructed from the quotient flow on our "balloon", depicted in Fig. 4. We only have to reinterpret that flow appropriately. The "balloon" is now obtained by reducing out the torus action represented by the combined flows of I and N_3 on a surface $I = \text{const.} > 0$ $N_3 = \text{const.} \geq 0$. Accordingly, the two equilibrium points in Fig. 3 now represent quasi-periodic solutions with two frequencies except in the case $N_3 = 0$ in which these solutions are again periodic with frequency $1 \pm \sqrt{\frac{2}{3}}\ C(\frac{I}{4})^{3/2}$ $(C = \sqrt{A^2 + B^2})$. Moreover the "basket of the balloon" $(x_3 = -1)$ now represents the invariant 3-manifold $z_1 = 0$, $I = \text{const.}$ which is filled with a two parameter family of periodic solutions of period π. Notice that, like the vertical mode in the elastic pendulum, this submanifold as a whole is unstable. The degeneracy of this subflow is only lifted after the fourth order term in the normal form K is "switched on". The latter can easily be seen to be a quadratic form $Q = Q(N_1, J, \vec{M})$ of the variables N_1, J, $\vec{M}$. Here, J, $\vec{M}$ are defined the same way as in Section 2 with z_2, z_3 taking over the role of z_1, z_2. The orbit manifold of the restriction of the flow to the submanifold $z_1 = 0$, $I = \text{const.} > 0$ is again a 2-sphere which in $\vec{M}$-space is given by the equation $\vec{M}^2 = J^2$. (Contrary to Section 2 we do not normalize the sphere by introducing the variables $\vec{x} = J^{-1}\vec{M}$.) Again the flow on this sphere is governed by Euler-type equations, namely by

$$\dot{\vec{M}} = \nabla_{\vec{M}}\, Q(0, J, \vec{M}) \times \vec{M} \quad .$$

Generically, 2, 4 or 6 periodic orbits on our invariant submanifold will survive after the Q-term has been "switched on". All of them are unstable since the submanifold itself has this property. Also, the two stable periodic orbits, found above before Q was "switched on", will continue to exist. However, the values N_1, N_2, N_3 on these solutions will undergo a displacement away from the original values: $N_1 = \frac{2}{3}\,I$, $N_2 = \frac{1}{6}\,I$, $N_3 = 0$. The situation with regard to the existence of periodic solutions is summarized in Fig. 8 (compare [10]).

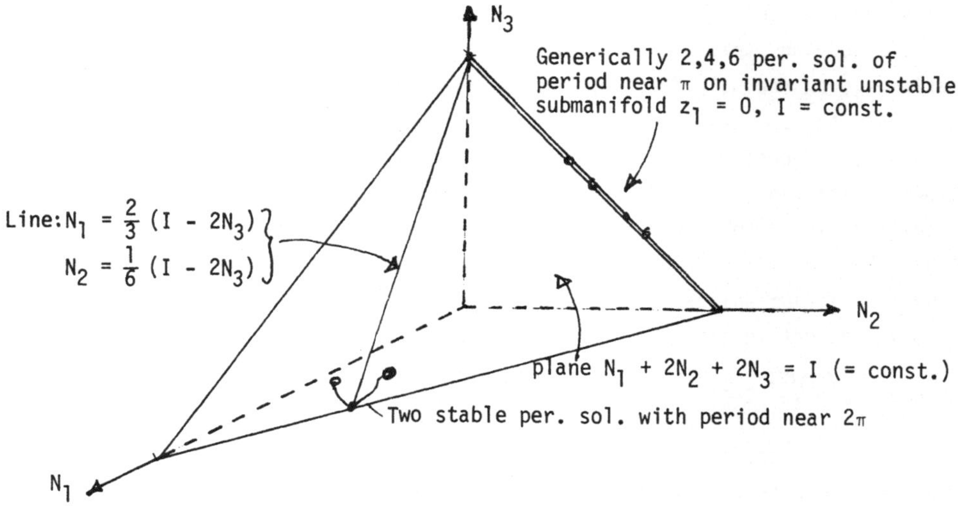

Fig. 8. Periodic solutions in the 1:2:2 resonance.

REFERENCES

[1] V.I. Arnold, Mathematical Methods of Classical Mechanics (Springer, New York, 1978), App. 8.

[2] J.J. Duistermaat, "On periodic solutions near equilibrium points of conservative systems," Arch. Rat. Mech. Anal. 45 143-160 (1972).

[3] D. Schmidt and D. Sweet, "A unifying theory in determining periodic families for Hamiltonian systems of resonance," J. Diff. Eq. 14, 597-609 (1978).

[4] M. Braun, "On the applicability of the third integral of motion," J. Diff. Eq. 13, 300-318 (1973).

[5] M. Kummer, "On resonant nonlinearly coupled oscillators with two equal frequencies", Comm. Math. Phys. 48, 53-79 (1976).

[6] M. Kummer, "On resonant classical Hamiltonians with two equal frequencies, "Comm. Math. Phys. 58, 85-112 (1978).

[7] R. Churchill, M. Kummer, and D. Rod, "On averaging, reduction, and symmetry in Hamiltonian systems, "J. Diff. Eq. 49, 359-414 (1983).

[8] M. Kummer, "On the stability of Hill's solutions of the plane restricted three body problem," Am. J. of Math. 101, 1333-1354 1979).

[9] M. Kummer, "A group theoretical approach to a certain class of perturbations of the Kepler problem," Arch. Rat. Mech. Anal. (to be published).

[10] E. Van der Aa and F. Verhulst, "Asymptotic integrability and periodic solutions of a Hamiltonian system in 1:2:2 resonance." Preprint, University of Utrecht, 1982.

[11] R. Churchill, G. Pecelli, and D. Rod," A survey of the Henon Heiles Hamiltonian with applications to related examples", Lecture Notes in Physics 93, 76-136 (1979).

[12] J. Moser, Lectures on Hamiltonian systems, Mem. Amer. Math. Soc. 81, 1-57 (1968).

[13] M. Kummer, "How to avoid "secular" terms in classical and quantum mechanics," Nuovo Cimento 1B, 123-148 (1971).

[14] M. Kummer, "On the three-dimensional lunar problem and other perturbation problems of the Kepler problem", J. of Analysis and Appl. 93, 142-194 (1983).

[15] J.J. Duistermaat, "Periodic solutions near equilibrium points of Hamiltonian systems," Analytical and numerical approaches to asymptotic problems in analysis 47, 27-33 (1981).

Lecture 2: REALIZATIONS OF THE REDUCED PHASE SPACE OF A HAMILTONIAN SYSTEM WITH SYMMETRY.

Martin Kummer
Department of Mathematics
University of Toledo
Toledo, Ohio 43606

1. REALIZATION OF THE REDUCED PHASE SPACE AS COADJOINT ORBIT

In connection with the 1:1 resonance we considered the Hopf map π: $\mathbb{C}^2 \setminus \{0\} \to \mathbb{R}^3 \setminus \{0\}$. As in Section 2 of my Lecture 1 we write it in nonnormalized form: $\pi(z) = M$, where $M_+ := M_1 + i M_2 = z_1\bar{z}_2$ and $M_3 = \frac{1}{2}(N_1 - N_2)$ $(N_\nu = |z_\nu|^2;\ \nu = 1, 2)$. Using the vector of Pauli-matrices:

$$\vec{\sigma} = \left[\begin{pmatrix} 0 & 1 \\ 1 & 0 \end{pmatrix}, \begin{pmatrix} 0 & i \\ -i & 0 \end{pmatrix}, \begin{pmatrix} 1 & 0 \\ 0 & -1 \end{pmatrix}\right] \tag{1}$$

and the standard inner product in $\mathbb{C}^2$

$$< z,w > \ = \ \bar{z}_1 w_1 + \bar{z}_2 w_2$$

we can also write it in the form

$$\pi(z) = \vec{M} = \frac{1}{2} < z, \vec{\sigma} z >. \tag{2}$$

As already pointed out in Lecture 1, π reduces out the circle action (= action of U(1)) : $(z_1, z_2) \to (\kappa z_1, \kappa z_2)$ $(|\kappa| = 1)$ which in turn agrees with the flow of the Hamiltonian $J := \frac{1}{2} <z, z>$. Restricting the map π to the 3-sphere $S^3 = J^{-1}(1)$ shows that S^3 is a circle bundle over the unit 2-sphere S^2 (compare Fig. 2 of Lecture 1). One can show that π is symplectic in the sense that if ω is an appropriate multiple of the area element on S^2 (see (6)) and if $d\Theta$ with $\Theta = \text{Im} < z, dz >$ is the canonical 2-form on $P := \mathbb{C}^2 \setminus \{0\}$, then we have

$$\pi^* \omega = i^* d\Theta. \tag{3}$$

Here $i: S^3 \to \mathbb{C}^2$ is the inclusion. In the language of Marsden-Weinstein (see [1], [2], [3] for a general introduction and [4] for a lucid discussion of nontrivial examples) this means that S^2 is the reduced phase space $J^{-1}(1)/U(1)$, where $J: \mathbb{C}^2 \setminus \{0\} \to \mathbb{R}$ is a moment map associated with the action of U(1) on $\mathbb{C}^2 \setminus \{0\}$.

We digress briefly to recall the general language used in connection with the reduced phase space of a Hamiltonian system. Let $\{P, \omega\}$ be a smooth symplectic manifold (ω = non-degenerate closed 2-form on even dimensional manifold P) on which the Lie group G acts via an action Φ that leaves ω invariant. This means that we are given a smooth map $\Phi : G \times P \to P$ such that for all $g, h \in G$, $x \in P$ the relations $\Phi(g, \Phi(h,x)) = \Phi(gh, x)$, $\Phi(e, x) = x$ and $\Phi_g^* \omega = \omega$ hold. Here $e \in G$ is the unit element of G and $\Phi_g : P \to P$ is defined as a "partial map" : $\Phi_g(x) = \Phi(g,x)$. Obviously, $g \to \Phi_g$ is a homomorpnism of G into the group of diffeomorphisms over P. For $x \in P$ we also define the partial map $\Phi^x : G \to P$ by holding x constant, i.e. $\Phi^x(g) = \Phi(g,x)$. This map is used to associate with each element X of the Lie algebra $G = T_e G$ (= tangent space to the unit element) a vector field X_P on P, called the fundamental vector field associated with $X \in G$ by setting

$$X_{P,x} = d\,\Phi^x\,(e)\,X \tag{4}$$

($d\,\Phi^x\,(e)$ = differential of Φ^x evaluated at e). We say that the action Φ of G on P has moment map $\Psi : P \to G^*$ (= dual of G) if every fundamental vector field X_P derives from a Hamiltonian in the sense that $\omega(X_P,\ ) = -d < \Psi(\ ), X >$. (Here $< , >$ denotes the pairing between G and G^*). In particular, if $\omega = d\Theta$, and the action Φ is exact symplectic, i.e. $\Phi_g^* \Theta = \Theta$, the action has moment map $\Psi : P \to G^*$, defined by $<\Psi(\ ),X> = \Theta(X_P)$.

We illustrate these concepts by showing that the Hopf map π is itself the moment map of an action of the Lie group SU(2) on $\mathbb{C}^2 \setminus \{0\}$. Indeed, it is well known from the theory of spin in quantum mechanics that each element of SU(2) has a representation $U = \exp [\ i \frac{s}{2} (\vec{\sigma} \cdot \vec{a})]$, where $s \in \mathbb{R}$, $|\vec{a}| = 1$. This representation allows an identification of the Lie algebra su(2) with $\mathbb{R}^3$ (the Lie bracket being the cross product (!)). The fundamental vector field $(\vec{a})_P$ on P: = $\mathbb{C}^2 \setminus \{0\}$ corresponding to $\vec{a} \in \mathbb{R}^3$ is $(\vec{a})_{P,z} = \frac{i}{2} (\vec{\sigma} \cdot \vec{a})z$. It derives from the Hamiltonian

$$\Theta((\vec{a})_P) = \frac{1}{2} \ <z, \sigma z> \ \cdot \ \vec{a} \ = \pi(z) \cdot \vec{a} \ . \tag{5}$$

Here, we recall that $\theta = \mathrm{Im} < z, dz >$. Also we have identified $(su(2))^*$ with $\mathbb{R}^3$ as well, and the pairing between su(2) and $(su(2))^*$ becomes the usual dot product in $\mathbb{R}^3$. Formula (5) explicitly exhibits the Hopf map in the disguise of a moment map of the canonical action of SU(2) on $\mathbb{C}^2 \setminus \{0\}$.

In general the moment map associated with an exact symplectic action Φ of a group G on a space $\{P, d\Theta\}$ is equivariant w.r. to the action Φ on P and the action $Ad^\#$ on G^*. Here $Ad^\#$ is defined via $Ad^\#_g := Ad^*_{g^{-1}} := [d\, I_{g^{-1}}\, (e)]^*$ $(g \in G)$, where in turn $I_g : G \to G$ is defined as follows: $I_g(h) = g\, h\, g^{-1}$ $(g, h \in G)$.

In our specific example the equivariance of π is illustrated by the formula $\pi\,(\exp\ \ i \frac{s}{2} (\vec{\sigma} \cdot \vec{a})\ \ ) = R_{\vec{a}}\,(s)\,\pi(z)$, where $R_{\vec{a}}\,(s)$ is a rotation about $\vec{a}$ through s. Again this formula follows from well known facts in the theory of spin in quantum mechanics. In our context it expresses the equivariance of π. The fact that SU(2) acts on $su(2)^*$ via rotations is a consequence of our identification of $(su(2))^*$ with $\mathbb{R}^3$. Accordingly, the orbits of this action (the so called coadjoint orbits) are concentric 2-spheres. According to the general theory of Kostant-Kirillov-Souriau these orbits carry a natural symplectic structure which in the present example is given by

$$\omega_{\vec{M}}(\vec{X}_{\vec{M}}, \vec{Y}_{\vec{M}}) = -J^{-2}\vec{M}\cdot(\vec{X}_{\vec{M}} \times \vec{Y}_{\vec{M}});\ \vec{X}_{\vec{M}}\ \vec{Y}_{\vec{M}} \text{ tang. to } S^2: = \{\vec{M}: |\vec{M}| = J\}. \tag{6}$$

For J = 1 this is precisely the symplectic structure which enters [3]. In fact, formula [11] of Lecture 1 which in the non-normalized form reads

$$\dot{\vec{M}} = \nabla_{\vec{M}} K \times \vec{M} \tag{7}$$

can be justified, using (6) (and (3)). Namely for $\vec{Y}_{\vec{M}}$ tangential to the sphere, we find (dropping subscripts $\vec{M}$):

$$\omega(\nabla K \times \vec{M}, \vec{Y}) = -J^{-2}\vec{M}\cdot[(\nabla K \times M) \times \vec{Y}]$$

$$= -J^{-2}\vec{M}\ \ [(\nabla K \cdot \vec{Y})\vec{M} - (\vec{M}\cdot\vec{Y})\ \nabla K] = -(\nabla K \cdot \vec{Y}).$$

This relation shows that the vector field $\nabla K \times \vec{M}$ derives from the Hamiltonian K by means of the symplectic form ω. (A less sophisticated way to derive the equations (7) is presented in [5] and [9] of Lecture 1).

Summarizing, we note that in our specific example we have realized the reduced phase space $J^{-1}(1)/U(1)$ as a coadjoint orbit of SU(2). In order to formulate a generalization of the situation of the example we shall use the symbol Ψ_G to denote a moment map of an action of a group G. Employing this notation in our example we have: $\pi = \Psi_E$ and $J = \Psi_F$ with E = SU(2), F = U(1). Indeed, G = E x F acts on $\mathbb{C}^2 \setminus \{0\}$ and π, J are components of the moment of this action. We also note that the map Ψ_E in the diagram:

$$S^3 = \Psi_F^{-1}(1) \xrightarrow[\Psi_E]{} \mathcal{O}: = S^2 \subset su(2)^*: = \mathbb{R}^3$$

drops to the quotient to define a symplectomorphism (e.g., a canonical diffeomorphism between $\Psi_F^{-1}(1)/U(1)$ and S^2). It is therefore plausible that the following Theorem gives the correct generalization of the situation at hand:

<u>Theorem</u>: Let $\{P, \omega\}$ be a smooth symplectic manifold on which the Lie group G = E x F E, F = connected Lie groups) acts by means of an action with equivariant moment map

$$\Psi_G = \Psi_E \times \Psi_F : P \to G^* = E^* \oplus F^* .$$

Assume $\mu_0 \in F^*$ is a regular value of Ψ_F with isotropy group $F_o \subset F$ (or $E \times F_o \subset G$) and assume that $E \times F_0$ acts transitively on $\Psi_F^{-1}(\mu_0)$. Then the image of $\Psi_F^{-1}(\mu_0)$ under the map Ψ_E is an orbit of the action of E in E^*. In fact, the map

$\Psi_E : \Psi_F^{-1}(\mu_0) \to \mathcal{O} \subset E^*$ drops to the quotient to define a symplectic covering map

$\Psi_E : \Psi_F^{-1}(\mu_0)/F_0 \longrightarrow \mathcal{O} \subset E^*$. In particular, if $\mathcal{O}$ is simply connected Ψ_E is a symplectomorphism.

A proof of this theorem follows from the general theory of homogeneous symplectic spaces [5].

As a second application consider the situation E = SU(2,2), F = U(1) with symplectic action on $\mathbb{C}^4\setminus\{0\}$ (= parametrized by $\eta, \zeta \in \mathbb{C}^2$), where the canonical 1-form is given by $\Theta = \mathrm{Im} < \eta, d\eta > - \mathrm{Im} < \zeta, d\zeta >$. (Here $U \in SU(2,2)$ acts by sending $\binom{\eta}{\zeta}$ to $U\binom{\eta}{\zeta}$, and $\kappa \in \mathbb{C}$ with $|\kappa| = 1$ acts by sending (η, ζ) to $(\kappa\eta, \kappa\zeta)$). Identifying $(su(2,2))^*$ with $\mathbb{R}^{15}$ we find $\Psi_E = (J, \vec{L}, \vec{A}, Q, P)$, $\Psi_F = I$, where $\binom{J}{I} = \frac{1}{2}(< \eta, \eta> \pm < \zeta, \zeta >)$, $Q + iP = (- i < \eta, \zeta >, < \eta, \vec{\sigma}\zeta >)$, $\binom{\vec{L}}{\vec{A}} = \vec{M} \pm \vec{N}$ with $M = \frac{1}{2} < \eta, \vec{\sigma}\eta >$, $\vec{N} = -\frac{1}{2} < \zeta, \vec{\sigma}\zeta >$. Reducing out the circle flow on the null-quadric $I^{-1}(0)$ our theorem realizes the reduced space $I^{-1}(0)/U(1)$ as the image of Ψ_E in $\mathbb{R}^{15}$ which simultaneously is an orbit of E. Using the explicit form of our map Ψ_E it is not difficult to show that this orbit is symplectomorphic to T^+S^3, i.e. to the cotangent bundle of the 3-sphere with its 0-section deleted. Since T^+S^3 and $I^{-1}(0)/U(1)$ are the phase spaces on which the regularized Kepler problem lives, depending on whether one uses Moser's [6] or Kustaanheimo-Stiefel's [7],[8], regularization, the foregoing constructions can be used to relate the two regularizations to each other [9].

2. THE REALIZATION OF THE REDUCED PHASE SPACE AS A COTANGENT BUNDLE

Again our point of departure is the Hopf map π: M: = $\mathbb{C}^2\setminus\{0\} \to$ N: = $R^3\setminus\{0\}$, defined by $\pi(z) = \frac{1}{2} < z, \vec{\sigma} z >$. We recall: π represents $\mathbb{C}^2\setminus\{0\}$ as a circle bundle over $\mathbb{R}^3\setminus\{0\}$, or stated otherwise, $\mathbb{R}^3\setminus\{0\}$ is the orbit space with respect to the action ϕ of U(1) on M, where $\phi(\kappa, z) = \kappa z$, $|\kappa| = 1$, $z \in M$. A lift Φ of this action to the contangent bundle P: = T*M = M x $\mathbb{C}$ is given by $\Phi(\kappa; (z,w)) = (\kappa z, \kappa w)$. Since this action leaves the canonical 1-form on P, namely $\Theta = \mathrm{Im} < w, dz >$ invariant, it has a moment map, given by the expression $\Psi(z, w) = \Theta(X_P) = \mathrm{Re} < w, z >$. Here X_P is the fundamental vector field, associated with the circle action:

$$X_{P,(z,w)} = (iz, iw).$$

Our goal is to find a realization of the reduced phase space $\Psi^{-1}(\mu)/U(1)$ for $\mu \in \mathbb{R}$. Such a realization is of crucial importance in the study of Hamiltonians of type:

$$H = \frac{1}{2} < w,w > + W(z), \ z \in \mathbb{C}^2\setminus\{0\}, \tag{8}$$

where the potential function $W(z)$ is assumed to be $U(1)$-invariant. For example the Hamiltonian of the planar 3-body problem in Jacobi coordinates is of this form.

It turns out that the theory of principal G-bundles with connections (see e.g. [10]) is particularily well adapted to the problem at hand. Intuitively, a principal G-bundle $\pi: M \to N$ is a fiber bundle whose fibers are copies of G (i.e. M can be thought of being made up of "strips" $U \times G$ where U runs through an open cover of N, the strips being glued together in a smooth way). The action ϕ of G on M consists of (left) translation along the fibers. ϕ has an exact symplectic lift Φ to $P := T^*M$ with moment

$$\Psi(\alpha) = [d\,\phi^x\,(e)]^*\,\alpha, \quad \alpha \in T_x^*\,M \tag{9}$$

Here ϕ^x for $x \in M$ is again the partial map $\phi^x: G \to M$ defined by $\phi^x(g) = \phi(g,x)$ so that the fundamental vector field X_M corresponding to $X \in \mathcal{G}$ is $X_M = d\,\phi^x(e)\,X$. Formula (9) then is a consequence of the following calculation: $< \Psi(\alpha), X> = \Theta_\alpha(X_P) = < \alpha, X_M > = < \alpha, d\,\phi^x(e)\,X > = < [d\,\phi^x(e)]^*\,\alpha, X >$. Since $\mathrm{Im}\, d\,\phi^x(e) = \mathrm{Ker}\, d\,\pi(X)$ we have $\mathrm{Im}\, d\,\pi\,(x)^* = \mathrm{Ker}\,(d\,\phi^x(e))^*$ and since $d\,\pi\,(x)$ is surjective $(d\,\pi\,(x))^*$ is injective. Hence there exists a map $\tilde{\pi}_0: \Psi^{-1}(0) \to T^*N$ which inverts $(d\,\pi(x))^*$ on its image for each x. The crucial fact is now that $\tilde{\pi}_0: \Psi^{-1}(0) \to T^*N$ is again a principal G-bundle and that moreover $\tilde{\pi}_0$ is symplectic in the sense that $\pi_0^*\,\Theta_N = \Theta_M|_{\Psi^{-1}(0)}$. ($\Theta_N$, Θ_M = canonical 1-forms of T^*N, T^*M). In particular, T^*N with its canonical 2-form $d\,\Theta_N$, is a realization of the reduced phase space $\Psi^{-1}(0)/G$. In order to get a corresponding result for $\mu \in \mathcal{G}^* \setminus \{0\}$ we introduce the concept of a connection ω on a principal G-bundle: ω is a smooth map $TM \to \mathcal{G}$ which is equivariant w.r. to the actions $d\phi$ and Ad of G on TM and on $\mathcal{G}$, respectively, and which moreover has the property that its restriction ω_x to T_xM is a left inverse of $d\,\phi^x\,(e)$: $\omega_x \cdot d\,\phi^x\,(e) = \mathrm{id}_{\mathcal{G}}$. By means of the connection we define the smooth retraction: $\rho_\omega: T^*M \to \Psi^{-1}(0)$ by setting $\rho_\omega(\alpha) = \alpha - \omega_x^*\,\Psi(\alpha)$ for $\alpha \in T_x^*M$. Its restriction $\rho_{\omega,\mu}$ to $\Psi^{-1}(\mu)$ defines a diffeomorphism $\Psi^{-1}(\mu) \to \Psi^{-1}(0)$. Hence, if for simplicity we assume that $\mu \in \mathcal{G}^*$ is G-invariant, the restriction $\tilde{\pi}_{\omega,\mu}$ of the canonical lift $\tilde{\pi}_\omega = \pi_0 \circ \rho_\omega$ of π to $\Psi^{-1}(\mu)$ defines a principal G-bundle $\tilde{\pi}_{\omega,\mu}: \Psi^{-1}(\mu) \to T^*N$ so that in particular T^*N is a realization of $\Psi^{-1}(\mu)/G$. However, the symplectic structure on T^*N is no longer the canonical one. In fact, one shows

$$\pi_{\omega,\mu}^*(d\,\Theta_N + \tau_N^*\,\Omega_\mu) = d\,\Theta_M|_{\Psi^{-1}(\mu)} \tag{10}$$

so that the canonical 2-form $d\,\Theta_N$ on T^*N has to be modified by the "magnetic" term $\tau_N^* \Omega_\mu$. Here τ_N is the canonical projection τ_N: $T^*N \to N$ and Ω_μ is the μ-component of the curvature of the connection ω, viewed as a two form on N (i.e. if ω_μ is the one form on M defined by the association

$$x \to \omega_x^* \mu \text{ , then } d\,\omega_\mu = \pi^* \Omega_\mu \text{ .)}$$

According to a theorem of A. Weil the cohomology class of Ω_μ turns out to be independent of the connection ω (i.e., an intrinsic property of the principal G-bundle) and its non-vanishing constitutes the obstruction to $\{T^*N,\, d\,\Theta_N\}$ being a realization of $\Psi^{-1}(\mu)/G$.

Summarizing, we see that if we start from a principal G-bundle π: $M \to N$ and $\mu \in G^*$ is G-invariant then each connection ω on that bundle defines a lift $\tilde{\pi}_\omega$: $T^*M \to T^*N$ of π with the property that its restriction $\tilde{\pi}_{\omega,\mu}$ to $\Psi^{-1}(\mu)$ defines a symplectomorphism between $\Psi^{-1}(\mu)/G$ and T^*N, provided T^*N is equipped with the symplectic structure $d\,\Theta_N + \tau_N^* \Omega_\mu$. The relationship between ω and the lift $\tilde{\pi}_\omega$ can be expressed by means of the formula

$$< \Psi(\alpha), \omega_x(\zeta) > + \quad < \tilde{\pi}_\omega(\alpha),\; d\,\pi(\zeta) > \; = \; < \alpha,\; \zeta >, \tag{11}$$

which is valid for all $\alpha \in T_x^* M$, $\zeta \in T_x M$.

We illustrate this construction by taking up the example in the Introduction of my first lecture. Starting with the relation $tr(\sigma_\nu \sigma_\mu) = 2\,\delta_{\mu\nu}$ $(\mu,\nu = 0,1,2,3)$ where $\sigma_\nu = (\sigma 0, \vec{\sigma})$ (σ_0 = unit matrix) we find $\sum_{\nu=0}^{3} \sigma_\nu^{\alpha\beta}\, \sigma_\nu^{\gamma\epsilon} = 2\delta^{\alpha\epsilon}\, \delta^{\beta\gamma}$.

Multiplying this relation by u_α, z_β, w_ϵ and enforcing Einstein's summation convention we obtain

$$< u, \sigma_\nu z > \;\; \sigma_\nu w = 2 < u, w > z$$

$$\text{or} \;\; < u, \vec{\sigma}\; z > \; \vec{\sigma}\; w = 2 < u, w > z - < u, z > w. \tag{12}$$

After some algebra the relation $\mathrm{Re} < z, w > \mathrm{Im} < z, u > + \mathrm{Im} < w, \vec{\sigma} z > \mathrm{Re} < z, \vec{\sigma} u > =$ $\mathrm{Im} < w, u > < z, z >$ results. Substituting in it $u = dz$, it can be written in the form:

$$\Psi(z)\; \mathrm{Im} < z, dz > + \; \mathrm{Im} < w, \vec{\sigma}\; z > d\,\pi(z)\; u = \; < z, z > \Theta_{Mz}. \tag{13}$$

Comparison of this relation with (11) shows that the connection $\omega_z = <z, z>^{-1}\, \mathrm{Im} < z,dz >$

on our S^1-bundle $\pi: \mathbb{C}^2 \setminus \{0\} \to \mathbb{R}^3 \setminus \{0\}$ leads to the following canonical lift $\tilde{\pi}$ of π
$\tilde{\pi}$: $\vec{x} = \frac{1}{2} < z, \vec{\sigma} z >$, $y = < z, z >^{-1}$ Im $< w, \vec{\sigma} z >$. For $(z,w) \in \Psi^{-1}(\mu)$ relation (13) becomes

$$\mu \tau_M^* \omega + \tilde{\pi}_\mu^* \Theta_N = \Theta_M|_{\Psi^{-1}(\mu)}, \tag{14}$$

where $\Theta_N = \vec{y} \cdot d\vec{x}$ is the canonical 1-form on $\mathbb{R}^2 \setminus \{0\}$ and $\tilde{\pi}_\mu = \tilde{\pi} \mid_{\Psi^{-1}(\mu)}$.

It is straightforward to verify that $d\omega = \pi^* \Omega$, $\Omega = * \vec{B} \cdot d\vec{x}$, where $\vec{B} = - \frac{1}{2} \nabla (r^{-1}) = \frac{1}{2} r^{-3} \vec{x}$ is the magnetic field of a magnetic monopole with magnetic charge 2π sitting at the origin of $\mathbb{R}^3$. (The star operator is as usually defined via the relations $* dx_1 = dx_2 \wedge dx_3$ and cyclically.) Exterior differentiation of the relation (14) yields precisely (10) with $\Omega_\mu = \mu\Omega$. Moreover, using the relation

$$< z,z > \vec{y} \cdot \vec{y} = < w,w > - (\text{Re} < z,w >)^2,$$

which also follows from (12), we obtain for the reduced Hamiltonian (8) the expression

$$H = r(\frac{1}{2} \vec{y}^{\,2} + \frac{1}{8} \mu^2 r^{-2} + \tilde{W} (\vec{x})), \tag{15}$$

where $\tilde{W} (\vec{x}) = W(\pi^{-1}(\vec{x}))$. The change of time variable $dt = r ds$ finally brings the Hamiltonian into the form

$$H = \frac{1}{2} \vec{y}^{\,2} + \frac{1}{8} \mu^2 r^{-2} + \tilde{W} (\vec{x}),$$

with corresponding differential equations

$$\frac{d\vec{x}}{ds} = \vec{y} , \quad \frac{d\vec{y}}{ds} = - \nabla_{\vec{x}} H + \mu (\vec{B} \times \vec{y}).$$

Thus, we see that the reduced Hamiltonian describes a particle of mass 1 in $\mathbb{R}^3 \setminus \{0\}$ that is not only subjected to the "amended" potential $\frac{1}{8} \mu^2 r^{-1} + \tilde{W} (\vec{x})$ but also to the magnetic field of a magnetic monopole sitting at the origin of $\mathbb{R}^3$. Here the angular momentum μ plays the role of the electric charge of the particle.

In Section 2 of this Lecture we have reviewed that part of [11] which deals with the case $G_\mu = G$ (G_μ = isotropy group of μ). If G_μ is properly contained in G and if $\pi: M \to N$ is a principal G_μ-bundle then according to Theorem 3 of [11] the reduced phase space $\Psi^{-1}(\mu)/G_\mu$ is realized as a symplectic submanifold of T^*N. Here T^*N is again equipped with a symplectic structure that differs from the canonical one by an appropriate "magnetic term".

The subject of realizations of the reduced phase space of Marsden-Weinstein has recently been taken up in great generality by Montgomery [12], who considers also situations in which the group action is nonfree and $\tau_M(\Psi^{-1}(\mu))$ is allowed to be a proper submanifold of M. (Note that the assumption $\tau_M(\Psi^{-1}(\mu)) = M$ is implicit in previous work on the reduced space).

REFERENCES

[1] J. Marsden and A. Weinstein, "Reduction of symplectic manifolds with symmetry," Rep. Math. Phys. 5, 121-130 (1974).

[2] R. Abraham and J. Marsden, Foundations of Mechanics (Benjamin/Commings, Reading, Mass. 1978), p. 253-369.

[3] V. I. Arnold, Mathematical Methods of Classical Mechanics (Springer, New York, 1978), Appendix.

[4] J. Moser, "Various aspects of integrable Hamiltonian systems," Prog. Math. 8, 233-289 (1980).

[5] A. A. Kirillov, Elements of the theory of representations (Springer, New York, 1976), p. 234.

[6] J. Moser, "Regularization of Kepler's problem and the averaging method on a manifold," Comm. Pure Appl. Math. 23, 609-636 (1970).

[7] P. Kustaanneimo and E. Stiefel, "Perturbation theory of Kepler motion based on spinor regularization," J. Reine Angew. Math. 218, 204-219 (1965).

[8] E. L. Stiefel and G. Scheifele, Linear and Regular Celestial Mechanics (Springer, New York, 1971).

[9] M. Kummer, "On the regularization of the Kepler problem", Comm. Math. Phys. 84, 133-152 (1982) (see also ref. [9] of Lecture 1).

[10] W. Greub, S. Halperin and R. Vanstone, Connections, Curvature and Cohomology, Vols. I, II (Academic, New York 1973).

[11] M. Kummer, "On the construction of the reduced phase space of a Hamiltonian system with symmetry," Indiana Univ. Math. 30, 281-290 (1981) (see also references in this paper).

[12] R. Montgomery, "The structure of reduced cotangent phase spaces for nonfree group actions," Preprint, University of California, Berkeley, 1983.

KAM TODAY

Richard Barrar
Mathematics Department
University of Oregon
Eugene, Oregon 97403

In a four page announcement in 1954 [10] Kolmogorov stated the following amazing theorem, which solved a very old and puzzling problem:

<u>Theorem 1</u>

Let

$$H(p,q) = H_0(p) + H_1(p,q), \quad \text{with } p = (p_1,\ldots,p_n), \quad q = (q_1,\ldots,q_n). \tag{1}$$

Further, let

$$\lambda_i = (\partial H_0/\partial p_i)|_{p_j} = (p_j)_0 \quad \text{(a constant)},$$
$$|\Sigma\, \lambda_i\, h_i| \geq \frac{\varepsilon}{(\|h\|)^s}, \text{ for integer } h_i, \text{ with } \|h\| = \sum_{i=1}^{n} |h_i| > 0,$$
$$s \text{ an integer } \geq n.$$

Then, if $H_1(p,q)$ is sufficiently small and if the Hessian of $H_0(p)$ does not vanish at $p_j = (p_j)_0$, there exists a canonical change of variable $(p,q) \leftrightarrow (P,Q)$

$$p_i = P_i + g_i(Q) + \Sigma P_j g_{ij}(Q), \quad q_i = f_i(Q) + Q_i, \tag{2}$$

with f_i, g_i, g_{ij} periodic of period 2π in the Q's, such that for $P_i = (P_i)_0$ (a constant), $Q_i = \lambda_i(t - \tau_i)$, (2) is a conditionally periodic solution of equation (1).

Kolmogorov never gave a proof of the theorem. Barrar [3] later gave a proof of Kolmogorov's result. Kolmogorov had assumed the Hamiltonian was analytic. However, Moser [11] showed that this restriction was not necessary. Arnold [1] generalized Kolmogorov's result to the case where $H_0(p)$ does not depend on all the variables p and Barrar [4] also treated this case.

If one restricts the problem to the surface of section $q_1 = 0$ on the energy surface $H = h$ and observes two successive crossings of the surface of section by orbits near a periodic orbit, then this theorem becomes a theorem about mapping a surface onto itself. For the case $n = 2$, this results in Moser's theorem on the invariant curves of area preserving mappings [12]. However, Kolmogorov's and Moser's formulation of this problem are not equivalent. For a discussion of this point see Arnold and Avez [2, Section 21.E].

These results, under the acronym KAM theory, have become a standard building block in treating mechanical problems.

The one unsettling part of the theory that remains is the phrase "for sufficiently small $H_1(p,q)$" in the statement of the theorem. The mathematical estimates one gets in the proofs of the theorems are very small for the magnitude of $H_1(p,q)$. The estimates one gets from numerical experiments are much larger (see Hénon and Heiles [9]). It would be very useful to have more precise and realistic estimates of the validity of the KAM procedure than are available at the present time.

In the present paper, we wish to initiate a program to obtain this extension. We will restrict attention to Hamiltonians of the form (1). However, it is clear that by using the techniques introduced by Arnold and Moser that our methods will extend to Hamiltonians where $H_0(p)$ does not depend on all the p_i's and to nonanalytic Hamiltonians.

As an analogy for the procedure we are about to suggest, let us briefly look at the way Riesz and Sz-Nagy [15, p. 143] treat linear integral equations.

Given a linear integral equation

$$f(x) + \int_a^b K(x,y)f(y) = g(x), \tag{3}$$

where $[a,b]$ is a bounded interval on the real line and $f(x)$ and $g(x) \in L_2[a,b]$, Riesz and Sz.-Nagy point out there are two easy cases where (3) can be solved, and the general case can be broken up into these easy cases.

First they show that if

$$\int_a^b\int_a^b |K(x,y)|^2 \, dxdy < 1, \tag{4}$$

then (3) can be solved by a Neumann series.

A second easy case is when the kernel $K(x,y)$ is of finite rank, i.e.,

$$K(x,y) = \sum_{i=1}^{r} \phi_i(x)\, \overline{\psi}_i(y), \tag{5}$$

for then substitution of (5) into (3) yields a set of linear algebraic equations.

Finally, they treat the general case of a linear integral equation by approximating a given kernel arbitrarily closely by a kernel of finite rank (which can always be done), and then solving the resulting equation by a Neumann series.

In the case of the KAM theory, the Kolmogorov procedure corresponds to the Neumann series. To treat the general case, we need some analogue of the kernel of finite rank in the theory of integral equations, i.e., if we have a general Hamiltonian $H(p,q) = H_0(p) + H_1(p,q)$ we need a way of reducing the magnitude of $H_1(p,q)$ before applying the Kolmogorov procedure.

I would like to suggest in this article (and will further develop in future papers) that this second easy case be the method of Delaunay.

Delaunay [5] developed his method to treat the moon's orbit. He devoted twenty years of work to do the actual numerical computation. His solution as Smart [16, p. 176] says "is the most complete literal solution ever achieved in this complicated problem."

Consider

$$H_1(p,q) = \Sigma A_\alpha(p)\, e^{i\alpha q},$$

$$\alpha = (\alpha_1,\ldots,\alpha_n) \quad \text{all } \alpha_i \text{ integers.}$$

The method of Delaunay is based on using canonical transformations to eliminate the larger $A_\alpha(p)e^{i\alpha q}$ from this Hamiltonian. In the lunar theory, the p_i's represent quantities such as eccentricities and inclinations of orbits. Delaunay eliminated 320 periodic terms, and in each term he eliminated all of the p_i's at least to the seventh order and several to the eighth order.

So although Poincaré showed that this type of procedure was only asymptotically convergent, we can see that in this practical case it worked to give very good numerical answers.

Further, Poincaré devoted the whole Chapter XIII of [14] to discuss the principle of the method of Delaunay. In summary, his conclusion is the following. If the perturbation is developed in powers of μ, then the general term in the answer will be of the form $At^m \cos(\nu t + h)$. If α is the degree of μ in A, and if m' is the sum of the exponents of the small divisors in any term, then Poincaré called the class of this term $\alpha - \frac{1}{2}(m + m')$. He showed (a) $\alpha - \frac{m}{2} - \frac{m'}{2} \geq \frac{1}{2}$, (b) that the method of Delaunay eliminates terms of class $\frac{1}{2}$, and (c) that the terms of low class have the greatest effect on the behavior of the system for long time periods.

In studying the method of Delaunay in any of the standard texts such as Hagihara [8, §9.1, 9.2] or Plummer [13, §144, 145], one notes that they are not mathematically rigorous and that they fail to treat canonical changes of variables $(p,q) \leftrightarrow (P,Q)$ that transform Hamiltonians periodic in $q_1,\ldots,q_n$ into ones that are periodic in $Q_1,\ldots,Q_n$, a condition which is required for our purposes.

We will now show how this can be accomplished.

To begin, we prove a result of Wintner [17, §181].

Lemma 1: Consider a Hamiltonian

$$H(p,q) = h. \tag{6}$$

If locally $\partial H/\partial p_n \neq 0$ so $p_n = K(p^*,q)$ on the manifold (6) with $p^* = (p_1,\ldots,p_{n-1},h)$, define

$$\tilde{H}(p^*,q) = p_n - K(p^*,q) = 0. \tag{7}$$

Then geometrically the orbits of

$$dq_i/dt = \partial H/\partial p_i, \quad dp_i/dt = -\partial H/\partial q_i, \tag{8}$$

and

$$dq_i/d\bar{t} = \partial\tilde{H}/\partial p_i, \quad dp_i/d\bar{t} = -\partial\tilde{H}/\partial q_i, \tag{9}$$

are the same, with $q_n = \bar{t} + c$.

Proof:

One has $\partial H/\partial p_n \neq 0$, $dq_n/dt \neq 0$ on an orbit of (8) on the manifold (6). So $t = f(q_n)$ on this orbit. Then on this orbit

$$dq_i/dq_n = (dq_i/dt)/(dq_n/dt) = (\partial H/\partial p_i)/(\partial H/\partial p_n)$$
$$= -\partial p_n/\partial p_i = -\partial K/\partial p_i = \partial\tilde{H}/\partial p_i.$$

On the other hand, from (9) $dq_n/d\bar{t} = 1$, or $q_n = \bar{t} + c$. A similar analysis applies to dp_i/dq_n, q.e.d.

Note that in the special case where

$$\tilde{H} = p_n - K(p^*, q_n) = 0,$$

i.e., K does not depend on $q_1, \ldots, q_{n-1}$, the system is integrable since

$$dq_n/d\bar{t} = 1, \quad dp_i/d\bar{t} = -\partial\tilde{H}/\partial q_i = 0, \quad i = 1, \ldots, n-1,$$
$$dq_i/d\bar{t} = \partial\tilde{H}/\partial p_i = -\partial K/\partial p_i \quad i = 1, \ldots, n-1.$$

As an even more special case, consider the Hamiltonian

$$H = \frac{p^2}{2} - \cos q = h,$$

i.e., the motion of a pendulum. Then

$$p = \sqrt{2(h + \cos q)}, \quad \tilde{H} = p - \sqrt{2(h + \cos q)} = 0,$$
$$dq/dt = \partial H/\partial p = p, \text{ so } dt/dq = 1/p = 1/\sqrt{2(h + \cos q)}$$
$$\text{or} \quad t = \int dq/\sqrt{2(h + \cos q)},$$

Hence, p and t are functions of the uniformizing variable q.

In this case, since we want $\frac{\partial H}{\partial p} = p \neq 0$, we restrict to $h > 1$, i.e., a rotating pendulum.

Returning to the method of Delaunay, assume we have a Hamiltonian of the form

$$H = p_n - K_1(p^*, q), \tag{10}$$

where $K_1(p^*,q)$ is periodic of period 2π in each of the q_i. Further, assume that by a linear change in the $q_1, \ldots, q_n$ the term $K_0(p^*, q_n)$, is the largest term in $K_1(p^*,q)$, i.e.,

$$K_1(p^*,q) = \Sigma A_\alpha (p^*) e^{i\alpha q},$$
$$\alpha = (\alpha_1, \ldots, \alpha_n), \text{ all } \alpha_i \text{ integers},$$

and

$$K_0(p^*, q_n) = \Sigma A_m(p^*) e^{imq_n}, \quad A_m(p^*) = A_{(0,\ldots,0,m)}(p^*),$$

is the largest term in $K_1(p^*,q)$.

Theorem 2:

By use of the generating function

$$S(p,q') = -\int_0^{q_n'} K_0(p^*,\xi)d\xi + (q_n'/2\pi)\int_0^{2\pi} K_0(p^*,\xi)d\xi + \sum_{i=1}^{n} p_i q_i' \tag{11}$$

the Hamiltonian (10) is transformed to one of the form

$$\tilde{H} = p_n' - K_0(p^{*\prime},q_n') + K_1(p^{*\prime},q') + (1/2\pi)\int_0^{2\pi} K_0(p^{*\prime},\xi)d\xi \tag{12}$$

with $p^{*\prime} = (p_1',\ldots,p_{n-1}',h)$, $q' = (q_1',\ldots,q_n')$, and is periodic of period 2π in each of the q_i'. Hence, we have eliminated the term K_0 from the Hamiltonian.

Proof:

Using the generating function (11), we find

$$\begin{aligned}
p_i' &= \partial S/\partial q_i' = p_i \qquad i = 1,\ldots,n-1,\\
p_n' &= \partial S/\partial q_n' = -K_0(p^*,q_n') + (1/2\pi)\int_0^{2\pi} K_0(p^*,\xi)d\xi + p_n\,,\\
q_n &= \partial S/\partial p_n = q_n'\,,\\
q_i &= \partial S/\partial p_i = -\int_0^{q_n'} (\partial K_0/\partial p_i)(p^*,\xi)d\xi\\
&\qquad + (q_n'/2\pi)\int_0^{2\pi} (\partial K_0/\partial p_i)(p^*,\xi)d\xi + q_i'\,,
\end{aligned} \tag{13}$$

or

$$q_i = \emptyset_i(p^{*\prime},q_n') + q_i'\,, \qquad i = 1,\ldots,n-1.$$

By construction, $\emptyset_i(p^{*\prime},q_n')$ is periodic of period 2π in q_n'. Hence, substitution of (13) into (10) yields the result.

To iterate this process one difficulty remains. To explain and resolve this difficulty let us write the Hamiltonian (12) as

$$\tilde{H} = p_n' - K_2(p^{*\prime},q') = 0\,. \tag{14}$$

Assume the largest term in K_2 is the one containing multiples of

$$q_n'' = \alpha_1 q_1' \dots + \alpha_n q_n' . \tag{15}$$

To apply Theorem 2 to this case, we would first have to make a canonical change of variables from $(p',q') \to (p'',q'')$ so that (15) held. We can do this by means of the generating function $S_1(p'',q') = \Sigma a_{ij} p_i'' q_j'$ with all a_{ij} integers, and in particular $a_{nj} = \alpha_j$, for then

$$q_i'' = \partial S_1 / \partial p_i'' = \Sigma a_{ij} q_i' , \tag{16}$$

$$p_j' = \partial S_1 / \partial q_j' = \Sigma a_{ij} p_i'' . \tag{17}$$

To invert (16), so that $q_j' = \Sigma\, b_{jh} q_h''$ with b_{jh} integers (so that the Hamiltonian is periodic in the q_i'' with period 2π), we would need $\det|a_{ij}| = 1$. We can always assume this to be true. Thus, substituting (16) and (17) into (14) and then using the implicit function theorem to solve for p_n'', we can assume that (14) is equivalent to

$$p_n'' - K_3(p^{*''},q'') = 0 \tag{18}$$

and the largest term in K_3 is of the form $K_4(p^{*''},q_n'')$. We can now apply Theorem 2 to (14) and keep iterating.

As an example consider the following Hamiltonian, which is similar to one considered by Escande and Doveil in [6] and [7],

$$H = \frac{p_2^2}{2} - M \cos q_2 - N \cos 2(q_1 - q_2) + p_1 = h. \tag{19}$$

When either M or N is zero, this reduces to the pendulum problem. However, if $M > N > 0$ the Hamiltonian is nonintegrable. Solving for p_2, we find

$$p_2 = K_1(p_1,h,q_1,q_2) = \sqrt{2(h + M \cos q_2 + N \cos 2(q_1 - q_2) - p_1} .$$

In this case, $K_0 = \sqrt{2(h + M \cos q_2 - p_1)}$.

So as long as $h + M \cos q_2 - p_1 > 0$, we can apply Theorem 2 to reduce the Hamiltonian to one of the form:

$$\tilde{H} = p_2' + \frac{1}{2\pi}\int_0^{2\pi} K_0(p_1',h,\xi)d\xi - (K_1 - K_0)$$

$$K_1 - K_0 = \Sigma\, A_n(p_1')\, e^{i(2q_1' - 2nq_2')} .$$

Then, if $A_m(p_1')$ is the largest of the $A_n(p_1')$'s in the region of p_1' that we are interested in, we can set $q_2'' = 2q_1' - 2mq_2'$ and proceed as in (16) and (17) to eliminate $A_m(p_1')\, e^{i(2q_1'-2mq_2')}$ from the Hamiltonian. Proceeding in this fashion, we would hope to reduce the Hamiltonian (19) for general M and N to one for which the Kolmogorov-Arnold-Moser theory would apply.

REFERENCES

[1] V. I. Arnold, "Small denominators and problems of stability of motion in classical and celestial mechanics," Russ. Math Surv. 18, 85 - 193 (1963).

[2] V. I. Arnold and A. Avez, Ergodic Problems of Classical Mechanics (Benjamin, 1968).

[3] R. B. Barrar, "Convergence of the von Zeipel procedure," Celestial Mechanics 2, 494 - 504 (1970).

[4] R. B. Barrar, "A Proof of the convergence of the Poincaré - Von Zeipel procedure in celestial mechanics, Am. J. Math. 88, 206 - 220 (1966).

[5] C. E. Delaunay, Mem. Acad. Sci. Paris 28, 29 (entire volumes), 1860 - 1867.

[6] D. F. Escande and F. Doveil, "Renormalization method for the onset of stochasticity in a Hamiltonian system", Phys. Lett. 83A, 307 - 310 (1981).

[7] D. F. Escande and F. Doviel, "Renormalization method for computing the threshold of the large-scale stochastic instability in two degrees of freedom Hamiltonian systems," Stat. Phys. 26, 257 - 283 (1981).

[8] H. Hagihara, Celestial Mechanics, Vol. II, Part 1 (MIT Press, 1972).

[9] M. Hénon and C. Heiles, "The applicability of the third integral of motion: Some numerical experiments," Astron. J. 69, 73 - 79 (1964).

[10] A. N. Kolmogorov, "The conservation of conditionally periodic motions with a small change in the Hamiltonian," Dokl. Akad. Nauk SSSR 98, 527 - 530 (1954).

[11] J. Moser, "A rapidly convergent iteration method and nonlinear differential equations: II, "Ann. Scuola Norm. Sup. Pisa 20, 499 - 533 (1966).

[12] J. Moser, "On invariant curves of area-preserving mappings of an annulus, Nachr. Akad. Wiss. Göttingen, 1 (1962).

[13] H. C. Plummer, An Introductory Treatise on Dynamical Astronomy (Dover, New York, 1960).

[14] H. Poincaré, Leçons de Mécanique Céleste, Vol. 1 (Gauthiers-Villars, Paris, 1905).

[15] F. Riesz and B. Sz-Nagy, Functional Analysis (Frederick Ungar, New York, 1955).

[16] W. M. Smart, Celestial Mechanics (Longmans, Green, New York, 1953).

[17] Aurel Wintner, The Analytical Foundations of Celestial Mechanics (Princeton University Press, 1941).

NOTE ON THE EVOLUTION OF THE LIE-DEPRIT TRANSFORM

Ahmed A. Kamel

Ford Aerospace & Communications Corporation

Palo Alto, CA 94303

1. INTRODUCTION

Deprit's introduction of the Lie transform early in 1969 [8] stirred great interest in the following three areas:

a. Equivalence of Deprit's formulation to that of Hori [14] and von Zeipel (Schniad [33], Kamel [17], Mersman [27], [28], Campbell and Jeffereys [6], Hori [15], Henrard and Roels [12]).

b. Improvement of the computational aspects of Deprit's algorithm (Kamel [17, 18]) and the development of similar recursive algorithms for Hori's Lie Series (Kamel [17], Mersman [27, 28], Campbell and Jeffereys [6]).

c. Generalization of the Lie-Deprit transform and Lie-Hori series to non-Hamiltonian systems, (Kamel [19, 20], Henrard [10, 11], Hori [15]).

This note will be limited to a summary o. the evolution of the Lie-Deprit transform and its equivalence to the well known method of averaging (Bogoliubov ınd Mitropolsky [3]) and the method of multiple scales (Nayfeh [30], Kevorkian and Cole [24]).

2. DEPRIT'S FUNDAMENTAL ALGORITHM

In the spirit of Lie's ideas [25], Deprit introduced the Lie transform $(x, X) \rightarrow (y, Y)$ using the canonical differential equations

$$\frac{dx}{d\epsilon} = W_X(x, X; \epsilon) \tag{1a}$$

$$\frac{dX}{d\epsilon} = -W_x(x, X; \epsilon) \tag{1b}$$

with

$$x = y,\ X = Y, \text{ when } \epsilon = 0.$$

In Eq. (1) the independent variable is the small parameter ϵ and the function W is the generator of the canonical transformation $(x, X) \rightarrow (y, Y)$. The solution of Eq. (1) can be represented by a Taylor series in ϵ. More generally, for any function F, the Taylor series can be written as

$$F(x, X; \epsilon) = \sum_{n=0}^{\infty} \frac{\epsilon^n}{n!} F_n(x, X) = \sum_{k=0}^{\infty} \frac{\epsilon^k}{k!} F^{(k)}(y, Y), \tag{2}$$

with

$$F_n(x, X) = \left[\frac{\partial^n}{\partial \epsilon^n} F(x, X; \epsilon) \right]_{\epsilon=0}, \; n \geq 0, \tag{3a}$$

$$F^{(k)}(y, Y) = \left[\frac{d^k}{d\epsilon^k} F(x, X; \epsilon) \right]_{\epsilon=0}, \; k \geq 0. \tag{3b}$$

The total derivative operator $d/d\epsilon$ is related to the partial derivative operator $\partial/\partial\epsilon$ and it has the following properties

$$\frac{dF}{d\epsilon} = \frac{\partial F}{\partial \epsilon} + F_x \frac{dx}{d\epsilon} + F_X \frac{dX}{d\epsilon} \tag{4a}$$
$$= \frac{\partial F}{\partial \epsilon} + L_W F,$$

$$\frac{d^k F}{d\epsilon^k} = \frac{d}{d\epsilon} \frac{d^{k-1} F}{d\epsilon^{k-1}}, \; k \geq 2, \tag{4b}$$

where L_W is the Lie-Deprit operator. The use of Eqs. (1) and (4a) leads to the usual Poisson bracket

$$L_W F = (F; W) = F_x \cdot W_X - F_X W_x. \tag{5}$$

Deprit selected the generating function W in the form

$$W(x, X; \epsilon) = \sum_{m=0}^{\infty} \frac{\epsilon^m}{m!} W_{m+1}(x, X), \tag{6}$$

and substituted Eqs. (2) and (6) in Eqs. (4a) and (4b) to obtain

$$\frac{d^k}{d\epsilon^k} F = \sum_{n=0}^{\infty} \frac{\epsilon^n}{n!} F_n^{(k)}(x, X), \; k \geq 1, \tag{7}$$

with

$$F_n^{(k)} = F_{n+1}^{(k-1)} + \sum_{m=0}^{n} C_m^n L_{m+1} F_{n-m}^{(k-1)}, \; k \geq 1, \; n \geq 0, \tag{8}$$

where C_m^n is the usual binomial coefficient

$$C_m^n = \frac{n!}{(n-m)! \, m!} \tag{9}$$

and L_p is the Lie-Deprit operator defined by the Poisson bracket of Eq. (5):

$$L_p F = (F;\ W_p),\ \ p \geqslant 1. \tag{10}$$

Note that in view of Eqs. (2), (3), (7), and (8), the Poisson bracket operations can be performed in terms of a dummy set of variables, e.g., $(z,\ Z)$, which are selected at the end of the calculations to be either $(x,\ X)$ or $(y,\ Y)$ as appropriate to calculate the coefficients of ϵ in Eq. (2). These are given by

$$F_n = F_n^{(0)} \text{ and } F^{(k)} = F_0^{(k)}. \tag{11}$$

To show how Eq. (8) is used to generate $F^{(k)}$ from F_n, Deprit used the forward triangle of Fig. 1 similar to Pascal's triangle. The forward flow from left to right leads to the required $F^{(k)}$.

Now, to obtain the inverse transformation $(y,\ Y) \rightarrow (x,\ X)$, or more generally to obtain F_n from $F^{(k)}$, Deprit proposed to perform this inversion using two additional triangles similar to that of Fig. 1 (see Eqs. (32) to (34) of Deprit [8]). The first triangle is used to calculate the inverse generating function V from W of Eq. (6), and the selected triangle is used to calculate F_n from $F^{(k)}$. This procedure for the calculation of the inverse was in marked contrast to Hori's method in which no new calculations are needed for the inverse (Mersman [27]). However, it was found (Kamel [17], [18]) that Deprit's algorithm of Eq. (8) can be simplified such that, like Hori's method, no new calculations are required for the inverse. This is summarized in the next section.

3. SIMPLIFICATION OF DEPRIT'S ALGORITHM

The simplification of Deprit's algorithm was performed in two steps. In the first step, Deprit's Eq. (8) was rearranged to eliminate the need for the calculation of the inverse generating function V. This was achieved by observing that Deprit's triangle can be constructed with backward flow from the right to the left, as shown in Fig. 2, to obtain F_n from $F^{(k)}$. The general backward algorithm was simply obtained by replacing k by $k + 1$ and $n + 1$ by n in Eq. (8) to get

$$F_n^{(k)} = F_{n-1}^{(k+1)} - \sum_{m=0}^{n-1} C_m^{n-1}\, L_{m+1} F_{n-m-1}^{(k)},\ n \geqslant 1,\ k \geqslant 0. \tag{12}$$

The second step in the simplification of Deprit's algorithm was to avoid any new operations for the inverse. This was achieved by observing that successive substitution of Eq. (12) into itself from $n = 1$ up leads to the form

$$F_n^{(k)} = -\sum_{j=0}^{n} C_j^n G_j F^{(k+n-j)}, \; n \geqslant 1, \; k \geqslant 0, \tag{13}$$

where G_j is a linear operator given by Kamel [17]-[19]:

$$G_0 = -1, \tag{14a}$$

$$G_j = -\sum_{m=1}^{j} C_{m-1}^{j-1} L_m G_{j-m}. \tag{14b}$$

The present author (Kamel [20]) later followed a suggestion by Dr. Henrard (1970) to use an operator $N_j = -G_j$ to eliminate the negative signs in Eqs. (13) and (14a). In a later publication, however, Dr. Henrard [11] offered a new derivation for the algorithm given by Eqs. (13) and (14) with $N_j = -G_j$ and referred to it as the algorithm of the inverse. Now, for $k = 0$ and $k = 1$, Eqs. (8), (11), and (13) yield

$$F_n = F^{(n)} - \sum_{j=1}^{n} C_j^n \, G_j F^{(n-j)}, \; n \geqslant 1, \tag{15a}$$

$$F^{(n)} = F_n + \sum_{j=1}^{n-1} \left[C_j^{n-1} G_j \, F^{(n-j)} + C_{j-1}^{n-1} L_j F_{n-j} \right] + L_n F_0, \; n \geqslant 1. \tag{15b}$$

The use of $F_{j,i} = -G_j F^{(i)}$ in Eqs. (14) and (15) leads to the recursive algorithm

$$F_n = F^{(n)} + \sum_{j=1}^{n} C_j^n F_{j,n-j}, \; n \geqslant 1, \tag{16a}$$

$$F^{(n)} = F_n - \sum_{j=1}^{n-1} \left[C_j^{n-1} F_{j,n-j} - C_{j-1}^{n-1} L_j F_{n-j} \right] + L_n F_0, \; n \geqslant 1, \tag{16b}$$

with

$$F_{0,0} = F^{(0)} = F_0 \, , \; F_{0,i} = F^{(i)}, \tag{16c}$$

$$F_{j,i} = -\sum_{m=1}^{j} C_{m-1}^{j-1} L_m F_{j-m,i}, \tag{16d}$$

which is the simplest algorithm to obtain F_n in terms of $F^{(n)}$ using Eq. (16a) or $F^{(n)}$ in terms of F_n using (16b). Except for the binomial coefficients, the same $F_{j,i}$ are used in both equations. This proves the fact that no new Poisson bracket operations are required to calculate the inverse (see also Eqs. (29) and (30) of Kamel, [18]). Formula (16b) is particularly useful in the calculation of the Hamiltonian function. At each order of perturbation ($n = 1, 2, \ldots$), the generating function W_n is selected such that $L_n F_0$ eliminates the short period terms in Eq. (16b) leaving the new Hamiltonian, $F^{(n)}$, with the long period and secular terms. In some cases, however, some reduction in the number of the Poisson bracket operations may be possible by choosing $F^{(n)}$ to be the old Hamiltonian instead of the new Hamiltonian. In this case $F = 0$ for all values of i as shown by Eq. (16d). Further reduction in the number of Poisson bracket operations can be achieved by observing that the use of the small parameter ϵ is only formal and, therefore, the functions $F^{(n)}$ can be redefined to contain other terms. For example, $F^{(n)} + \epsilon^m F^{(n+m)}$ can be used to replace $F^{(n)}$ in Eq. (16) in a similar manner as used by Howland [13].

Equation (16d) can be visualized using the triangle of Fig. 3. This triangle is simpler than the triangles of Figs. 1 and 2 because of the absence of any cross computations. This is due to the fact that each element $F_{j,i}$ in the ith column is computed in terms of the elements above it in the same column and, as a result, the ith column drops out from the computations when $F^{(i)} = 0$. Similarly, Eq. (16a) can be considered as the sum of certain combinations of the horizontal elements in Fig. 3.

Formula (16b) is essentially the same as formula (49) of Kamel [18], which was implemented by Char using the MACSYMA algebraic program (see McNamara [26]).

4. GENERALIZED LIE-DEPRIT TRANSFORM

The initial effort to extend the Lie-Deprit transform to non-Hamiltonian systems was published by the present author in a SUDAAR 389 report in Oct. 1969 (Kamel [19]). Private communication of this report to Drs. Deprit and Henrard and to Prof. Hori resulted in a more rigorous derivation by Henrard [10] and similar extension of Lie series by Hori [16]. Professor Hori's first presentation of his extension was at the Summer Institute in Orbital Mechanics at the University of Texas at Austin in May 1970. This, apparently, led Prof. Giacaglia [9] to conclude that Prof. Hori's extension was independent of the present author's development published in *Celestial Mechanics* (Kamel [20]) at about the same time that Prof. Hori presented his extension at Texas. However, it is important to point out that Prof. Hori referred to the present author's SUDAAR 389 report in February of 1970 (Hori [15]). Also, Prof. Hori's publication of his extension (Hori [16]) referred to the present author's two publications on the subject and never claimed that his finding was independent.

Now, the extension of the Lie-Deprit transform to non-Hamiltonian systems starts with the same basic idea of the Krylov-Bogoliubov method of averaging that applies a near identity transformation $y \rightarrow x$ in the form of power series in the small parameter ϵ,

$$y = x + \sum_{n=1}^{\infty} \frac{\epsilon^n}{n!} x_n(x), \tag{17}$$

to the N-dimensional system of differential equations,

$$\dot{y} = g(y; \epsilon) = \sum_{k=0}^{\infty} \frac{\epsilon^k}{k!} g^{(k)}(y), \tag{18}$$

so that the resulting system of differential equations

$$\dot{x} = f(x; \epsilon) = \sum_{k=0}^{\infty} \frac{\epsilon^k}{k!} f_k(x) \tag{19}$$

has a simple form. To obtain the relationship between f_k and $g^{(k)}$, a generalized Lie-transform of Eq. (1) was introduced (Kamel [19], [20]):

$$\frac{dx}{d\epsilon} = \tilde{W}(x; \epsilon) = \sum_{m=0}^{\infty} \frac{\epsilon^m}{m!} \tilde{W}_{m+1}(x), \tag{20}$$

with $x = y$ when $\epsilon = 0$.

Now, differentiating Eq. (17) with respect to time and using Eqs. (18) and (19) yields

$$g = [A]f, \tag{21}$$

where A is the $N \times N$ Jacobian matrix given by

$$A = \left[\frac{\partial y}{\partial x}\right] = I + \sum_{n=1}^{\infty} \frac{\epsilon^n}{n!} \left[\frac{\partial x_n}{\partial x}\right]. \tag{22}$$

Note that the sequence $g^{(k)}(y)$ of Eq. (18) is related to $g(y; \epsilon)$ by

$$g^{(k)}(y) = \left[\frac{d^k}{d\epsilon^k} g\right]_{\epsilon=0}, \ k \geqslant 0, \tag{23}$$

which can be obtained by successive differentiation of Eq. (21) with respect to ϵ. For $k = 1$

$$\frac{dg}{d\epsilon} = [A] \frac{dt}{d\epsilon} + \left[\frac{dA}{d\epsilon}\right] f, \tag{24}$$

where, using Eq. (20),

$$\frac{df}{d\epsilon} = \frac{\partial f}{\partial \epsilon} + \left[\frac{\partial f}{\partial x}\right] \tilde{W}, \tag{25}$$

and $dA/d\epsilon$ may be obtained in terms of $\tilde{W}$ using the fact that y is the initial condition of Eq. (20) and, therefore, is independent of ϵ. Thus, in view of Eqs. (22) and (25) with $f = y$,

$$\frac{dy}{d\epsilon} = \frac{\partial y}{\partial \epsilon} + [A]\,\tilde{W} = 0. \tag{26}$$

Also, since the right-hand side of Eq. (26) is zero, its partial derivative must equal its total derivative with respect to ϵ. This leads to

$$\frac{dA}{d\epsilon} = -[A]\left[\frac{\partial \tilde{W}}{\partial x}\right], \tag{27a}$$

where

$$A = \mathrm{I} \text{ when } \epsilon = 0. \tag{27b}$$

Substitution of Eqs. (25) and (27a) into Eq. (24) lead to

$$\frac{dg}{d\epsilon} = [A]\Delta f, \tag{28}$$

with

$$\Delta f = \frac{\partial f}{\partial \epsilon} + \tilde{L}_W f \tag{29}$$

where $\tilde{L}_W$ is the generalized Lie-Deprit operator similar to L_W of Eq. (5),

$$\tilde{L}_W f = \left[\frac{\partial f}{\partial x}\right]\tilde{W} - \left[\frac{\partial \tilde{W}}{\partial x}\right] f \tag{30}$$

and $[\partial f/\partial x]$, $[\partial W/\partial x]$ are the $N \times N$ Jacobian matrices of f and W.

Now, in view of Eqs. (21) and (28), the successive differentiation of Eq. (28) leads to

$$\frac{d^k g}{d\epsilon^k} = [A]\Delta^k f,\ k \geqslant 0, \tag{31}$$

where

$$\Delta^0 f = f, \tag{32a}$$

$$\Delta^k f = \Delta\Delta^{k-1} f,\ k \geqslant 1. \tag{32b}$$

In view of Eqs. (22), (23), and (31), the coefficients of Eq. (18) are given by

$$g^{(k)}\,(y) = [\Delta^k f(x;\,\epsilon)]_{\epsilon=0}. \tag{33}$$

Note the similarity between Eqs. (29), (32), (33) and (4a), (4b), (3b), respectively. This shows that recursive algorithms similar to those of Sections 2 and 3 can be obtained. In particular, an algorithm similar to Eq. (16b) is given by

$$g_{0,0} = g^{(0)} = f_0, \; g_{0,i} = g^{(i)}, \tag{34b}$$

$$g_{j,i} = -\sum_{m=1}^{j} C_{m-1}^{j-1} \tilde{L}_m \, g_{j-m,i}, \tag{34c}$$

$$\tilde{L}_j f = \left[\frac{\partial f}{\partial x}\right] \tilde{W}_j - \left[\frac{\partial \tilde{W}_j}{\partial x}\right] f, \tag{34d}$$

which is essentially the same as Eqs. (47) to (49) of Kamel [19] and Eq. (37) of Kamel [20]. In these equations, the generator $\tilde{W}_n$ is selected such that $\tilde{L}_n f_0$ eliminates the short-period terms in Eq. (34a). In the Krylov-Bogoliubov method of averaging, however, the x_n of Eq. (17) is used to eliminate the short-period terms. To obtain x_n in terms of $\tilde{W}_n$, substitute Eqs. (17), (20), and (22) in Eq. (26) to get

$$x_n = -\tilde{W}_n - \sum_{m=1}^{n-1} C_{m-1}^{n-1} L'_m \, x_{n-m}, \tag{35a}$$

with

$$L'_m F = \left[\frac{\partial F}{\partial x}\right] \tilde{W}_m . \tag{35b}$$

Also, for any function F of the form

$$F = \sum_{n=0}^{\infty} \frac{\epsilon^n}{n!} F_n(x) = \sum_{k=0}^{\infty} \frac{\epsilon^k}{k!} F^{(k)}(y), \tag{36}$$

the $F^{(k)}$ are related to F_n by a formula similar to (16a) where $F_{j,i}$ is defined by (also, see Eq. 26 of Kamel [20])

$$F_{j,i} = -\sum_{m=1}^{j} C_{m-1}^{j-1} L'_m \, F_{j-m,i}, \tag{37}$$

where L'_m is defined by Eq. (35b).

It should be pointed out that the use of $\tilde{W}_n$ instead of x_n to eliminate the short-period terms in Eq. (34a) allowed the development of the compact recursive formulation for the Krylov-Bogoliubov method of averaging. This is in marked contrast to Musen's method (Musen [29]). Also, since the Lie-Hori series is equivalent to the Lie-Deprit transform, which was shown here to be equivalent to the Krylov-Bogoliubov method, it is expected that the Lie-Hori series is also equivalent to the method of averaging as shown by Ahmed and Tapley [1].

Simple examples that demonstrate the use of Eq. (34) can be found in Kamel [20] and Nayfeh [30]. More sophisticated application can be found in Kamel and Hassan [22] and Brown [4].

The generalized Lie-Deprit transform can be shown to be also equivalent to the method of multiple scales whose basic idea is to solve Eq. (18) by using the near identify transformation of Eq. (17) and an expansion of the total derivative operator d/dt in terms of multiple time scales

$$\frac{d}{dt} = \sum_{k=0}^{\infty} \frac{\epsilon^k}{k!} \frac{\partial}{\partial t_k}, \tag{38a}$$

where

$$t_k = \frac{\epsilon^k}{k!} t, \; k \geqslant 0. \tag{38b}$$

At each order of the expansion, the vector $\partial x_n / \partial t_0$ ($n = 1,2, \ldots$) is selected to eliminate the short-period terms and the vector $\partial x / \partial t_k$ ($k \geqslant 0$) is selected to eliminate the secular terms. At the end of the process, the total derivative of the vector x is obtained from

$$\frac{dx}{dt} = \sum_{k=0}^{\infty} \frac{\epsilon^k}{k!} \frac{\partial x}{\partial t_k}. \tag{39}$$

To establish the equivalence to the generalized Lie-Deprit transform, compare Eq. (39) with Eq. (19) to get

$$f_k = \frac{\partial x}{\partial t_k}, \; k \geqslant 0, \tag{40}$$

and use Eq. (35a) to get the vector x_n from the generating vector $\tilde{W}_n$. Note that, as in the case of the method of averaging, the use of $\tilde{W}_n$ instead of x_n allowed the development of the compact recursive formulation of Eq. (34) for the method of multiple scales.

Based on the above analysis, the generalized Lie-Deprit transform is equivalent to both the method of averaging and the method of multiple scales. However, it should be pointed out that the near identity transformation of Eq. (17) is not the same for both methods. In the method of averaging, x is usually defined in terms of polar coordinates, i.e., radii and phase angles while in the method of multiple scales, x is given by the equation

$$\frac{\partial x}{\partial t_0} = g(0), \tag{41}$$

whose solution may be written as

$$x = x(x_a) \tag{42}$$

where x_a are the polar coordinates used as x in the method of averaging.

The choice of x in the method of multiple scales leads to fewer trigonometric terms than in the method of averaging (Kamel and Duhamel [23]). On the other hand, the integration of the $\tilde{W}_n$ equations become more complicated (Eq. 64, Kamel [20]). Also, some of the partial derivatives $\partial F/\partial x$ become more difficult to perform unless a change of variables from x to x_a is used. In this case, $\partial F/\partial x$ is replaced by

$$\frac{\partial F}{\partial x} = \left[\frac{\partial F}{\partial x_a}\right]\left[\frac{\partial x_a}{\partial x}\right], \tag{43}$$

where $[\partial x_a/\partial x]$ is the Jacobian of the inverse of Eq. (42),

$$\frac{\partial x_a}{\partial x} = \left[\frac{\partial x}{\partial x_a}\right]. \tag{44}$$

Finally, the equivalence between the method of averaging and Kevorkian's two-scale method was given by Sarlet [32].

5. THE HAMILTONIZATION OF NON-HAMILTONIAN SYSTEMS

A completely different approach to develop the recursive formulas of Eqs. (34) to (37), for the methods of averaging and multiple time scales, was given by Kamel [21]. In this approach, the non-Hamiltonian system was expressed in form of a Hamiltonian of twice the order of the original system as shown by Birkhoff (pp. 55-58, Birkhoff [2]).

This Hamiltonization process has been long known to researchers in optimization and control (see for example, Bryson and Ho [5]), Powers and Tapley [31]; Choi and Tapley [7], Kamel and Hassan [22]) and it is achieved by using adjoint vectors Y and X (also known as Lagrange multipliers). In this case, the Hamiltonian

F is defined by

$$F = Y \cdot g(y; \epsilon) = X \cdot f(x; \epsilon), \tag{45}$$

where the linearity in the adjoint variables is preserved by selecting the generating function W to be

$$W = X \cdot \tilde{W}(x). \tag{46}$$

In view of Eqs. (2), (6), (18), (19), and (20),

$$F^{(n)} = Y \cdot g^{(n)}(y), \; F_n = X \cdot f_n(x), \; W_{n+1} = X \cdot \tilde{W}_{n+1}(x), \quad n \geqslant 0. \tag{47}$$

Substitution of Eq. (47) in Eq. (16) leads to Eqs. (34) to (37). This is because the Lie-Deprit operators L_W, $\tilde{L}_W$, and L'_W of Eqs. (5), (30), and (35) are related as follows:

$$L_W[X \cdot F(x)] = X \cdot \tilde{L}_W F(x), \tag{48a}$$

$$L_W[F(x)] = L'_W F(x). \tag{48b}$$

Therefore, the canonical algorithm of Eq. (16) is also equivalent to the methods of averaging and multiple scales. Finally, the solution of the equations of motion of the adjoint vectors (X, Y) are not required. These equations, however, are given by

$$\dot{X} = -\frac{\partial F}{\partial x} = -\left[\frac{\partial f}{\partial x}\right] X, \tag{49a}$$

$$\dot{Y} = -\frac{\partial F}{\partial y} = -\left[\frac{\partial g}{\partial y}\right] Y, \tag{49b}$$

and in view of Eqs. (18) and (19), (X, Y) can be interpreted as the variations ∂x and ∂y propagated backward in time. These variations may be useful in the study of the dynamical stability about a given nominal trajectory (x, y).

6. CONCLUSION

The methods of averaging and multiple scales were shown to be equivalent to the Lie-Deprit transform. General recursive algorithms (Eqs. (34) and (35)) were obtained for these methods using the generating vector defined by Eq. (20). These algorithms are convenient for computerized algebraic manipulation using existing scientific computers or the ever increasing capability of personal computers. Finally, the extension of the Lie-Deprit transform to the perturbation solution of partial differential equations is an open subject for future research.

ACKNOWLEDGMENTS

The author gratefully thanks Drs. Cawley, Deprit, Sáenz, and Zachary for their invitation to the workshop and for the simulating discussions.

REFERENCES

[1] A. H. Ahmed and B. D. Tapley, "Equivalence of generalized Lie-Hori method and the method of averaging," *Celestial Mechanics* **33**, 1-20 (1984).

[2] G. D. Birkhoff, *Dynamical Systems* (Am. Math. Soc. Colloq. Pub., IX, Providence, Rhode Island, 1927).

[3] N. N. Bogoliubov and Y. A. Mitropolsky, *Asymptotic Methods in the Theory of Nonlinear Oscillations* (Gordon and Breach, New York, 1961).

[4] B. C. Brown, "Secular effects in the orbits of the Galilean satellites," *Celestial Mechanics* **23**, 203-221 (1981).

[5] A. E. Bryson and Y. C. Ho, *Applied Optimal Control* (Hemisphere, Washington, D.C., 1981), revised ed.

[6] J. A. Campbell and W. H. Jeffereys, "Equivalence of the perturbation theories of Hori and Deprit," *Celestial Mechanics* **2**, 467-473 (1970).

[7] J. S. Choi and B. D. Tapley, "An extended canonical perturbation method," *Celestial Mechanics* **7**, 77-90 (1973).

[8] A. Deprit, "Canonical transformations depending on a small parameter," *Celestial Mechanics* **1**, 12-30 (1969).

[9] G. E. O. Giacaglia, *Perturbation Methods in Non-Linear Systems*, (Springer, New York, 1972).

[10] J. Henrard, "On a perturbation theory using Lie transforms," *Celestial Mechanics*, **3**, 107-120 (1970).

[11] J. Henrard, "The algorithm of the inverse for Lie transforms," *Recent Advances in Dynamical Astronomy*,

Tapley and Szebehely, Eds. (Reidel, Dordrecht, Holland), pp. 250-259.

[12] J. Henrard and J. Roels, "Equivalence for Lie transforms," *Celestial Mechanics* **10**, 497-512 (1974).

[13] R. A. Howland, "An accelerated elimination technique for the solution of perturbed Hamiltonian systems," *Celestial Mechanics* **15**, 327-352 (1977).

[14] G. Hori, "Theory of general perturbations with unspecified canonical variables," *Publ. Astron. Soc. Japan* **18**, 287-296 (1966).

[15] G. Hori, "Comparison of two perturbation theories based on canonical transformations," *Publ. Astron. Soc. Japan* **22**, 191-198 (1970).

[16] G. Hori, "Theory of general perturbations for noncanonical systems," *Pub. Astron. Soc. Japan,* **23**, 567-587 (1971).

[17] A. A. Kamel, "Perturbation theory based on Lie transforms and its application to the stability of motion near Sun-Perturbed Earth-Moon triangular libration points," PhD Thesis, SUDAAR No. 391, Dec 1969, Dept. of Aeronautics and Astronautics, Stanford University. See also NASA CR-1622, August 1970.

[18] A. A. Kamel, "Expansion formulae in canonical transformation depending on a small parameter," *Celestial Mechanics* **1**, 190-199 (1969).

[19] A. A. Kamel, "Perturbation method in the theory of nonlinear oscillations," SUDAAR No. 389, Oct 1969, Dept. of Aeronautics and Astronautics, Stanford University. See also NASA CR-1554, March 1970.

[20] A. A. Kamel, "Perturbation method of the theory of nonlinear oscillations," *Celestial Mechanics,* **3**, 90-106 (1970).

[21] A. A. Kamel, "Lie transforms and the Hamiltonization of non-Hamiltonian systems," *Celestial Mechanics* **4**, 397-405 (1971).

[22] A. A. Kamel and S. D. Hassan, "A perturbation treatment for optimal slightly nonlinear systems with

linear control and quadratic criteria," *J. Optimization Theory and Applications* **2**, 386-404 (1973).

[23] A. A. Kamel and T. Duhamel, "A second order solution of the main problem of artificial satellites using multiple scales," to be published in *J. Guidance and Control*.

[24] J. Kevorkian and J. D. Cole, *Perturbation Methods in Applied Mathematics* (Springer, New York, 1981).

[25] M. S. Lie, *Theorie der Transformationsgruppen* (Teubner, Leipzig, 1888, Vol. 1, Chap. 1).

[26] B. McNamara, "Super-convergent adiabatic invariants with resonant denominators by Lie Transforms," *J. Math. Phys.* **19**, 2154-2164 (1978).

[27] W. A. Mersman, "A new algorithm for the Lie transformation," *Celestial Mechanics* **3**, 81-89 (1970).

[28] W. A. Mersman, "Explicit recursive algorithm for the construction of equivalent canonical transformation," *Celestial Mechanics* **3**, 384-389 (1971).

[29] P. Musen, "On the high order effects in the methods of Krylov-Bogoliubov and Poincaré," *J. Astr. Sciences* **12**, No. 4, 129-134 (1965).

[30] A. H. Nayfeh, *Perturbation Methods* (Wiley, New York, 1973).

[31] W. F. Powers and B. D. Tapley, "Canonical transformation applications to optimal trajectory analysis," *AIAA Journal* **7**, 394-399 (1969).

[32] W. Sarlet, "On a common derivation of the averaging method and the two-timescale Method," *Celestial Mechanics* **17** 299-312 (1978).

[33] H. Shniad, "The equivalence of von Zeipel mappings and Lie transforms," *Celestial Mechanics* **2**, 114-120 (1969).

LIE TRANSFORMS: A PERSPECTIVE

John M. Finn
Laboratory for Plasma and Fusion Energy Studies
University of Maryland
College Park, Maryland 20742

1. INTRODUCTION

Canonical transformations on Hamiltonian systems performed by means of Lie transforms have been used with success for about 15 years now in such diverse areas as celestial mechanics [1-3], accelerator design [4], geometrical optics [5], electron microscopes, and plasma physics [6-17]. In the last area, applications have included formulation of guiding center theory [6], motion of particles in waves [7-10], magnetic field line trajectories [11], and basic studies of nonlinear plasma theory [9, 12]. In addition to all of these applications, these techniques have proved their usefulness in studies of the basic properties of Hamiltonian systems, e.g., nonlinear resonances and ergodicity [13-15]. I have only included examples of such work in Refs. [1-17]; I decided early on that it was futile to attempt to be comprehensive.

I fear that some of us enthusiasts may have attempted to sell Lie transforms as blunt instruments that can be bought off the shelf (except in New York City) and used to batter any Hamiltonian into submission. One thing I hope to convince the reader of is that they are, on the contrary, exceedingly sharp instruments which, to be used to their potential, need to be custom built for the problem at hand. On the other hand, I should try to encourage the timid that use of these techniques does not require mastery of Lie algebras, Lie groups, group representation theory, and comparative Eastern theology.

Lie transforms were reintroduced to the modern world by Hori [1] in the late sixties. A few years later, Deprit [2] generalized this notion so that the most general canonical transformation near the identity could be represented by such transforms, and refinements have been made by the many practitioners that have used this method with success since then. Somewhat later, Dragt and I introduced a completely different type of Lie transform [13, 14] which consists of successive transformations described by simple Lie series of the Hori type, i.e., infinite product Lie transforms, and proved that they too could represent the most general canonical transformation near the identity.

At this date Lie transforms of both varieties are widely used, and a comparison of the efficiency of the two types of Lie transforms in use has been recently published [16]. In these pages I present and discuss some of the properties of both types of Lie transforms and show some manipulations that are useful in practice. I also compare the two methods and present a method of obtaining the generators of each type of transform from the generators of the other. At the least, this last exercise shows that the two formalisms are nontrivially related. This material, which follows a section (2) on the basics of Lie series, is contained in Sections 3 and 4. Section 5 contains some examples of direct applications of Lie transforms, i.e., applications where the actual time evolution of the system is described in terms of infinite product Lie transforms. Included is a compact, and I hope comprehensible, rederivation of the Campbell-Baker-Hausdorff formula. Normalization of Hamiltonian systems is the most widely used application of Lie transform techniques and the subject of the last section, Section 6. This material contains discussions of nonresonant and resonant cases, time dependent cases with either small amplitude or adiabatic evolution, cases with zero frequencies of the linearized motion, and Kolmogorov's superconvergent algorithm. All of these applications are discussed in terms of infinite product Lie transforms, although, as shown in Section 6, the basic framework for normalization (including superconvergence) is similar to that using Deprit Lie transforms.

2. FUNDAMENTALS

In this section I will establish notation and review briefly some background material that is necessary to understand Lie transform techniques.

First, we will assume to be dealing in canonical coordinates, so that the coordinates $q_1,\dots,q_n$ and momenta $p_1,\dots,p_n$ can be written in shorthand form $z_i = q_i$ for $i = 1,\dots,n$ and $z_i = p_{i-n}$ for $i = n+1,\dots,2n$. Thus the fundamental Poisson bracket relationship is

$$[z_i,z_j] = J_{ij} , \tag{1}$$

where

$$J = \begin{pmatrix} 0 & I_n \\ -I_n & 0 \end{pmatrix} . \tag{2}$$

From this, it is clear that the Poisson bracket of two functions f and g is

$$[f,g] = J_{ij} \frac{\partial f}{\partial z_i} \frac{\partial g}{\partial z_j} ; \tag{3}$$

here and elsewhere we employ the summation convention.

With any sufficiently differentiable function f on phase space we associate an operator F, $\hat{f}$ or ad(f) by the rule

$$Fg = [f,g] . \tag{4}$$

This operator, the Lie operator associated with f, has the following properties: it is linear and is a derivation with respect to multiplication, i.e.,

$$F(gh) = (Fg)h + g(Fh) . \tag{5}$$

In addition, it is a derivation with respect to the Lie (i.e., Poisson) bracket,

$$F[g,h] = [Fg,h] + [g,Fh] . \tag{6}$$

This relationship follows from the Jacobi identity $[f,[g,h]] + [g,[h,f]] + [h,[f,g]] = 0$. Also, by antisymmetry of the bracket we have $Fg = - Gf$. Finally we note the identity

$$(FG - GF)h = \mathrm{ad}[f,g]h , \tag{7}$$

which also follows from the Jacobi identity. This shows the homomorphism between the Lie algebra of functions, whose bracket is the Poisson bracket, and the Lie algebra of differential operators on that function space, whose bracket is the commutator; this is the usual adjoint representation. The sign on the right side of (4) was picked so that (7) would not be polluted with a minus sign.

Corresponding to the Lie algebra element H we can associate the one parameter subgroup of the Lie group

$$\exp(sH) = e^{sH} = \sum_{n=0}^{\infty} \frac{s^n}{n!} H^n , \tag{8}$$

called the Lie series generated by H, or by h. Except as noted we will ignore questions of convergence and treat infinite series as formal series. From (5) and (6) we find that e^{sH} has the following properties:

$$e^{sH}(fg) = e^{sH}f\, e^{sH}g , \tag{9a}$$

$$e^{sH}[f,g] = [e^{sH}f,\, e^{sH}g] , \tag{9b}$$

$$e^{sH}f(z) = f(e^{sH}z) . \tag{9c}$$

The last property holds for analytic functions, i.e., power series, by a repeated application of (9a). The Lie series associated with H has two important properties. First, the transformation defined by

$$w_i(s) = e^{sH}z_i \tag{10}$$

is canonical; this follows from property (9b) by $[w_i(s),w_i(s)] = [e^{sH}z_i,e^{sH}z_j] = e^{sH}[z_i,z_j] = e^{sH}J_{ij} = J_{ij}$. Second, the Lie series satisfies $w_i(0) = z_i$ and

$$\frac{d}{ds} e^{sH} = He^{sH} = e^{sH}H . \tag{11}$$

The latter property shows that the transformation (10) satisfies

$dw_i/ds = Hw_i(s)$, i.e., that it is generated by the time-independent Hamiltonian $-h$, with the parameter s taking the role of time. We also note from (9c) that $d/ds\, h(w(s)) = [h(w(s)),h(w(s))] = 0$, i.e., that h is an invariant of the transformation (10). Equation (9c) shows how a function of z transforms under (10). Also, it is easy to see that the inverse of (10) is given by e^{-sH}. (This formula is deceptively simple; its proof involves the fact that $h(w) = h(z)$ and that Poisson brackets with respect to w and with respect to z are equal.)

We should comment here that if h also depends explicitly on s, i.e., $h = h(z,s)$, the transformation (10) is still canonical, and h (with its s dependence frozen) is still an invariant of the transformation. These are the Lie transforms of Hori [1]; for more detail see the paper by Deprit in this volume. However, the operators $U(s) = \exp(sH)$ do not form a one-parameter subgroup $[U(s_1)U(s_2) \neq U(s_1 + s_2)]$, Eq. (11) does not hold, and $-h$ is not the Hamiltonian for the flow. In fact, this flow for $0 < s_1 < s$ is generated by the Hamiltonian $-h(z,s)$, i.e., with parameter s_1 fixed at the final value s. The conclusion that this class of transformations cannot encompass the whole class of canonical transformations connected to the identity is immediate, since not all members of the latter class are integrable.

3. THE LIE TRANSFORM OF DEPRIT

In his 1969 paper [2], A. Deprit introduced a Lie transform more general than the Lie series (10) or the Lie transform of Hori [1] (i.e., (10) with $h = h(z,s)$). This transform is written as

$$w_i(s) = \exp(s\Delta_d)z_i \ , \tag{12}$$

where the operator Δ_d corresponding to the generator $d(z,s)$ is defined by $\Delta_d f = Df + \partial f/\partial s$. We can show that every canonical transformation connected to the identity can be written in this form. First we prove the following.

Lemma 3.1: If the canonical transformation U(s) which maps z to w(s) is canonical for all sufficiently small s, then there exists a function $d(w,s)$ with

$$\frac{dw_i}{ds} = [d,w_i] = Dw_i \ . \tag{13}$$

Note that this implies that $-d$ is a time (s) dependent Hamiltonian for U(s).

Proof: By assumption, $[w_i(s),w_j(s)] = J_{ij}$, from which we obtain by differentiating $[f_i,w_j] + [w_i,f_j] = 0$, where $f_i(w,s) = dw_i/ds$. Defining new variables $\tilde{w}_i = J_{ij}w_j$, we find $\partial f_i/\partial\tilde{w}_j = \partial f_j/\partial\tilde{w}_i$, showing that the differential $f_i d\tilde{w}_i$ is exact. Therefore, a function d exists having $f_i = -\partial d/\partial\tilde{w}_i$, or $f_i = J_{ki}\partial d/\partial w_k = [d,w_i]$, which completes the proof.

Now, we can extend the phase space by defining q_{n+1} as the parameter s and p_{n+1} as its canonical conjugate. Then, defining $d_* = d(z,q_{n+1}) - p_{n+1}$ and an extended bracket by $\{f,h\} = [f,h] + \partial f/\partial q_{n+1}\partial h/\partial p_{n+1} - \partial f/\partial p_{n+1}\partial h/\partial q_{n+1}$, we find that the extended equations of motion corresponding to (13) are

$$\frac{dw_\mu}{ds} = \{d_*,w_\mu\} \equiv D_*w_\mu \tag{14}$$

for $\mu = 1, \ldots, 2n+2$. Now, since d_* does not depend upon s explicitly, we obtain

$$w_\mu(s) = e^{sD_*} z_\mu , \tag{15}$$

which, when applied for $i = 1, \ldots, 2n$ is exactly the Lie transform of Eq. (12).

At this point some observations are in order. First, this Lie transform is clearly a generalization of the Lie series (10), reducing to the latter if d does not depend explicitly on s. In fact, if the transform $w_i(s) = U(s)z_i$ defined by (12) satisfies the one parameter group property $U(s_1)U(s_2) = U(s_1 + s_2)$, then the generator g is independent of s [14]. Next, note that Eq. (13) implies

$$\frac{dU(s)}{ds} = D(w,s)U(s) . \tag{16}$$

Now, notice that $dw_i/ds = [d(w,s),w_i] = [d(U(s)z,s),U(s)z_i] = [U(s)d(z,s),U(s)z_i]$ by 9(c); this in turn equals $U(s)[d(z,s),z_i]$ because of (9a). Therefore we find

$$\frac{dU(s)}{ds} = U(s)D(z,s) , \tag{17}$$

where, it should be noted, d or D is evaluated at the initial phase space point z but at the final parameter s. It should be emphasized that by d(z,s) we do not mean d(w(s),s), expressed in terms of the initial z_i and s. Incidentally, in extended phase space, writing $w_\mu(s) = V(s)z_\mu$, we obtain from (15)

$$\frac{dV(s)}{ds} = D_*V(s) = V(s)D_* , \tag{18}$$

where d_*, being an invariant of motion, can be evaluated at any value of s. The second form in (18) reduces to (17) if we evaluate z_μ at $s = 0$ and notice that the operation of V(s) is to apply U(s) and to propagate the "time" from 0 to s. The first form of (18) is immediately seen to be equivalent to (16).

Equation (17) is in a particularly useful form, since we can iterate the integral form

$$U(s) = I + \int_0^s ds_1 U(s_1)D(z,s_1) \tag{19}$$

to obtain the time-ordered exponential formula

$$U(s) = I + \int_0^s ds_1 D(z,s_1) + \int_0^s ds_1 \int_0^{s_1} ds_2 D(z,s_2)D(z,s_1) + \ldots . \tag{20}$$

It is important to note the order of the operators in the $O(s^2)$ and higher terms.

This form is convenient because, with the generator D depending on the old variables z rather than the new variables w as in (16), it can be immediately applied to z_i.

The inverse of an operator U(s) satisfies $dU(s)^{-1}/ds = -U(s)^{-1}(dU(s)/ds)U(s)^{-1}$; therefore, we obtain from (17)

$$\frac{dU(s)^{-1}}{ds} = - D(z,s)U(s) \ . \tag{21}$$

Thus, the inverse of a Lie transform of this variety cannot be immediately written as a Lie transform of this variety, but this does not seem to be a limitation for applications. In fact, the Lie transforms of Dewar [17] satisfy relations like (21) rather than (17). The analogous iteration provides

$$U(s)^{-1} = I - \int_0^s ds_1 D(z,s_1) + \int_0^s ds_1 \int_0^{s_1} ds_2 D(z,s_1)D(z,s_2) + \dots , \tag{22}$$

i.e., $D \to -D$ and the operators are applied in reverse order. This property can also be obtained by reversing the order of the integrations in (20).

Another property that follows from the fact that this type of Lie transform is a simple Lie series in the extended phase space $w^{ext} = \{w_\mu\}_{\mu=1}^{2n+2}$ is the rule for evolution of functions. From (9c) we find

$$f(w(s),s) = f(w^{ext}(s)) = f(e^{sD_*} z^{ext}) = e^{sD_*} f(z^{ext}) = e^{sD_*} f(z,s) = e^{sD} f(z,s). \tag{23}$$

Here, we have written q_{n+1} as s to indicate that d_* no longer operates on it. This result has already been anticipated in arriving at (17).

The time-ordered exponential (20) gives the same formulas for a generator of the form

$$d(z,s) = \sum_{n=0}^{\infty} \frac{s^n}{n!} d_{n+1}(z) \tag{24}$$

as does the triangular algorithm of Deprit [2]. This algorithm, which is based directly upon (23), shows that for

$$f(z,s) = \sum_{m=0}^{\infty} \frac{s^m}{m!} f_m(z) , \tag{25}$$

we have

$$D_*^k f = \sum_{m=0}^{\infty} \frac{s^m}{m!} f_m^{(k)}(z) \tag{26}$$

with

$$f_m^{(k)}(z) = f_{m+1}^{(k-1)}(z) + \sum_{\ell=0}^{m} \binom{m}{\ell} D_{\ell+1} f_{m-\ell}^{(k-1)}(z) \ . \tag{27}$$

For future reference we write them explicitly out to fourth order:

$$w_i(s) = \sum_{n=0}^{\infty} \frac{s^n}{n!} w_i^{(n)}(z) , \tag{28}$$

$$w_i^{(0)}(z) = z_i , \tag{29a}$$

$$w_i^{(1)} = D_1 z_i , \tag{29b}$$

$$w_i^{(1)}(z) = (D_2 + D_1^2)\, z_i , \tag{29c}$$

$$w_i^{(3)}(z) = (D_3 + D_2 D_1 + 2D_1 D_2 + D_1^3) z_i , \tag{29d}$$

$$W_i^{(4)}(z) = (D_4 + D_3 D_1 + 3D_1 D_3 + D_2 D_1^2 + 2D_1 D_2 D_1 + 3D_1^2 D_2 + 3D_2^2 + D_1^4) z_i . \tag{29e}$$

4. INFINITE PRODUCT LIE TRANSFORMS

Some time after the Lie transform of Sec. 3 was introduced, Dragt and Finn [13, 14] presented an alternate representation of canonical transformations by means of Lie series. This was the infinite product representation

$$w_i(s) = U(s) z_i = e^{sF_1} e^{s^2 F_2} \ldots e^{s^n F_n} \ldots z_i , \tag{30}$$

where the functions $f_k(z)$ have no s dependence. Although it was originally introduced [13, 14] with the small parameter s labeling the distance from a fixed point in phase space, and therefore the functions $f_k(z)$ were polynomials homogeneous of degree k+2, this representation is trivially generalized to the more general situation where s represents any small parameter. In fact, in the special case where the f_k are homogeneous polynomials, the linear part of the transformation is not naturally associated with a small parameter and must be treated separately [13, 14]. Also, a more general form of infinite product Lie transform, which occurs often in practice, is $e^{A_0} e^{A_1} \ldots$, where $A_k = A_k(z,s)$. Each of the factors here is a Hori-type [1] Lie transform, i.e., its s dependence is passive.

This representation of a canonical transformation as a composition of simple Lie series is reminiscent of the Euler angle representation of rotations, with the following exception: the Euler angle representation is merely for convenience, since any rotation can be written as a single exponential. The simple Lie series, on the other hand, does not generate the most general canonical transformation, as we have noted in Section 2. That the infinite product does generate the general transformation is the content of our representation theorem.

Theorem: If $z_i \to w_i(s)$ is an analytic canonical transformation for all sufficiently small s and $w_i(0) = z_i$, there exist functions $f_1(z)$, $f_2(z)$, ... such that the transformation can be written in the form (30).

Proof: (This is a generalization of the proof of [14] to arbitrary small parameter s.) Expanding in s, we have $w_i(s) = z_i + sg_{i,1}(z) + O(s^2)$. Then the requirement $[w_i(s),w_j(s)] = J_{ij}$ shows $[z_i,g_{j,1}] = [z_j,g_{i,1}]$ and hence, as in (13), $g_{i,1} = [f_1,z_i] = F_1 z_i$. In this case, however, the $g_{i,1}$ do not depend upon s and therefore f_1 is a function of z alone. We therefore find $\exp(-sF_1)w_i(s) = z_i - sF_1 z_i + sg_{i,1} = z_i + O(s^2)$. Similarly, at each order, we write $\exp(-s^n F_n)\ldots\exp(-sF_1)w_i(s) = z_i + s^{n+1}g_{i,n+1}(z)$ and, since this transformation is also canonical obtain $g_{i,n+1} = F_{n+1}z_i$, showing $\exp(-s^{n+1}F_{n+1})\ldots\exp(-sF_1)w_i(s) = z_i + O(s^{n+2})$, proving the inductive step. Inverting, and as before ignoring issues of convergence, we finally obtain (30).

At this point a few comments are in order. We have already used the fact that the inverse of a Lie transform (30) is represented by the infinite product of inverse Lie series in reverse order,

$$U(s)^{-1} = \ldots e^{-s^n F_n} \ldots e^{-s^2 F_2} e^{-sF_1} . \tag{31}$$

This formula provides an existence theorem for Lie transforms in an order reverse to that of (30); that is, for a canonical transformation M(s) apply the theorem of this section to $M(s)^{-1}$. Also, using (9c) for each of the factors in (26), we see that a function transforms as

$$g(w(s)) = e^{sF_1} \ldots e^{s^n F_n} \ldots g(z) . \tag{32}$$

The inverse (31) is not of the same form as the original transformation (30). For most applications this is not a problem. However, the order of any product of Lie series can be reversed in the following manner. First write, for example,

$$M(s) = e^{s^2 F_2} e^{sF_1} = e^{sF_1} e^{-sF_1} e^{s^2 F_2} e^{sF_1}$$

$$= e^{sF_1} e^{-s\hat{F}_1} e^{s^2 F_2} , \tag{33}$$

where $\hat{F}_1 = \mathrm{ad}F_1$ belongs to the adjoint representation of the Lie algebra of operators, and $\hat{F}_1 G = F_1 G - GF_1$. Continuing, we find

$$M(s) = e^{sF_1} \exp(s^2 e^{-s\hat{F}_1} F_2)$$

$$= e^{sF_1} e^{s^2 A_2} , \tag{34}$$

where $a_2(z,s) = \exp(-sF_1)f_2$. Note, however, that this Lie transform is of the more general form mentioned in the first paragraph because the second factor of (34) is a Hori-type Lie transform rather than a simple Lie series.

The proof I have given for the above theorem is constructive, so that one can obtain the generators $h_i(z)$ from the explicit form of the transformation. It is also possible to obtain the infinite product generators f_i from the generator $d(z,s)$ of the inverse Deprit Lie transform or vice versa. Indeed, writing

$$U(s) = e^{sA_1} e^{s^2 A_2} \dots e^{s^n A_n} \dots \tag{35}$$

with $a_n = a_n(z)$, we find

$$\begin{aligned}\frac{dU(s)}{ds} &= A_1 U(s) + 2se^{sA_1} A_2 e^{s^2 A_2} \dots + \dots + ns^{n-1} e^{sA_1} \dots e^{s^{n-1}A_{n-1}} A_n e^{s^n A_n} \dots \\ &= A_1 U(s) + 2s(e^{sA_1} A_2 e^{-sA_1}) U(s) + \dots \\ &\quad + ns^{n-1}(e^{sA_1} \dots e^{s^{n-1}A_{n-1}} A_n e^{-s^{n-1}A_{n-1}} \dots e^{-sA_1}) + \dots \,.\end{aligned} \tag{36}$$

Using the identity $e^{sA_1} A_2 e^{-sA_1} = e^{s\hat{A}_1} A_2 = \mathrm{ad}(e^{sA_1} a_2)$, as in (33), we find

$$\frac{dU(s)}{ds} = \sum_{n=0}^{\infty} s^n T_{n+1}(z,s) U(s) \tag{37}$$

with

$$T_n(z,s) = ne^{s\hat{A}_1} e^{s^2\hat{A}_2} \dots e^{s^{n-1}\hat{A}_{n-1}} A_n \,.$$

Comparing with Eq. (21) it is clear that we can compare coefficients to obtain the generator for the inverse Deprit Lie transform

$$-\sum_{n=0}^{\infty} \frac{s^n}{n!} D_{n+1}(z) = \sum_{n=0}^{\infty} s^n T_{n+1}(z,s) \,, \tag{38}$$

or, explicitly for the first three orders

$$\begin{aligned} &A_1 = - D_1 \,, \quad A_2 = -\frac{1}{2} D_2 \,, \\ &A_3 = -\frac{1}{6} D_3 - \frac{2}{3} (A_1 A_2 - A_2 A_1) = -\frac{1}{6} D_3 - \frac{1}{3} (D_1 D_2 - D_2 D_1) \,. \end{aligned} \tag{39}$$

Demoting the operators to functions on phase space, we have

$$\begin{aligned} &a_1 = - d_1 \,, \quad a_2 = -\frac{1}{2} d_2 \,, \\ &a_3 = -\frac{1}{6} d_3 - \frac{2}{3} [a_1,a_2] = -\frac{1}{6} d_3 - \frac{1}{3} [d_1,d_2] \,. \end{aligned} \tag{40}$$

Clearly these relations could be used to find the $d_n(z)$ from the $a_n(z)$ or vice versa; this construction also provides an alternate proof of the representation theorem of this section. Also, it is clear from the construction that each d_n is obtained from repeated bracket of the a_n and vice versa, i.e., that the d_n are in the Lie algebra spanned by the a_n and vice versa. To obtain the analogous formulas for the Deprit Lie transform rather than its inverse, merely replace D_i by $-D_i$ and reverse the order of the D_n.

To gain more understanding of the relationship between the two Lie transform representations, let us write (30) in extended phase space as

$$w_\mu(s) = e^{F_{*0}} e^{F_{*1}} e^{F_{*2}} \dots z_\mu , \tag{41}$$

where $f_{*k}(w_\mu) = q_{n+1}^k f_k(w_i)$, and where $F_{0*} = - p_{n+1}$. The f_* notation indicates that the f_{*k} operators are obtained with extended Poisson brackets, i.e., the parameter s has an active role in the factor $\exp(F_{*k})$ for $k > 1$, that of advancing p_{n+1}. The original z_i are advanced exactly as in (30), and $\exp(sF_{*0})$ advances q_{n+1}. The exponentials in (37) can be combined into one exponential by the Campbell-Baker-Hausdorff (CBH) formula (see Section 5) exp $(sG*)$. Then $g*$ can be equated with $d*$ of (14). In fact, it is easy to see that the leading terms of $g*$ are $f_{*0} + f_{*1} = q_{n+1} f_1 - p_{n+1}$, and that p_{n+1} does not occur in any other terms, as in d_*. The fact that the two Lie transform representations are related by the CBH formula shows that they are nontrivially related and is consistent with the observation made after Eq. (40).

It is also of interest to compute the terms of (30) explicitly to fourth order. Writing $w_i(s) = U(s)z_i$, we easily obtain

$$\begin{aligned} U(s) &= I + sF_1 + s^2[\tfrac{1}{2} F_1^2 + F_2] + s^3[\tfrac{1}{3!} F_1^3 + F_1F_2 + F_3] \\ &\quad + s^4[\tfrac{1}{4!} F_1^4 + \tfrac{1}{2} F_1^2F_2 + F_1F_3 + \tfrac{1}{2} F_2^2 + F_4] + O(s^5) . \end{aligned} \tag{42}$$

Comparing with (29) we see that the main difference in form is that the terms of (42) are normal ordered in the sense that any combination F_kF_ℓ occurs only if $k < \ell$. (This relationship is, of course, reversed for the transforms of reverse order, as in (31). Thus we see [16] that the number of terms of order s^k is $p(k)$, the partition of k, whereas for the Deprit Lie transform the number of terms of order s^k is 2^{k-1}. As discussed in [16], p(k) is much smaller than 2^{k-1} even for fairly small k, so that explicit expansion of the infinite product Lie transform requires far fewer operations formally. I say formally because most of the extra operations required by the Deprit algorithm involve permutations of products of the D_k for small k. Therefore the actual savings in computational efficiency of the infinite product algorithm depends critically upon the problem being solved, in particular, on the degree of complexity of the higher order generators (D_k or F_k) relative to the lower order generators. Finally, I should mention that in actually performing a perturbation expansion by either method to high order, the manipulations (by machine, usually) are not performed by explicit expansion as in (29) and (42) but by recursive algorithms [2, 13]. The point here is that at a certain step of either recursion, combinations may occur that reduce the computation that would be required by (29) or (42). For example, in a power series expansion, at each recursive step, coefficients of monomials may add, so that the following recursive steps need be applied only to the combined monomials. Because of this, the increase in efficiency of algorithms based upon the infinite product Lie transforms is further dependent upon the particular problem.

5. DIRECT APPLICATIONS

In this section I will discuss some applications in which Lie transform techniques are applied directly to solutions of Hamiltonian systems, i,e., taking advantage of the fact that solutions of such systems can be written as Lie transforms. I find infinite product Lie transforms more useful for those applications because the small parameter s in (30) need not be associated with time; for these applications the Deprit Lie transform provides a Taylor series in time, which is not useful on long time scales.

First, let us establish some rules for dealing with exponentials of abstract operators. In order to express $\exp(A + \varepsilon B) = \exp(A_0)\exp(\varepsilon A_1)\ldots\exp(\varepsilon^k A_k)\ldots$, note that

$$U(t) = e^{t(A + \varepsilon B)} \tag{43}$$

satisfies $dU/dt = (A + \varepsilon B)U$. Letting $U_0 = e^{tA}$ and writing $U = U_0 V_1$, we find

$$dV_1/dt = \varepsilon C(t)V_1; \; C(t) = U_0^{-1} B\, U_0 = e^{-tA} B e^{tA} = e^{-t\hat{A}} B \,, \tag{44}$$

where again $\hat{A}B$ is the commutator $AB - BA$. Now, let us define

$$U_1 = \exp[\varepsilon D(t)]; \; D(t) = \int_0^t C(t_1)dt_1 = \frac{1}{\hat{A}} (1 - e^{-t\hat{A}})\varepsilon B \,. \tag{45}$$

By expanding this out, we obtain $dU_1/dt = \varepsilon C(t) + O(\varepsilon^2) = \varepsilon C(t)U_1 + O(\varepsilon^2)$, so that $U_1 = V_1 + O(\varepsilon^2)$. Thus, we obtain

$$U(t) = e^{tA} e^{\varepsilon D(t)} + O(\varepsilon^2) \,. \tag{46}$$

In order to continue, we need to find the differential equation satisfied by U_1. Writing dU_1/dt as the limit of $\{\exp[\varepsilon D(t)+\varepsilon\delta C(t)] - \exp[\varepsilon D(t)]\}/\delta$ and applying (46), we find

$$\frac{dU_1}{dt} = U_1 \frac{1}{\hat{D}(t)} [1 - e^{-\varepsilon\hat{D}(t)}]C(t) \,. \tag{47}$$

Now, writing $V_1 = U_1V_2$, we obtain

$$dV_2/dt = E(\varepsilon,t)V_2 \ ,$$

$$E(\varepsilon,t) = [\varepsilon e^{-\varepsilon\hat{D}} - \frac{1}{\hat{D}}(1 - e^{-\varepsilon\hat{D}})]C(t) = -\frac{1}{2}\varepsilon^2[D(t),C(t)] + O(\varepsilon^3) \ . \tag{48}$$

Clearly, we can repeat the procedure, defining U_2 in a way analogous to U_1. Specifically, we can write $U_2 = \exp\ [\int_0^t E(\varepsilon,t_1)dt_1]$, giving $U_2 = V_2 + O(\varepsilon^4)$; if it is more convenient, the approximate expression for $E(\varepsilon,t)$ in (48) may be used, but the error will be of order ε^3.

We summarize the results we have obtained so far:

Lemma 5.1: The differential equation $dV/dt = \epsilon C(t)V$ has solution $V(t) = \exp[\epsilon \int_0^t C(t_1)dt_1] + O(\epsilon^2)$.

Lemma 5.2: The exponential $\exp[A + \epsilon B]$ can be written in the form $\exp(A_0)\exp(A_1)\ldots\exp(A_n)\ldots$, where $A_0 = A$ and $A_{k+1} = O(\epsilon^{2^k})$. Alternatively, simpler forms for the A_k can be obtained with $A_k = O(\epsilon^k)$.

The advantage of such expansions for canonical transformations is that the truncated infinite product is canonical to all orders in ε. Lemma 5.2 shows that the infinite product can be made superconvergent in ε.

The results obtained so far can be used to give an easy derivation of the Campbell-Baker-Hausdorff expansion. Using (46) as we did to obtain (47), it is easy to show that

$$\frac{d}{ds} e^{H(s)} = e^{H(s)} \frac{1}{\hat{H}}(1 - e^{-\hat{H}}) \frac{dH(s)}{ds} \ . \tag{49}$$

Now, writing $e^F e^{sG} = e^{H(s)}$, and differentiating both sides with respect to s, we find

$$G = \frac{1}{\hat{H}}(1 - e^{-\hat{H}}) \frac{dH(s)}{ds} \tag{50}$$

or

$$\frac{dH(s)}{ds} = \frac{\hat{H}}{1 - e^{-\hat{H}}} G \ . \tag{51}$$

As far as I know, this is the simplest and most transparent derivation of the Campbell-Baker-Hausdorff or CBH formula (51). Summarizing:

Theorem 5.3: The operator $e^F e^{sG}$ can be written in the form $e^{H(s)}$, where $H(s)$ satisfies the differential equation (51) with initial condition $H(0) = F$.

From (51) we can obtain the successive terms in the Taylor series of H

$$H'(0) = \frac{\hat{F}}{1 - e^{-\hat{F}}} G \ , \tag{52}$$

$$H''(0) = \hat{H}'(0) \frac{1}{1 - e^{-\hat{F}}} G - \frac{\hat{F}}{1 - e^{-\hat{F}}} \hat{G} e^{-\hat{F}} \frac{1}{1 - e^{-\hat{F}}} G , \tag{53}$$

and so forth. Also note that (51) can be written

$$\frac{dH(s)}{ds} = - \frac{1}{1 - e^{-s\hat{G}} e^{-\hat{F}}} \hat{G} H(s) , \tag{54}$$

which is of the same form as the equation for V_1, i.e., (44), so that this series can be made superconvergent also. That is, letting $H(s) = W(s)A$, with $W(0) = 1$, we find that $W(s)$ satisfies Eq. (54). The algorithm summarized in Lemmas 1 and 2 gives $W(s) = \exp[W_1(s)]\ldots\exp[W_k(s)]\ldots$, where $W_{k+1} = O(s^{2^k})$. The advantage of superconvergent series is that they avoid the small denominator problems of series such as (52,53). We will return to these issues later.

To illustrate these methods, let us consider a harmonic oscillator with slowly varying frequency

$$h = \frac{1}{2} p^2 + \frac{1}{2} \omega^2(\varepsilon t) q^2 . \tag{55}$$

Performing a transformation $y = p/\sqrt{\omega}$, $x = \sqrt{\omega}\, q$, we obtain the new Hamiltonian

$$\tilde{h} = \frac{1}{2} \omega(\varepsilon t)(y^2 + x^2) + \frac{\varepsilon\omega'(\varepsilon t)}{2\omega} xy . \tag{56}$$

The equations of motion are $dZ_i/dt = M_{ij}(t)Z_j$, or $dU/dt = MU$, where $M(t)$ is

$$\begin{bmatrix} \delta(\varepsilon t) & \omega(\varepsilon t) \\ -\omega(\varepsilon t) & -\delta(\varepsilon t) \end{bmatrix} \tag{57}$$

$Z_i(t) = U_{ij}(t)Z_i(0)$, and $\delta = \varepsilon\omega'(\varepsilon t)/2\omega$. We can write this in terms of Pauli matrices as $M = i\omega\sigma_2 + \delta\sigma_3$. Now, first write $U_0 = \exp[i\theta_0(t)\sigma_2]$, where $\theta_0(t) = \int_0^t \omega(\varepsilon t_1)dt_1$ and $U = U_0 W$. From this we easily see

$$dW/dt = \delta(\varepsilon t)U_0^{-1}\sigma_3 U_o W \;\; = \delta(\varepsilon t)[\cos 2\theta_0(t)\sigma_3 + \sin 2\theta_0(t)\sigma_1]W , \tag{58}$$

giving

$$W = \exp[\sigma_3 \int_0^t \delta(\varepsilon t_1)\cos 2\theta_o(t_1)dt_1 + \sigma_1 \int_0^t \delta(\varepsilon t_1)\sin 2\theta_0(t_1)dt_1] , \tag{59a}$$

$$= \exp[\sigma_3 \int_0^t \delta(\varepsilon t_1)\cos 2\theta_0(t_1)dt_1]\exp[\sigma_1 \int_0^t \delta(\varepsilon t_1)\sin 2\theta_0(t_1)dt_1] , \tag{59b}$$

both to within $O(\delta^2)$. The first equality follows from Lemma V-1 and the second from Theorem 5.3. Thus we have shown that it is possible (to order δ) to factor the solution U(t) into three exponentials each involving a single Pauli matrix. In fact, since the Pauli matrices form a closed Lie algebra, this is possible to all orders. Note that U_0W, using (59a), is of the form (30) with the parameter s representing the time rate of change of ω rather than time itself. Bakshi [18] has used similar techniques to perform plasma kinetic theory calculations. Also, Dragt and coworkers [4,5,19] have developed and applied direct infinite product Lie techniques to orbits of particles in accelerators, light rays in optical systems, and electron microscopes.

The CBH formula (51) or (54) has proved to be useful in studying properties of area preserving two-dimensional maps, studied because they simulate the surface of section return map for an autonomous Hamiltonian system with two degrees of freedom. For example, the class of maps

$$x' = \lambda[x + f(x+y)],$$

$$y' = \lambda^{-1}[y - f(x+y)], \tag{60}$$

with a hyperbolic fixed point at the origin is the composition of two maps $x_1 = x + f(x+y)$, $y_1 = y - f(x+y)$, and $x' = \lambda x_1$, $y' = \lambda^{-1}y_1$. Because of this property, it can easily be written as a product of two Lie series $z_i' = e^{F_0}e^{G}z_i$, where $g = -xy\ln\lambda$ and

$$f_0(x,y) = -\int_0^{x+y} f(\zeta)d\zeta \ . \tag{61}$$

(Note that the order of the Lie series is in the opposite order to the function composition rule [13]. We can use the rule (34) for reversing order to obtain

$$z_i' = e^{G}e^{F_1}z_i \ , \tag{62}$$

where $f_1(x,y) = e^{-G}f_0(x,y) = f_0(x/\lambda,\lambda y)$.

Specifically, for the quadratic Cremona map [13] with $f(x+y) = \alpha(x+y)^2/4$, we obtain $f_0(x,y) = -\alpha(x+y)^3/12$ and $f_1(x,y) = -\alpha(x/\lambda + \lambda y)^3/12$. The product Lie transform in (62) is now in a form to which CBH series can be applied. We obtain $z_i' = e^{H(\alpha)}z_i$ where $h(\alpha)$, obtained by (52), (53) has

$$h(0) = g = -xy\ \ln\lambda \ , \tag{63a}$$

$$h'(0) = -\frac{1}{4}\ln\lambda\left[\frac{x^3 + \lambda^3y^3}{\lambda^3 - 1} + \frac{x^2y + \lambda xy^2}{\lambda - 1}\right] , \tag{63b}$$

$$h''(0) = \frac{1}{16}\ln\lambda\left[\frac{(\lambda-1)(\lambda^2-1)}{(\lambda^3-1)(\lambda^4-1)}x^4 + \frac{2(\lambda-1)}{\lambda^3-1}x^3y + \frac{2\lambda^2(\lambda-1)}{\lambda^3-1}xy^3 + \frac{\lambda^4(\lambda-1)(\lambda^2-1)}{(\lambda^3-1)(\lambda^4-1)}y^4\right]$$
$$+\ldots$$

$$+ \frac{1}{16} x^2y^2 \{-3 \ln \lambda [\frac{\lambda^3}{(\lambda^3-1)^2} + \frac{\lambda}{(\lambda-1)^2}] + \frac{(\lambda+1)(2\lambda^2+\lambda+2)}{\lambda^3-1}\} . \tag{63c}$$

This series, when applied to the elliptic form of the map [13] with $\lambda = e^{i\phi}$, shows the small denominators $[e^{ik\phi} - 1$ in $h^{(k-1)}(0)]$ alluded to in the CBH series (52), (53). For the hyperbolic form of the map, i.e., for λ real, this series converges for α small enough (in a small enough neighborhood of the origin) and gives an exact invariant $h(\alpha)$. It has been used as a magnifying glass to detect homoclinic angles as small as 10^{-15} degrees in the map for $\lambda \neq 1$ [13]. (At $\lambda = 1$ the map reduces to e^{F_1} and hence is integrable.)

Another well-studied map [14]

$$x' = x + y \quad (\text{Mod } 2\pi)$$

$$y' = y - \varepsilon \sin(x + y) \quad (\text{Mod } 2\pi) \tag{64}$$

can also be factored: $x_1 = x + y$, $y_1 = y$, $x' = x_1$, $y' = y_1 - \varepsilon \sin x_1$, and can therefore immediately be cast in the form $z_i' = e^G e^{\varepsilon F} z_i$ with $g = -y^2/2$ and $f = \cos x$. The CBH series gives $z_i' = e^{H(\varepsilon)} z_i$ with

$$h(\varepsilon) = -\frac{y^2}{2} + \frac{\varepsilon y}{2} \frac{\sin(x+y) - \sin x}{1 - \cos y} + O(\varepsilon^2) . \tag{65}$$

Note that (65) is not an expansion for small x,y but only for small ε.

6. NORMALIZATION

By far the most common use of Lie transform techniques is to perform canonical transformations on Hamiltonian systems to bring them to a more usable form. These methods are generalizations of the venerable Poincaré-Von Zeipel-Birkhoff method of normalization. As originally formulated, this technique employed generating functions of mixed variables of the form

$$\sum_i q_i P_i + \varepsilon W(q,P) .$$

To first order in ε, this produces the identical transformation as the Deprit Lie transform with $D = -\varepsilon W$ or the infinite product Lie transform with $F_1 = -W$. Any normalization that can be performed by mixed variable generating functions can also be performed by either type of Lie transform, and vice versa. The advantages of Lie transforms over mixed variable generating functions are: (i) Obtaining the transformation from the generator requires only Poisson bracket operations and no inversions; it is therefore simpler to perform and preserves analyticity.
(ii) The inverse transformation may be obtained easily (see (21), (22), (31).)

the first and second terms on the left in (71) into these spaces, respectively. It is not necessary to define H_0^+; indeed, as long as $H_0 f_k$ belong to $R(H_0)$, Γ_k is merely $r_k - H_0 f_k$. However, it is often useful to use H_0^+ in describing the properties of Γ. As discussed by Prigogine [20], any operator $H_0 = [h_0, \cdot]$ is antihermitian, and hence has $N(H_0) = N(H_0^+)$, with respect to the inner product formed by integration over all phase space. However, the functions that we deal with are usually unbounded (e.g., Taylor series expansions about an equilibrium orbit) and therefore this inner product is not defined. Finally, we should mention that in general it is possible to demand that $H_0 f_k$ belong to a proper subspace of $R(H_0)$. For an example, see subsection D. However, for most applications it turns out to be best to put as little as possible into Γ_k and therefore to put all the range space part of r_k into $H_0 f_k$.

I have attempted to impress upon the reader that there is some degree of flexibility in reducing a Hamiltonian to normal form. However, one cannot remove any offending term from the remainder Hamiltonian r_k at order k; as we have discussed, only terms in the range of H_0 may be removed. In Subsections A through E we will present some examples of this normalization procedure.

A. Nonresonant Cases

Consider the Hamiltonian with n degrees of freedom

$$h = \sum_{i=1}^{n} \omega_i J_i + \tilde{h}(\phi, J) , \tag{73}$$

where the unperturbed system of n uncoupled oscillators is in action-angle variables $(\phi, J) \equiv (\phi_1, \ldots, \phi_n, J_1, \ldots, J_n)$. Also assume that the frequencies $\omega_i \neq 0$ are incommensurate, i.e., that there do not exist integers m_i, not all zero, such that $\sum_i m_i \omega_i = 0$. The operator H_0 is given by

$$H_0 = - \sum_{i=1}^{n} \omega_i \frac{\partial}{\partial \phi_i} . \tag{74}$$

If h is analytic and $\tilde{h}$ represents all nonlinearities (i.e., $\tilde{h}$ consists of terms of degree 3 and higher in the z_i), it can be written as

$$\tilde{h} = \sum_{k=3}^{\infty} \sum_{|m| \leq k} h_{\underset{\sim}{m}}^{(k)} (J) e^{i m_1 \phi_1 + \ldots + i m_0 \phi_n} , \tag{75}$$

where $\underset{\sim}{m} = (m_1, \ldots, m_n)$ and $|\underset{\sim}{m}| = \Sigma |m_i|$. The coefficients $h_{\underset{\sim}{m}}^{(k)}$ are homogeneous of degree k/2 with respect to the J_i, and may for convenience be labelled with ε^{k-2}. The terms for k = 3 that are in $R(H_0)$ are those with $|m| \neq 0$. The rest, i.e., functions of the J_i alone, are in $N(H_0)$. (Note that H_0 is antihermitian with respect to any inner product involving integration over angles $\phi_1, \ldots, \phi_n$.) Therefore, at the first step [i.e., Eq. (70)], Γ_1 contains terms independent of the ϕ_i (is "averaged"

(iii) Functions are transformed easily according to (23) or (32).

A normalization procedure, like any perturbation theory, begins with a Hamiltonian expanded in a small parameter ε,

$$h = \sum_{k=0}^{\infty} \varepsilon^k h_k(z) \ . \tag{66}$$

We treat time dependence by extending the phase space to include time and its canonically conjugate momentum. I have in mind removing objectionable terms from h order by order; to that end we will construct a sequence of canonical transformations each of which normalizes h to a given order. Also, we will assume that the unperturbed Hamiltonian h_0 needs no transformation. So the total canonical transformation is naturally written in the form

$$z_i' = e^{-\varepsilon F_1} e^{-\varepsilon^2 F_2} \ldots z_i \ . \tag{67}$$

The new Hamiltonian satisfies $\Gamma(z') = h(z)$ or

$$\Gamma = \ldots e^{\varepsilon^n F_n} \ldots e^{\varepsilon^2 F_2} e^{\varepsilon F_1} h \ . \tag{68}$$

Substituting (66) into (68) and expanding, we find to first order in ε

$$F_1 h_0 + h_1 = \Gamma_1 \tag{69}$$

or

$$H_0 f_1 + \Gamma_1 = h_1 \ . \tag{70}$$

At order ε^k we obtain

$$H_0 f_k + \Gamma_k = r_k \ , \tag{71}$$

where the remainder r_k consists of all terms of order ε^k in the expression

$$e^{\varepsilon^{k-1} F_{k-1}} \ldots e^{\varepsilon F_1} (h_0 + \varepsilon h_1 + \ldots \varepsilon^k h_k) \ . \tag{72}$$

Clearly, the first term on the left in (71) must be the part of r_k in the range of the operator H_0, and Γ_k must be everything else.

In many cases the range and null spaces of H_0, $R(H_0)$ and $N(H_0)$, respectively, cover the whole space F of functions, and putting Γ_k in $N(H_0)$ is advantageous. For example, see subsections A and B. However, this property $R(H_0) + N(H_0) = F$ does not hold in general, for example if H_0 is nilpotent. An example is shown in subsection C. In such cases it may be useful to define an inner product in F and an adjoint of H_0 with respect to this inner product, H_0^+. Then we can use the fact that the whole space F is the direct sum of $R(H_0)$ and $N(H_0^+)$, the null space of H_0^+, and put with respect to the ϕ_1) and each term of f_1 of the form

$$\varepsilon f_{m}^{(3)}(J)e^{im_1\phi_1 + \ldots + im_n\phi_n} \tag{76}$$

has

$$\varepsilon f_{\underset{\sim}{m}}^{(3)}(J) = \frac{i\varepsilon h_{\underset{\sim}{m}}^{(3)}(J)}{m_1\omega_1 + \ldots + m_n\omega_n} . \tag{77}$$

From the form of (77) it is clear that the remainder r_k of (71) can also be expressed in the form like (75) and that at order ε^{k-2} we have

$$\varepsilon^{k-2} f_{\underset{\sim}{m}}^{(k)}(J) = \frac{i\varepsilon^{k-2} r_{\underset{\sim}{m}}^{(k)}(J)}{m_1\omega_1 + \ldots + m_n\omega_n} . \tag{78}$$

Several comments are in order. First, by the incommensurability assumption, the denominators in (78) are never zero, so the process never terminates. However, small denominators are present, i.e., these denominators can become arbitrarily small at some high order. The second comment is that any method of normalization, whether based on mixed variable generating functions, Deprit Lie transforms, or our infinite product formalism, produces the same formal recursive step (71). (See the comments preceeding Eq. (66).) The difference between the approaches is in the method of computation of the remainder, r_k. Finally, in (78) we have tacitly assumed that the f_k contains no null space terms (i.e., functions of the J_i alone). This is not necessary. Some exploration of the effects of including such null space terms in the generators has been done for a specific class of Hamiltonians [13], but in general it is not known how much real flexibility this adds to a normalization calculation.

Since the normal form Hamiltonian $\Gamma = \Sigma\Gamma_k$ is a function of the action variables alone, this procedure produces the transformation to action-angle variables $J_i' = \exp(-\varepsilon F_1)\exp(-\varepsilon^2 F_2)\ldots J_i$, $\phi_i' = \exp(-\varepsilon F_1)\exp(-\varepsilon^2 F_2)\ldots\phi_i$, assuring (except for questions of convergence) that the J_i' are n independent integrals in involution, i.e., $[J_i', J_j'] = 0$. Also, note that secularities are automatically avoided.

For time-dependent problems with Hamiltonians of the form

$$h = h_a(z) + \varepsilon_t h_t(z,t) , \tag{79}$$

where h_t is periodic in time with a fundamental frequency ω_0 of the same order as the unperturbed frequencies in h_a, and where ε_t is small, the same procedure can be used. We merely extend the phase space and write

$$k = h_a(z) + \varepsilon_t h_t(z, q_{n+1}/\omega_0) + \omega_0 p_{n+1} \tag{80}$$

and treat $k_0 \equiv h_0 + \omega_0 p_{n+1}$ as the new unperturbed Hamiltonian in action-angle variable form. This case is nonresonant if there are no integers $m_0, m_1, \ldots, m_n$

such that $m_0\omega_0 + m_1\omega_1 + \dots + m_n\omega_n = 0$. In this case the normalization procedure follows exactly as in Eqs. (73)-(78). Strictly speaking, in this case the range of k_0 is the whole function space, because the differential equation (71) can be solved as an initial value problem in time. Describing the range space as in (78) is equivalent to removing secularities.

B. Resonance cases

When relations of commensurability exist in the unperturbed frequencies of motion, the process outlined in the last section terminates because of zero denominators (see Eq. (78)). This difficulty is traced to an incorrect prescription of the range and null spaces of H_0. The correct treatment for such resonant problems was given by Gustavson [21], using mixed variable generating functions. To make the theory more transparent, we first present it in two degrees of freedom and, of course, with Lie transforms. If $\tilde{m}_1\omega_1 + \tilde{m}_2\omega_2 = 0$ is the relation of commensurability for the Hamiltonian of the form (73), the null space of H_0 consists of function of the action variables J_1, J_2 plus the space spanned by functions of the form

$$g_{m_1,m_2}(J_1,J_2)e^{im_1\phi_1 + im_2\phi_2}$$

with $m_1\omega_1 + m_2\omega_2 = 0$, i.e., $m_1/m_2 = \tilde{m}_1/\tilde{m}_2$. The range space is the complementary subspace of functions. Thus the normal form Hamiltonian is of the form

$$\Gamma = \Gamma(J_1,J_2,\tilde{m}_1\phi_1 + \tilde{m}_2\phi_2) \ , \tag{81}$$

which is not in action-angle form. However, (81) represents an integrable system, because Γ itself and $I = \tilde{m}_2 J_1 - \tilde{m}_1 J_2$ are independent commuting integrals. The terms that must be left in Γ are exactly those resonant with the unperturbed motion; in the surface of section given by Γ = const., $\phi_1 = 0$ and $J_1 = J_1(\Gamma, J_2, \phi_2)$ [by (81)], the level curves of $I = I(\phi_2, J_2, \Gamma)$ show islands due to this resonance.

The generalization of this procedure to n degrees of freedom [21] is a follows. Suppose there exist k independent relations of commensurability, i.e., an integer matrix M of rank k with

$$M_{ij}\omega_j = 0 \ . \tag{82}$$

Then any basis function of the form $g_m(J)\exp(im_1\phi_1 + \dots + im_n\phi_n)$, where the vector $\tilde{m}$ is a linear combination of the rows of M with integer coefficients, is in the null space of H_0, as are functions of the J_i alone. The normalized Hamiltonian again belongs to the null space and the range of H_0 is the complementary subspace. Now take a linear combination $\alpha_\ell J_\ell$ of action variables. This commutes with Γ if it commutes with $M_{ij}\phi_j$ for $i = 1,\dots,k$. This gives $M_{i\ell}\alpha_\ell = 0$, showing that the vector

$\underset{\sim}{\alpha}$ is in the null space of M, which has dimension n-k. Therefore, the normalized system possesses n - k + 1 independent commuting integrals of motion, including the Hamiltonian itself. If the series converges, the same can be said for the original Hamiltonian.

Time dependent resonant systems are treated in the same manner, by extending the phase space as usual.

C. Cases with zero frequencies

Up to this point we have considered systems whose linear part about an equilibrium point has all normal mode frequencies nonzero. Normalization of systems with zero frequencies was first described in [13, 15]. Examples of systems with a zero frequency include the motion of charged particles in a magnetic field; the linearized motion along the field has no restoring force and hence zero frequency. Mirroring of particles because of inhomogeneities is due to nonlinear terms.

As an example, consider a Hamiltonian system with two degrees of freedom whose linear part is given by

$$h_0 = \frac{1}{2} p_1^2 + \omega_2 J_2 , \tag{83}$$

that is, $\omega_1 = 0$ and so (q_1, p_1) are essentially in action-angle form as is $J_2 = (p_2^2 + q_2^2)/2$. Then the Lie operator associated with h_0 is

$$H_0 = - p_1 \frac{\partial}{\partial q_1} - \omega_2 \frac{\partial}{\partial \phi_2} . \tag{84}$$

The first term in H_0 is nilpotent: any function of p_1 alone $g(p_1) = O(p_1)$ is in its null space and is also the image of $-q_1 g(p_1)/p_1$. Alternatively, we can characterize the operator $J_+ = - p_1 \partial/\partial q_1$ as a raising operator, as in quantum mechanics. The homogeneous polynomials of degree k, $|m\rangle = q_1^{k-m} p_1^m$, satisfy $J_+|m\rangle = -(k-m)|m+1\rangle$. Therefore, if the first term on the left of (71) is in the range of H_0, as it must be, we cannot characterize Γ_k as being in the null space of H_0. However, we recall the theorem that for any operator such as H_0, the whole vector space (of functions, here) is the direct sum of the range of H_0 and the null space of its adjoint, defined with respect to some inner product [13, 15]. As we have discussed, it is not necessary to introduce an adjoint in order to carry out the normalization, but it helps in interpreting what the normal form Hamiltonian does for us. Let me now try to motivate my choice of inner product [13, 15]. If our basic expansion is that of power series in q,p about the equilibrium point $q_1 = q_2 = p_1 = p_2 = 0$, then terms of order ε^k in the expansion label the subspace V_{k+2} of homogeneous polynomials of degree k+2. Since the lowering operator L (and H_0 also) maps V_k to itself, we are motivated to define all monomials of different degree as orthogonal and the monomials

$$|j_1m_1j_2m_2\rangle = \left[\frac{(2j_1)!(2j_2)!}{(j_1-m_1)!(j_1+m_1)!(j_2-m_2)!(j_2+m_2)!}\right]^{1/2} q_1^{j_1-m_1} p_1^{j_1+m_1} q_2^{j_2-m_2} p_2^{j_2+m_2} \tag{85}$$

for $-j_1 < m_1 < j_1$ and $j_1 + j_2 = (k+2)/2$ as an orthonormal basis. For $J_+^{(1)} = \mathrm{ad}(p_1^2/2) = -p_1\partial/\partial q_1$ and $J_-^{(1)2} = \mathrm{ad}(q_1^2/2) = q_1\, \partial/\partial p_1$ we find $J_+^{(1)}|j_1m_1j_2m_2\rangle = -\sqrt{(j_1-m_1)(j_1+m_1+1)}\,|j_1,m_1+1,j_2,m_2\rangle$ and $J_-^{(1)}|j_1m_1j_2m_2\rangle = \sqrt{(j_1+m_1)(j_1-m_1+1)}\,|j_1,m_1-1,j_2,m_2\rangle$. Furthermore, the adjoint of $J_+^{(1)}$ is $-J_-^{(1)}$ and vice versa. [The coefficients in (85) were picked to make this last relation come out without grubby factors, but are in no way necessary.] Defining similar operators $J_+^{(2)}, J_-^{(2)}$ on (q_2,p_2) we find again $(\mathrm{ad}\ J_2)^+ = (J_+^{(1)} + J_-^{(2)})^+ = -J_-^{(2)} - J_+^{(2)} = -\mathrm{ad}\ J_2$ so that ad J_2 is antihermitian. From all this we find

$$H_0^+ = -q_1\frac{\partial}{\partial p_1} + \omega_2\frac{\partial}{\partial\phi_2}$$

$$= -\ \mathrm{ad}\left[\frac{q_1^2}{2} + \omega_2 J_2\right] . \tag{86}$$

Therefore the null space of H_0^+ consists of all functions of q_1 and J_2; the range of H_0 is the complementary subspace. Since the normal form terms Γ_k belong to $N(H_0^+)$ for $k > 1$, we find

$$\Gamma = \frac{1}{2}p_1^2 + \omega_2 J_2 + \tilde{\Gamma}(q_1,J_2) . \tag{87}$$

Again, this is not in action-angle form, but it is very useful. Obviously J_2 and Γ are commuting integrals of motion and (87) can be solved by quadratures (or reduced to normal form for the one remaining degree of freedom.) This procedure has been used [13, 15] to normalize the Störmer problem (dipole magnetic field) and, from this normal form Hamiltonian, compute periodic orbits and find minute regions of nonintegrability for fairly low energy [13, 15].

D. Adiabatic invariants

Another type of time dependent problem is one in which the time dependent terms are not small but instead vary slowly. As an example, pick

$$h = \frac{1}{2}p^2 + \frac{1}{2}\omega^2(\varepsilon t)q^2 , \tag{88}$$

as in Section 5. After a preparatory transformation, we find (56), and defining $x_0 = \varepsilon t$, obtain a new Hamiltonian

$$k = \frac{1}{2}\omega(x_0)(y^2 + x^2) + \frac{\varepsilon\omega'(x_0)}{2\omega(x_0)}xy + \varepsilon y_0 . \tag{89}$$

We have introduced the time as a coordinate in this manner in order to place the y_0

term in k_1. Defining action angle variables (J,ϕ) by $y = -\sqrt{2J}\sin\phi$, $x = \sqrt{2J}\cos\phi$ we have

$$k = \omega(x_0)J - \frac{\varepsilon\omega'(x_0)}{2\omega(x_0)} J\sin 2\phi + \varepsilon y_0 , \tag{90}$$

giving

$$K_0 = -\omega(x_0)\frac{\partial}{\partial\phi} + J\omega'(x_0)\frac{\partial}{\partial y_0} . \tag{91}$$

The second term in K_0 is ugly and I deal with it as follows: in the first order normalization equation (70), let Γ_1 contain the terms of h_1 independent of ϕ (none, because $\sin 2\phi$ averages to zero) plus εy_0. Then

$$f_1 = -\frac{\omega'(x_0)J\cos 2\phi}{4\omega^2(x_0)} . \tag{92}$$

Using this, it can be seen that the remainders r_k for $k > 2$ will not contain y_0 (although the εy_0 term in k_1 contributes to all the r_k), so that only the first term in K_0 need be kept in (71). That is, the f_k can be picked without y_0 dependence; $K_o f_k$ is chosen to belong to a proper subset of $N(K_0)$ for $k > 1$. Then the null space terms in r_k are those terms which, when expanded in the $e^{im\phi}$ basis, depend only on J and x_0, i.e., do not average to zero with respect to ϕ. (Note: Since ϕ is transformed at each step, averaging to zero with respect to ϕ is not equivalent to averaging to zero with respect to the original ϕ.)

Since $\Gamma_1 = 0$, we find that J commutes with $\Gamma = \omega(x_0)J$ (in the sense of extended Poisson brackets). This is the transformed action variable; expressing it in the old variables used in (90) we obtain

$$J = e^{-\varepsilon F_1} J_{old} = J_{old} - \frac{\varepsilon\omega'(x_0)}{2\omega^2(x_0)} J_{old}\sin 2\phi_{old} = \frac{1}{2\omega} p^2 + \frac{1}{2}\omega q^2 + \frac{\dot{\omega}}{2\omega^2} qp = k/\omega , \tag{93}$$

which is in agreement with the adiabatic invariant computed by other methods.

E. Superconvergent series

Kolmogorov's improvement of the Poincaré-Von Zeipel scheme, which enabled Arnold and Moser to prove their famous KAM theorem [22], has been described in terms of Deprit Lie transforms [23]. My point here is to show how simply this can be performed, or at least described, using infinite product Lie transforms.

Superconvergence, or renormalization, is achieved by using the normalized Hamiltonian Γ_k as part of the unperturbed operator, i.e., $H_0 \to H_0 + \hat{\Gamma}_1 + \hat{\Gamma}_2 + \ldots + \hat{\Gamma}_k$, as soon as Γ_k is available. Thus, we first transform $h = h_0 + \varepsilon h_1 + \ldots$ to form

$$h^{(1)} = e^{\varepsilon F_1} h \ . \tag{94}$$

The first order term gives

$$h_1^{(1)} = F_1 h_0 + h_1 \tag{95}$$

or

$$H_0 f_1 + \Gamma_1 = h_1 \ , \tag{96}$$

as in Sec. VI. Thus we pick $h_1^{(1)} = \Gamma_1$ to be in normal form as before. Next, apply the next transform to obtain

$$h^{(2)} = e^{\varepsilon^2 F_2} e^{\varepsilon F_1} h \ . \tag{97}$$

The zeroth and first order terms are h_0 and Γ_1, and the second and third order terms give

$$h_2^{(2)} = F_2 h_0 + h_2^{(1)} \ , \tag{98}$$

$$h_3^{(2)} = F_2 \Gamma_1 + h_3^{(1)} \ . \tag{99}$$

The higher order terms involve F_2^2 and therefore will not be useful in obtaining F_2. Adding and rearranging, we find

$$(H_0 + \hat{\Gamma}_1) F_2 + \Gamma_2 + \Gamma_3 = h_2^{(1)} + h_3^{(1)} \ . \tag{100}$$

That is, $h_2^{(2)}$ and $h_3^{(2)}$ can be put into normal form by considerations of the range of $H_0 + \hat{\Gamma}_1$. This process can be continued and, in computing F_k, we find $\Gamma_k + \ldots + \Gamma_{2k-1}$ with our "unperturbed" operator $H_0 + \hat{\Gamma}_1 + \ldots + \hat{\Gamma}_{k-1}$ because, as before, the process need only terminate at the order where F_k^2 comes in. Therefore, each F_k is of order 2^k in ε (but has all orders up to $2^k - 1$ order in it.) That is, the successive corrections are quadratic with respect to their antecedent, as in Newton's method.

This technique, which has without a doubt shown its usefulness in mathematical proofs, may or may not be advantageous in actual perturbation calculations because, in spite of quadratic convergence, terms of all orders in ε are still contained in the generators and in Γ. Also, inverting the operators $H_0 + \hat{\Gamma}_1 + \hat{\Gamma}_2 + \ldots$ on their range spaces may be difficult and not easy to automate for a machine. However, this method is in principle attractive because, for example, the distinction between resonant and nonresonant cases described in subsections A and B is somewhat artificial. I say this because this distinction is based upon a property of the unperturbed Hamiltonian, namely commensurability, that can be disturbed by an arbitrarily small perturbation. This distinction does not occur in the superconvergent algorithm because the "unperturbed" operator is successively updated.

ACKNOWLEDGEMENTS

I wish to thank André Deprit, Alex Dragt, Etienne Forrest, Celso Grebogi, and Curtis Menyuk for stimulating and enlightening discussions.

REFERENCES

[1] G. Hori, Publ. Astron. Soc. Japan 18, 287 (1966).

[2] A. Deprit, Celestial Mech. 1, 12 (1969); A. Deprit, J. Henrard, J. Price, and A. Rom, Celestial Mech. 1, 222 (1969).

[3] A. Kamel, Celestial Mech. 3, 90 (1970).

[4] A. Dragt, IEEE Trans. Nucl. Sci. NS-26, 3601 (1979); D. Douglas and A. Dragt, IEEE Trans. Nucl. Sci. NS-30, 2442 (1983).

[5] A. Dragt, J. Opt. Soc. Am. 72, 372 (1982).

[6] R. Littlejohn, Phys. Fluids 24, 1730 (1981); J. Plasma Phys. 29, 111 (1983).

[7] R. Dewar, J. Phys. 11, 9 (1978).

[8] C. Grebogi and A. Kaufman, "Multi-dimensional canonical symplectic maps for gyro-resonance crossings," in Long-Time Prediction in Dynamics, L.E. Reichl and A.G. Szebehely, Eds. (Wiley, New York, 1983).

[9] C. Grebogi and R. Littlejohn, Phys. Fluids 27, 1996 (1984).

[10] C. Menyuk, "Particle motion in the field of a modulated wave," University of Maryland, Plasma Preprint 83-058 (1983).

[11] J. Cary and R. Littlejohn, Ann. Phys. 151, 1 (1983).

[12] S. Johnston and A. Kaufman, Phys. Rev. Lett. 40, 1266 (1978).

[13] J.M. Finn, Doctoral Thesis, University of Maryland, 1974 (unpublished).

[14] A. Dragt and J.M. Finn, J. Math. Phys. 17, 2215 (1976).

[15] A. Dragt and J.M. Finn, J. Math. Phys. 20, 2649 (1979); A. Dragt and M.J. Finn, J. Geophys. Res. 81, 2327 (1976).

[16] J. Cary, Phys. Rep. 79, 131 (1981).

[17] R. Dewar, J. Phys. A 9, 2043 (1976).

[18] P. Bakshi, Boston College Report COO-2714-12, 1976 (unpublished).

[19] A. Dragt and E. Forrest, J. Math. Phys. 24, 2734 (1983).

[20] I. Prigogine, Non-Equilibrium Statistical Mechanics (Interscience, New York, 1962), p. 17.

[21] F. Gustavson, Astron. J. 71, 670 (1964).

[22] V.I. Arnold and A. Avez, Ergodic Problems of Classical Mechanics (Benjamin, New York, 1968); A. Kolgomorov, Dokl. Akad. Nauk. 98 (1954); J. Moser, Nachr. Akad. Wiss. Göttingen, Math. Phys. Kl. II (1962); V.I. Arnold, Russian Mathematical Surveys 18, 85 (1963).

[23] B. McNamara, J. Math, Phys. 10, 2154 (1978).

THE COVARIANT LIE-TRANSFORMED PLASMA ACTION PRINCIPLE

Allan N. Kaufman
Physics Department and Lawrence Berkeley Laboratory
University of California
Berkely, CA 94720

The Lie Transform is a systematic technique [1] for obtaining the nonlinear quasi-static effects of high-frequency phenomena. We consider, in particular, a plasma of charged particles in a self-consistent electromagnetic wave. Each particle oscillates about an "oscillation center" which undergoes nonoscillatory motion [2,3]. We obtain the so-called "ponderomotive" Hamiltonian for the oscillation center in relativistically covariant form and demonstrate, via the action principle, that it determines the dielectric susceptibility and wave propagation.

We begin with the (eight) canonical variables $z = (r,p)$, which satisfy the covariant Hamiltonian equations

$$dr^{\mu}/d\tau = \partial H/\partial p_{\mu}, \qquad dp_{\mu}/d\tau = -\partial H/\partial r^{\mu}, \tag{1}$$

with the invariant Hamiltonian function $H(z;A)$, a functional of the Maxwell potential $A(x)$. The correct particle evolution equations are obtained by the choice[†] (c=1):

$$H = (p - eA(r))^2/2m, \tag{2}$$

noting that

$$A_{\mu}(r) = \int d^4x \; \delta^4(x-r) \; A_{\mu}(x). \tag{3}$$

Consider the family of all phase-space trajectories $z(\tau)$, each parametrized smoothly by its proper time τ. The seven-dimensional time-like surface $\tau = 0$ will be called the "initial-condition" surface. We introduce arbitrary coordinates η on this surface, and let $g(\eta)d^7\eta$ denote the number of particles (of a given species) in $d^7\eta$. With the trajectory $z(\tau;\eta)$ considered as an eight-dimensional field on the eight-dimensional phase-space (τ,η) [it's actually a mapping of phase space onto itself], we construct the action functional for the system

$$S[z(\tau,\eta),A(x)] = \int d^7\eta \; g(\eta) \int d\tau \; [p_{\mu}(\tau,\eta)dr^{\mu}(\tau,\eta)/d\tau - H(p,r;A)]$$

$$- \int d^4x \; F_{\mu\nu}F^{\mu\nu}/16\pi, \tag{4}$$

where $F_{\mu\nu} = \partial_{\mu}A_{\nu} - \partial_{\nu}A_{\mu}$ is the Maxwell field.

Variation of S with respect to $z(\tau;\eta)$ yields the Hamiltonian equations (1), while variation with respect to A(x) yields the Maxwell equation

$$F^{\mu\nu}{}_{,\nu} = 4\pi j^{\mu}, \tag{5}$$

with

$$j^{\mu}(x) = \int d^7\eta \; g(\eta) \int d\tau \; j^{\mu}(x;z) \tag{6}$$

and

$$j^{\mu}(x;z) = -\; \delta H(z)/\delta A_{\mu}(x). \tag{7}$$

Using (2) and (7), we obtain:

$$j^{\mu}(x;z) = (e/m)\; \delta^4(x-r)\; (p^{\mu} - eA^{\mu}). \tag{8}$$

We now define the (invariant) Vlasov distribution

$$f(\bar{z}) = \int d^7\eta \; g(\eta) \int d\tau \; \delta^8(\bar{z} - z(\tau,\eta)) \tag{9}$$

and derive the Ignatiev equation [4]

$$\{f(z),\; H(z)\} = 0 \tag{10}$$

in terms of the covariant canonical Poisson bracket:

$$\{a,b\} = (\partial a/\partial r^{\mu})(\partial b/\partial p_{\mu}) - (\partial a/\partial p_{\mu})(\partial b/\partial r^{\mu}). \tag{11}$$

To derive (10), we introduce the intermediate distribution

$$\bar{g}(\bar{z};\tau) = \int d^7\eta \; g(\eta)\; \delta^8(\bar{z} - z(\tau,\eta)) \tag{12}$$

and obtain by (1):

$$\partial g/\partial\tau = -\{g,H\}. \tag{13}$$

Integration over τ then yields (10), since $g = 0$ at infinite τ.

We now restrict the Maxwell potential to represent a single wave of eikonal form:

$$A_{\mu}(x) = \tilde{A}_{\mu}(x)\; \exp\; i\theta(x)/\epsilon + \text{c.c.}, \tag{14}$$

where the amplitude $\tilde{A}$ and the gradient of the phase $k_{\mu}(x) = \partial_{\mu}\theta(x)$ are slowly varying fields. We substitute (14) into (2) and then invoke the Lie Transform to eliminate rapidly oscillating terms.

The new Hamiltonian K is related to the old H by the formula [1]

$$K = [\exp\; i\; \{w,\cdot\}]\; H, \tag{15}$$

where w(z) is a suitably chosen generating function. Expanding both H and K in powers of the amplitude $\tilde{A}$, we have $K^{(0)} = H^{(0)}$ and

$$K^{(1)} = H^{(1)} + \{w,\; H^{(0)}\}. \tag{16}$$

We can require $K^{(1)}$ to vanish if we ignore the problems of resonant denominators. Solving (16) for w(z), and proceeding to second order, we obtain $K^{(2)}(z)$, which we denote by $\Psi(z;A)$, the relativistically invariant ponderomotive potential:

$$\Psi(z;A) = \int d^4x\, A^*_\mu(x)\, \Psi^\mu{}_\nu(x;z) A^\nu(x), \tag{17}$$

where

$$\Psi^\mu{}_\nu(x;z) = \delta^4(x-r)(e^2/m)(p\cdot k)^{-2}\, [k^2 p^\mu p_\nu + \delta^\mu_\nu (k\cdot p)^2 - (k\cdot p)(p^\mu k_\nu + k^\mu p_\nu)]. \tag{18}$$

This may be expressed more concisely as

$$\Psi = m\,|d\tilde{u}/d\tau|^2/(d\Theta(r)/d\tau)^2, \tag{19}$$

and reduces in the rest-frame to the familiar

$$\Psi = e^2\,|\tilde{E}|^2/m\omega^2. \tag{20}$$

Invariance of the phase-space Lagrangian under a canonical transformation:

$$\int d\tau\, [p_\mu dr^\mu/d\tau - H(p,r)] = \int d\tau\, [\bar{p}_\mu d\bar{r}^\mu/d\tau - K(\bar{p},\bar{r})], \tag{21}$$

where the overbar denotes oscillation-center variables, converts the action to

$$S = \int d^7\eta\, g(\eta) \int d\tau\, [\bar{p}_\mu d\bar{r}^\mu/d\tau - \bar{p}^2/2m - \Psi(z;A)] - \int d^4x\, F^*_{\mu\nu} F^{\mu\nu}/8\pi. \tag{22}$$

Defining the invariant oscillation-center distribution $F(z) = \int d^7\eta\, g(\eta) \int d\tau\, \delta^8(z-\bar{z}(\tau,\eta))$, we obtain, in analogy to the steps leading to (10), the corresponding Ignatiev equation:

$$\{F(z),\ p^2/2m + \Psi(z;A)\} = 0. \tag{23}$$

In order to vary S with respect to $\tilde{A}_\mu(x)$ and $\Theta(x)$, we first substitute (17) into (22), obtaining for the terms bilinear in A

$$S^{(2)} = \int d^4x\, \tilde{A}^*_\mu(x) D^\mu{}_\nu(x;F) \tilde{A}^\nu(x), \tag{24}$$

where the dielectric matrix $D^\mu{}_\nu$ is the sum of its vacuum part

$$D_{vac}{}^\mu{}_\nu(k(x)) = (k^\mu k_\nu - k^2 \delta^\mu_\nu)/4\pi \tag{25}$$

and the susceptibility

$$\chi^\mu{}_\nu(x;F) = -\int d^8z\, F(z)\, \Psi^\mu{}_\nu(x;z). \tag{26}$$

The relation (26) is the "K-χ theorem" [5-8] which is seen to be the essential ingredient of the action functional coupling F(z) to $A_\mu(x)$.

We now express the dielectric matrix in terms of its local eigenvalues $D_\alpha(x)$ and eigenvectors $e_\alpha(x)$:

$$D^\mu{}_\nu(x) = \sum_\alpha D_\alpha(x) e^\mu_\alpha(x) e^{\alpha *}_\nu(x), \tag{27}$$

whence

$$S^{(2)} = \int d^4x\, D_\alpha(x)\, |A_\alpha(x)|^2, \tag{28}$$

with $A_\alpha(x) = e^\alpha{}_v{}^*(x)\tilde{A}^v(x)$, the projection of A on the α eigenvector.

We now vary $S^{(2)}$ with respect to $A_\alpha(x)$, obtaining the eikonal equation for the phase

$$D_\alpha(x,k = \partial\theta/\partial x) = 0 \tag{29}$$

associated with polarization e_α. This yields the covariant ray equations

$$dx^\mu/d\sigma = -\,\partial D_\alpha/\partial k_\mu, \qquad dk_\mu/d\sigma = \partial D_\alpha/\partial x^\mu . \tag{30}$$

Variation with respect to $\theta(x)$ yields the wave-action conservation law [9] $\partial J^\mu(x)/\partial x^\mu = 0$, where the wave-action density four-vector is

$$J^\mu(x) = -\,A_\alpha(x)^2 \partial D_\alpha(x,k)/\partial k_\mu . \tag{31}$$

Since the eigenvalues $D_\alpha(x)$ are functionals of the oscillation-center distribution F(z) by (26), we have thus obtained a closed self-consistent set of coupled equations for F(z) and the wave amplitude and phase.

D. D. Holm was instrumental in the formulation of the principles presented here at the Aspen Center for Physics. This research was supported by the Office of Energy Research of the U.S. Department of Energy.

REFERENCES

[1] J. R. Cary, Phys. Rep. 79, 131 (1981).

[2] R. L. Dewar, Phys. Fluids 16, 1102 (1973).

[3] R. L. Dewar, Aust. J. Phys. 30, 533 (1977).

[4] Yu. G. Ignatiev, Zh. Eksp. Teor. Fiz. 81, 3 (1981) [Sov. Phys. JETP 54, 1 (1981)].

[5] J. R. Cary and A. N. Kaufman, Phys. Rev. Lett. 39, 402 (1977).

[6] S. Johnston and A. N. Kaufman, Phys. Rev. Lett. 40, 1266 (1978).

[7] J. R. Cary and A. N. Kaufman, Phys. Fluids 24, 1238 (1981).

[8] A. N. Kaufman and B. M. Boghosian, Contemporary Math. 28, 169 (1984).

[10] R. L. Dewar, Astrophys. J. 174, 301 (1972).

† Ed. note: The "classical" (i.e., noncovariant) particle Hamiltonian is $h \equiv -\,p_0 = p^0$ (space-favoring metric: (+1,+1,+1,-1)). Using eq. (2), together with $H = -\,\frac{1}{2}m$, which holds when τ is the invariant (particle) proper time, the zero-component of the first of eqs (1) gives $h = m\,\frac{dr^0}{d\tau} + ea^0 = [(\underset{\sim}{p} - e\underset{\sim}{a})^2 + m^2]^{\frac{1}{2}} + ea^0$.

GEOMETRIC HAMILTONIAN STRUCTURES AND PERTURBATION THEORY*

Stephen Omohundro
Lawrence Berkeley Laboratory and Physics Department
University of California
Berkeley, CA 94720

1. INTRODUCTION

In this lecture we discuss the ideas in [25] intuitively and heuristically. That paper assumes a background in geometric mechanics and has detailed proofs. Here we will give the flavor of the structures and develop the needed background material. We will state the results and indicate why they are true without detailed proof. We begin with some introductory remarks, discuss a geometric picture for nonsingular perturbation theory, introduce the needed Hamiltonian mechanics including the crucial process of reduction in the presence of symmetry, describe the Hamiltonian structure of nonsingular perturbation theory, and close with some discussion of these ideas in connection with the method of averaging.

It is of interest to list the seminal ideas that form the background of the present work. In 1808 Lagrange introduced the description of the dynamics of celestial bodies in terms of what we today call Hamilton's equations [1,2]. His motivation was the reduction of the enormous labor involved in a straightforward perturbation analysis, which required tedious computations to be performed on each component of the dynamical vector field, to manipulations of a single function: the Hamiltonian. The description in terms of Lagrange brackets led to several other benefits: Lagrange showed that the value of the Hamiltonian and the structure of the brackets were both invariant under the dynamics, leading to a useful check of the complex calculations (which at that time were done by hand). In addition, he was able to show that the invariance of the Hamiltonian could be used to prove stability of certain equilibria. As the century progressed, Hamiltonian mechanics was refined and the connections with variational principles and optics were made. By the turn of the century Poincaré [3] had developed very powerful Hamiltonian perturbation methods utilizing generating functions, introduced the notion of asymptotic expansion, and begun the geometric and topological approach to dynamics. In 1918 Emmy Noether [4] made the connection between symmetries and conserved quantities. The development of quantum mechanics rested heavily on the Hamiltonian framework [5] by analogy with optics and served to put it firmly at the center of the modern formulation of fundamental physics [6]. During the 1960s the coordinate free description of Hamiltonian structures in terms of symplectic geometry was developed [7,8]. About this time the Hamiltonian method of Lie transforms greatly simplified Hamiltonian perturbation theory [9]. The 1970s saw enormous developments in the geometric approach to mechanics and largely as a result of these, an

*This work was supported by the Office of Basic Energy Sciences of the U.S. Department of Energy under Contract No. DE-AC03-76SF00098.

ever wider range of physical systems have been described in Hamiltonian terms. Some examples are: quantum mechanics [10], fluid mechanics [11,12,13], Maxwell's equations [15,16], the Maxwell-Vlasov and Poisson-Vlasov [14,15] equations of plasma physics, elasticity theory [17,18], general relativity [18], magnetohydrodynamics [11,17], multi-fluid plasmas [17,19], chromohydrodynamics [20], superfluids and superconductors [21], the Korteweg de Vries equation [22], etc. These developments have shed light on the underlying symmetry structure of these theories, have yielded improved stability results based on Arnol'd's stability method [23], and have given insight into the reasons for the integrability of certain systems [24].

For the most part, however, these structures describe fundamental underlying models in the various fields. In actual applications we almost always make numerous approximations which may or may not respect the underlying Hamiltonian structure. It is folklore within the particle physics community and elsewhere that perturbation methods which respect the underlying symmetries and conservation laws yield much better approximations to the actual system than those which do not. It is of interest, then, to try to do perturbation theory within the Hamiltonian framework and to obtain structures relevant to the approximate system. One may thus hope to understand the relation between the structures of systems which are limiting cases of known systems (e.g., does the KdV Poisson bracket arise naturally from that of the Boussinesq equations?) [31]. The history of Hamiltonian mechanics is inextricably tied to perturbation methods. For the most part, though, the Hamiltonian structure was used to simplify the perturbation method and the geometric structure of the perturbation method itself was not explored. We have found in several examples that taking this structure into account leads to simplifications (as in the problem of guiding center motion discussed later in this paper) and to deeper insight into the approximate system (as in modulational equations for waves in the eikonal limit) [28].

We have therefore been engaged in a program of investigating the Hamiltonian structure of the various perturbation theories used in practice. In this paper we describe the geometry of a Hamiltonian structure for nonsingular perturbation theory applied to Hamiltonian systems on symplectic manifolds and the connection with singular perturbation techniques based on the method of averaging.

2. GEOMETRIC PERTURBATION THEORY

The modern setting for describing an evolving system is a dynamical system. The state of the system is represented by a point in a manifold M. *A manifold* is a space which locally looks like Euclidean space and in which there is a notion of derivative. Globally a manifold may be connected together in a funny way, as in a sphere or a torus. If you know where you are, the dynamics tells you where you're going. Thus dynamics is represented by a vector field on the manifold of states. A *dynamical system* is a manifold with a vector field defined on it. In coordinates the dynamics gives a set of first order O.D.E.s, one for each coordinate. Typical dynamical systems with state spaces of three dimensions or greater have chaotic dynamics with extremely complicated trajectories for which one can prove there is no closed form exact description. If the dynamics simplifies then there is usually some physically relevant special feature such as a symmetry which causes the simplification.

In important physical applications, we often find ourselves close to a system which simplifies and we are interested in the effect of our deviation from it. We express this deviation in terms of the small parameter ϵ. If we are given dynamics in the form

$$\dot{x} = X_0 + \epsilon X_1 + \frac{\epsilon^2}{2} X_2 + \ldots \tag{1}$$

in terms of the vector fields X_i with initial conditions described by $x(\epsilon, t = 0) = y(\epsilon)$, we may attempt to express the solution in an asymptotic series in ϵ:

$$x(t) = x_0(t) + \epsilon x_1(t) + \frac{\epsilon^2}{2} x_2(t) + \ldots . \tag{2}$$

Choosing coordinates x^a $(1 \leqslant a \leqslant N)$ in a local patch and plugging this assumed asymptotic form into the equation of motion gives:

$$\begin{aligned} \dot{x}_0^a + \epsilon \dot{x}_1^a + \frac{\epsilon^2}{2!} \dot{x}_2^a + \ldots = {} & X_0^a(x_0 + \epsilon x_1 + \frac{\epsilon^2}{2!} x_2 + \ldots) + \\ & + \epsilon X_1^a(x_0 + \epsilon x_1 + \frac{\epsilon^2}{2!} x_2 + \ldots) + \\ & + \frac{\epsilon^2}{2!} X_2^a(x_0 + \epsilon x_1 + \frac{\epsilon^2}{2!} x_2 + \ldots) + \ldots . \end{aligned} \tag{3}$$

Since asymptotic expansions are unique, we may equate coefficients of equal powers of ϵ to get equations for $x_0, x_1, \ldots$:

$$\begin{aligned} \dot{x}_0^a &= X_0^a(x_0), \\ \dot{x}_1^a &= \sum_{b=1}^{N} \frac{\partial X_0^a}{\partial x^b}(x_0) \cdot x_1^b + X_1^a(x_0), \\ \dot{x}_2^a &= \sum_{b,c=1}^{N} \frac{\partial^2 X_0^a}{\partial x^b \partial x_c}(x_0)\, x_1^b x_1^c + \sum_{b=1}^{N} \frac{\partial X_0^a}{\partial x^b}(x_0) \cdot x_2^b \\ &\quad + 2 \sum_{b=1}^{N} \frac{\partial X_1^a}{\partial x^b}(x_0) \cdot x_1^b + X_2^a(x_0), \\ &\ldots . \end{aligned} \tag{4}$$

If $y(\epsilon) = y_0 + \epsilon y_1 + \frac{\epsilon^2}{2} y_2 + \ldots$ is an asymptotic expansion for the initial condition $y(\epsilon)$, then the initial conditions for these equations are: $x_0(t = 0) = y_0$, $x_1(t = 0) = y_1$, $\ldots$.

These equations immediately raise a number of questions. They are defined in terms of physically irrelevant coordinates; is the perturbation structure independent of these coordinates? If the original equations are Hamiltonian, are these equations? In Jth order perturbation theory, how are we to interpret this evolution of many variables $x_0, x_1, \ldots, x_J$? The goal of this work is to answer these questions.

Let us turn to the geometric interpretation of these equations. It is easiest to understand the first order perturbation equations:

$$\dot{x}_0^a = X_0^a(x_0)$$

$$\dot{x}_1^a = \sum_{b=1}^{N} \frac{\partial X_0^a}{\partial x^b}(x_0) \cdot x_1^b + X_1^a(x_0) \tag{5}$$

$$x_0(t=0) = y_0, \quad x_1(t=0) = y_1.$$

We would like to determine the geometric nature of the quantities x_0 and x_1. To understand what we mean by this let us recall the relationship between geometric quantities and coordinates. A function on a manifold is an intrinsically defined thing, it assigns a real number of each point of the manifold. A coordinate system on a region of an N dimensional manifold is a collection of N real valued functions $x^1, \ldots, x^N$ defined on that region, whose differentials are linearly independent at each point. In these coordinates, the gradient of a function is a collection of N numbers, the derivatives with respect to each of the x^a. Geometrically, however, it is wrong to think of these as just real numbers because they change if we change our coordinate system. For example, if we choose coordinates whose values at each point of the region are twice those of $x^1, \ldots, x^N$ then the components of the gradient of a function are halved. We introduce a geometric object whose relationship to the manifold at a given point is like that of the gradient of a function and we call it a *covector* or *one-form.* The collection of all covectors at a point is defined to be the *cotangent space* at that point and the collection of all cotangent spaces taken together form the *cotangent bundle.* Similarly the components of a *vector* at a point double with the coordinates. All vectors at a point taken together form the *tangent space* at that point and all tangent spaces taken together form the *tangent bundle TM* of *M.* Vectors and covectors are thus different objects when we consider more than one coordinate system, even though they both have N components in any given system. Our interest here will be to find out whether the quantities $x_0^a, \ldots, x_J^a$ for $1 \leqslant a \leqslant N$ have any geometric structure that is independent of a given coordinate system.

Intuitively, the first order quantity x_1 represents a small deviation from the unperturbed quantity x_0. Because x_0 can vary over the whole manifold M, we expect it to represent a point in the manifold. As ϵ gets smaller $x_0 + \epsilon x_1$ approaches the point x_0 and x_1 measures the first order rate of approach to x_0. Two different paths in the manifold approaching the point x_0 as ϵ approaches zero have the same x_1 if and only if they are tangent at x_0. This, however, is the defining criterion for a vector at the point x_0. We thus expect x_1 to lie in the tangent space to M over the point x_0. The x_0, x_1 dynamics then takes place on the tangent bundle TM. We will describe this dynamics on TM intrinsically in terms of vector fields derived from $X(\epsilon)$ on M.

The solution of a system of O.D.E.s tells us the state at each time t of a system which began with each initial condition. Geometrically, this is a mapping of M to itself for each t. If the solution doesn't run off of the manifold, then the uniqueness and smoothness of solutions with given initial conditions tells us that this map is a *diffeomorphism* (i.e. a smooth, 1-1, onto map with smooth inverse). This one-parameter family of diffeomorphisms labelled by t is called the *flow* of the dynamical vector field. As ϵ varies, the corresponding flows of $X(\epsilon)$ will vary. Perturbation theory describes that variation. Any time we have a mapping f from one manifold to another, we may define its *differential* Tf. This is a map that takes the tangent bundle of the first manifold to the tangent bundle of the second. It describes how infinitesimal perturbations at a point are sent to infinitesimal perturbations at the image point. In coordinates, it acts on the tangent space at a point via the Jacobian matrix of f at that point.

Through abuse of notation, let us denote the flow of the unperturbed vector field X_0 by $x_0(t)$. $x_0(t, y_0)$ is the point to which y_0 has flowed in time t under X_0. A small perturbation in M from a given orbit will evolve under X_0 according to the derivative of this flow $Tx_0(t)$. This is a flow on the manifold TM, and the vector field of which it is the flow may be written

$$\tilde{X}_0 \equiv \left.\frac{d}{dt}\right|_{t=0} Tx_0(t). \tag{6}$$

$\tilde{X}_0$ is a vector field on TM defined without recourse to coordinates that represents the effect of the unperturbed flow on perturbed orbits. In coordinates $\tilde{X}_0$ has components

$$\begin{aligned} \dot{x}_0^a &= X_0^a(x_0), \\ \dot{x}_1^a &= \sum_{b=1}^{N} \frac{\partial X_0^a}{\partial x^b}(x_0) \cdot x_1^b. \end{aligned} \tag{7}$$

This dynamics is exactly that part of the perturbation dynamics (5) which depends on X_0. The part which depends on X_1 may also be defined intrinsically. If we define

$$\tilde{X}_1(x, \mathrm{v}) \equiv \left.\frac{d}{dt}\right|_{t=0} (\mathrm{v} + t\, X_1(x)), \tag{8}$$

then the entire first order perturbation dynamics on TM is given by

$$\tilde{X}_0 + \tilde{X}_1. \tag{9}$$

3. THE GEOMETRY OF JTH ORDER PERTURBATION THEORY

We have succeeded in finding a geometric, coordinate-free interpretation for first order perturbation theory. We now would like to extend this to higher orders. The geometric object that arises is called a *jet*. To understand the setting, we discuss a number of relevant spaces.

How are we to think of the exact equation for the evolution of an ϵ dependent point $x(\epsilon)$ under ϵ-dependent evolution equations $\dot{x}(\epsilon) = X(\epsilon, x)$ with ϵ-dependent initial conditions $y(\epsilon)$? It is useful to think of the ϵ-dependent point $x(\epsilon)$ as a curve in the space $I \times M$, where I is the interval (say $[0,1]$) in which ϵ takes its values. If we think of $x(\epsilon)$ as a map from I to M, then the curve is the graph of this map. The dynamical vector field $X(\epsilon)$ naturally lives on $I \times M$ and its I component is zero everywhere. The flow of $X(\epsilon)$ on $I \times M$ takes paths to paths by letting each point of a path move with the flow. Our initial conditions are represented by paths (if they are independent of ϵ then they are straight lines). The true dynamics takes paths to paths. Even if the initial conditions are ϵ independent, the dynamics bends the path over. Thus we really should think of our dynamics as living on the infinite dimensional *path space*

$$P_1M \equiv \left\{\text{space of all paths } p\colon\ I \to I \times M \text{ of the form } p\colon\ \epsilon \mapsto (\epsilon, x(\epsilon))\right\}, \tag{10}$$

where as before $I = [0,1]$.

This projects naturally onto

$$P_0M \equiv \left\{\text{equivalence classes in } P_1M \text{ where } p_1 \sim p_2 \text{ iff } p_1(0) = p_2(0)\right\}. \tag{11}$$

P_0M is naturally isomorphic to M and represents the domain of the unperturbed dynamics. The equivalence classes forget all perturbation information and only remember behavior at $\epsilon = 0$. We are interested in spaces through which this projection of real to unperturbed dynamics factors. Perturbation theory tries to study behavior infinitesimally close to $\epsilon = 0$ without actually getting there. For each $0 \leqslant \alpha \leqslant 1$ we may define

$$P_\alpha M \equiv \left\{\text{equivalence classes in } P_1M \text{ where } p_1 \sim p_2 \text{ iff } p_1(\epsilon) = p_2(\epsilon) \text{ for all } 0 \leqslant \epsilon \leqslant \alpha\right\}. \tag{12}$$

These allow us to consider more and more restricted domains of ϵ, but there is always a continuum of ϵ's to traverse before reaching $\epsilon = 0$. For each $1 \geqslant \alpha_1 \geqslant \alpha_2 \geqslant 0$ we have the natural maps

$$P_1M \to P_{\alpha 1}M \to P_{\alpha 2}M \to P_0M. \tag{13}$$

We are interested in structure between even the smallest $P_\alpha M$ with $\alpha \neq 0$ and P_0M. We may introduce *germs* of paths:

$$GM \equiv \{\text{equivalence classes in } P_1M \text{ where } p_1 \sim p_2 \text{ iff there exists } \alpha > 0 \text{ such that } p_1(\epsilon) = p_2(\epsilon) \text{ for all } 0 \leqslant \epsilon \leqslant \alpha\} \tag{14}$$

and for any $\alpha > 0$ we have $P_\alpha M \to GM \to P_0M$. The germs capture behavior closer to $\epsilon = 0$ than any given ϵ, but still contain much more information than perturbation theory gives us (germs depend on features of functions in a little neighborhood that are not captured in a Taylor series).

Finally we may introduce spaces of *jets* of paths at $\epsilon = 0$ with integer $1 \leqslant J \leqslant \infty$:

$$JM \equiv \left\{ \text{equivalence classes in } P_1M \text{ where } p_1 \sim p_2 \text{ iff} \right.$$

for all C^∞ functions f on $I \times M$ we have : (15)

$$\left. \frac{\partial^i}{\partial \epsilon^i}\right|_{\epsilon=0} f(p_1(\epsilon)) = \left. \frac{\partial^i}{\partial \epsilon^i}\right|_{\epsilon=0} f(p_2(\epsilon)) \quad \text{for } 0 \leqslant i \leqslant J \Bigg\}.$$

Thus the space of J-jets gives the first J terms in a Taylor expansion of the curve around $\epsilon = 0$ in any coordinate system. Clearly,

$$GM \to \infty M \to IM \to JM \to P_0M \quad \text{for } I > J.$$

Thus the jets focus on information closer to $\epsilon = 0$ than even the germs.

If x^a for $1 \leqslant a \leqslant N$ are coordinates on $M \approx P_0M \approx OM$, then we may introduce coordinates $\{x_0^a, x_1^a, \ldots, x_J^a\}$ for $0 \leqslant J \leqslant \infty$ on JM to represent the equivalence class of the curve

$$x_0^a + \epsilon x_1^a + \frac{\epsilon^2}{2!} x_2^a + \ldots + \frac{\epsilon^J}{J!} x_J^a$$

in $I \times M$ (near $\epsilon = 0$ this will not leave the chart on which the x^a are defined).

The claim here is that JM represents geometrically the perturbation quantities $x_0, \ldots, x_J$. It may seem strange to go through the infinite dimensional space P_1M to get to it, but we shall see (especially when looking at the Hamiltonian structure) that it organizes and simplifies the structures of interest. It is a completely intrinsic and natural (category theorists would say functorial) operation to go from the original dynamical manifold M to the path space P_1M to the jet space JM. We shall now show that the dynamics on M also induces natural dynamics on P_1M and then projects from there down to JM where it is the perturbation dynamics we are interested in. Later we will show that a Hamiltonian structure on M leads to Hamiltonian structures on P_1M and JM. The dynamics $\dot{x} = X(\epsilon, x)$ takes elements of P_1M to other elements of P_1M and in fact takes equivalence classes to equivalence classes for each of $P_\alpha M$, GM, ∞M, JM, and M. This is what allows us to obtain an induced dynamics on each of these spaces. To determine this dynamics explicitly, we must understand what a tangent vector on each space is.

Intuitively, a vector represents a little perturbation to a point. We define it precisely as an equivalence class of tangent curves, where the curve represents the direction of perturbation and the equivalence class ensures that only the first order motion is reflected in the tangent vector. A point in the path space P_1M is a path in $I \times M$. A small perturbation thus gives a nearby path. Each point of the path is perturbed a little bit and we are interested in the first order perturbation. Thus we expect a tangent vector to a point in path space to be a vector field in $I \times M$ along the corresponding path. A curve $p(\gamma)$ in P_1M parameterized by γ defines a curve $p(\epsilon, \gamma)$ for each ϵ through $p(\epsilon, \gamma = 0)$ in $I \times M$. The equivalence class of curves in P_1M defining a vector thus reduces to an equivalence class of curves in M for each ϵ. We may therefore identify a tangent vector to p in P_1M with a field of vectors over p in $I \times M$ such that each vector has no $\frac{\partial}{\partial \epsilon}$ component. For $p \in P_1M$ a vector $\tilde{V} \in T_p(P_1M)$ is a map

$$\tilde{V} : I \rightarrow I \times TM \tag{16}$$

taking $\tilde{V} : \epsilon \mapsto (\epsilon, V(\epsilon))$ where $V(\epsilon) \in T_{p(\epsilon)}M$.

The tangent spaces to the quotient spaces are defined by taking the derivatives of the natural projections. Because $P_0M \approx M$, we see that $TP_0M \approx TM$. Because $1M \approx TM$ we see that $T1M \approx TTM$. Thus the first order perturbation space $1M$ is naturally TM and the dynamics is a vector field on TM as we saw earlier.

As with all tangent bundles, TJM has a natural coordinate chart, derived from the coordinates $\{x_0^a, \ldots, x_J^a\}$, $1 \leqslant a \leqslant N$ on JM defined earlier. We obtain coordinates $\{x_0^a, \ldots, x_J^a, \mathrm{v}_0^a, \ldots, \mathrm{v}_J^a\}$ by writing the corresponding vector as

$$\sum_{a=1}^{N} \left\{ \mathrm{v}_0^a \frac{\partial}{\partial x_0^a} + \ldots + \mathrm{v}_J^a \frac{\partial}{\partial x_J^a} \right\}.$$

We would like to know to which set of components $\{\mathrm{v}_0^a, \ldots, \mathrm{v}_J^a\}$ the equivalence class of a vector $V(\epsilon)$ on P_1M corresponds.

To the path $x^a(\epsilon)$ representing a point in P_1M corresponds the point coordinatized by

$$x_k^a = \left.\frac{\partial^k}{\partial \epsilon^k}\right|_{\epsilon=0} x^a(\epsilon), \quad 1 \leqslant a \leqslant N, \quad 0 \leqslant k \leqslant J, \tag{17}$$

in JM. To the curve of paths $x^a(\epsilon, \gamma)$ in P_1M corresponds the curve

$$x_k^a(\gamma) = \left.\frac{\partial^k}{\partial \epsilon^k}\right|_{\epsilon=0} x^a(\epsilon, \gamma), \quad 1 \leqslant a \leqslant N, \quad 0 \leqslant k \leqslant J, \tag{18}$$

in JM. The vector tangent to this curve in TP_1M has coordinates

$$V^a(\epsilon) = \left.\frac{\partial}{\partial \gamma}\right|_{\gamma=0} x^a(\epsilon, \gamma).$$

In TJM this corresponds to

$$\begin{aligned} \mathrm{v}_k^a &= \left.\frac{\partial}{\partial \gamma}\right|_{\gamma=0} \left\{ \left.\frac{\partial^k}{\partial \epsilon^k}\right|_{\epsilon=0} x^a(\epsilon, \gamma) \right\} \\ &= \left.\frac{\partial^k}{\partial \epsilon^k}\right|_{\epsilon=0} \left\{ \left.\frac{\partial}{\partial \gamma}\right|_{\gamma=0} x^a(\epsilon, \gamma) \right\} \end{aligned} \tag{19}$$

$$= \frac{\partial^k}{\partial \epsilon^k}\bigg|_{\epsilon=\epsilon} V^a(\epsilon).$$

Let us now consider the effect of the dynamics $\dot{x} = X(\epsilon, x)$ on paths. This lifts to a vector field on P_1M given by

$$\tilde{X} \text{ where } \tilde{X}(p) : \epsilon \mapsto X(\epsilon, p(\epsilon)). \tag{20}$$

This is the path space dynamical vector field. In coordinates, the corresponding vector field on JP is

$$V_k^a(x_0, \ldots, x_J) = \frac{\partial^k}{\partial \epsilon^k}\bigg|_{\epsilon=0} X^a(\epsilon, x_0 + \epsilon x_1 + \ldots + \frac{\epsilon^J}{J!} x_J)$$

$$= \frac{\partial^k}{\partial \epsilon^k}\bigg|_{\epsilon=0} X^a(\epsilon, x_0) + k \frac{\partial^{k-1}}{\partial \epsilon^{k-1}}\bigg|_{\epsilon=0} \sum_{m=1}^{2N} \frac{\partial}{\partial x_0^m} X^a(\epsilon, x_0)\, x_m^a + \ldots, \tag{21}$$

which is exactly the perturbation dynamics up to order J obtained in equations (4)!

We have thus found the natural geometric setting for Jth order perturbation theory in a certain jet bundle. The picture of the dynamics of paths in $I \times M$ is an extremely fruitful one. One can prove that the solution of the perturbation equations (4) really is the asymptotic expansion of the true solution just by noting that they are the equations of evolution of the jets of the paths evolving under the true dynamics. The coordinates in which the dynamics are expressed are irrelevant as regards the perturbation dynamics and therefore we can do perturbation theory on manifolds and in infinite dimensions as is required for many physical systems. Next we will review modern Hamiltonian mechanics and then show that the perturbation dynamics is Hamiltonian in a natural way if the unperturbed dynamics is.

4. GEOMETRIC HAMILTONIAN MECHANICS

The evolution of mechanical systems is traditionally described in terms of generalized coordinates q_i and their conjugate momenta p_i. One introduces the Hamiltonian function

$$H(q_1, \ldots, q_n, p_1, \ldots, p_n) \tag{22}$$

and the Poisson bracket

$$\{f, g\} = \sum_{i=1}^{n} \left(\frac{\partial f}{\partial q_i} \frac{\partial g}{\partial p_i} - \frac{\partial f}{\partial p_i} \frac{\partial g}{\partial q_i} \right) \tag{23}$$

of two functions of q_i and p_i. Any observable f evolves according to the evolution equation

$$\dot{f} = \{f, H\}. \tag{24}$$

For a detailed description of the modern approach see [7], [8], and [30]. The modern perspective regards the particular coordinates p_i and q_i as physically irrelevant. Just as general relativity isolates the physically relevant essence of local coordinates in a metric tensor, modern classical mechanics views the Poisson bracket structure (not necessarily expressed in any coordinate system) as the physical entity. Just as physics in space-time is invariant under transformations that preserve the metric, physics in phase space is invariant under the *canonical transformations* which preserve the Poisson bracket. In the modern viewpoint one proceeds axiomatically and does not require canonical coordinates. Dynamics occurs on a *Poisson manifold.* This is a manifold of states with a Poisson bracket defined on it. From this viewpoint a Poisson bracket is a bilinear map from pairs of functions to functions which makes the space of functions into a Lie algebra and acts on products like a derivative does:

I. Bilinearity: $\{af_1 + bf_2, g\} = a\{f_1, g\} + b\{f_2, g\}$

II. Anti-symmetry: $\{f, g\} = -\{g, f\}$

III. Jacobi's identity: $\{f,\{g, h\}\} + \{g,\{h, f\}\} + \{h,\{f, g\}\} = 0$ (25)

IV. Derivation property: $\{f, gh\} = \{f, g\}h + \{f, h\}g$.

The *Hamiltonian* is a function on the Poisson manifold. The evolution of local coordinates z^i is obtained from a Hamiltonian H and the Poisson bracket $\{,\}$ via:

$$\dot{z}^i = X_H \cdot z^i = \{z^i, H\}. \tag{26}$$

X_H is the *Hamiltonian vector field* associated with H and defines a dynamical system. The fourth property of a Poisson bracket implies the useful expression:

$$\{f, g\} = \sum_{i,j} \frac{\partial f}{\partial z^i} \{z^i, z^j\} \frac{\partial g}{\partial z^j}. \tag{27}$$

Thus the Poisson bracket is equivalent to an antisymmetric contravariant two-tensor:

$$J^{ij} \equiv \{z^i, z^j\}. \tag{28}$$

If this is nondegenerate, its inverse $\omega \equiv J^{-1}$ is a closed, nondegenerate two-form called a *symplectic structure.* In this case our Poisson manifold is known as a *symplectic manifold.* The terminology is due to Herman Weyl. If one works in canonical coordinates, the matrix ω has a square equal to minus the identity matrix. In this sense, ω was thought of as a complex structure. Intrinsically, however, ω is a two-form and thus takes a vector and returns a one-form, preventing us from squaring ω. One needs a metric to "lower an index" and obtain a complex structure. To eliminate this confusion, Weyl took the Latin roots *com* and *plex* and converted them to their Greek equivalents *sym* and *plectic.*

Because we do not require nondegeneracy, a Poisson manifold is a more general notion than a symplectic manifold. If J is degenerate, then there are directions in phase space in which no Hamiltonian vector field can point. The available directions lie tangent to submanifolds which fill out the Poisson manifold and on which J

is nondegenerate. The highest dimensional of these form a foliation of their union and so are known as *symplectic leaves*. The only usage of the term symplectic in English is to describe a small bone in the head of a fish. Because Poisson is French for fish, the lower dimensional symplectic submanifolds are known as *symplectic bones* (the terminology is due to Alan Weinstein) [29]. Together the symplectic leaves and the symplectic bones fill out the Poisson manifold and any Hamiltonian dynamics is restricted to lie on a single bone or leaf. Any function which is constant on each bone and leaf Poisson commutes with every other function. Any function which Poisson commutes with every function is automatically a constant of the motion, regardless of the Hamiltonian and is called a *Casimir function*.

A natural symplectic manifold arises from Lagrangian mechanical systems on a configuration space C. The Lagrangian L lives on the tangent bundle TC (velocities being tangent to the curves of motion in configuration space are naturally tangent vectors). Hamiltonian mechanics takes place on the cotangent bundle T^*C (momenta, being derivatives of L with respect to velocity, are naturally dual to velocities and thus are covectors). T^*C has a natural symplectic structure $\omega = -d\theta$ where θ is an intrinsically defined one-form. It acts on tangent vectors v to T^*C at the point (x, α) by first pushing them down to the base C by the natural projection π which gives the basepoint of a covector and then inserting the result into the one-form α on C. Thus:

$$\theta(v) = \alpha(\pi_* v). \tag{29}$$

In coordinates q^a on C, $\theta = p_a dq^a$ and $\omega = dq^a \wedge dp_a$. This generalizes the usual structure in terms of canonical p's and q's to configuration spaces which are manifolds. Symmetry is responsible for most of the simplified systems about which we perturb and plays an intimate role in our geometric theory. We therefore introduce some key modern ideas and basic examples relating to Hamiltonian symmetry.

5. HAMILTONIAN SYSTEMS WITH SYMMETRY

Perhaps the central advantageous feature of systems with a Hamiltonian structure is a generalization of *Noether's theorem* relating symmetries to conserved quantities. Noether considered symmetries of the Lagrangian under variations of configuration space. One may introduce generalized coordinates $q_1, \ldots, q_n$ where $q_2, \ldots, q_n$ are constant under the symmetry transformation and q_1 varies with the transformation. For example, we might take the configuration space to be ordinary Euclidean 3-space where the action of the symmetry is translation in the x direction, and utilize the coordinates $q_1 = x$, $q_2 = y$, $q_3 = z$. That L is invariant means that it does not depend on q_1, i.e., q_1 is an ignorable coordinate. The Euler-Lagrange equations:

$$\frac{d}{dt}\left(\frac{\partial L}{\partial \dot{q}}\right) - \frac{\partial L}{\partial q} = 0 \tag{30}$$

show that in this case the momentum $p_1 = \dfrac{\partial L}{\partial \dot{q}_1}$ conjugate to q_1 is actually a constant of the motion.

By going to a Hamiltonian description in terms of Poisson brackets we may extend Noether's theorem in a fundamental way. We may consider a one-parameter symmetry transformation of the whole phase space as opposed to just configuration space. If this transformation preserves the Hamiltonian and the Poisson bracket (i.e., is a canonical transformation) then it is associated with a conserved quantity. We will see that this extension of Noether's theorem is essential in the case of gyromotion and in other examples.

One-parameter families of canonical transformations of this type may be represented as the "time" s evolution of some function J, treated momentarily as a Hamiltonian. Parametrizing our transformation by s and labelling points in phase space by $\underline{z}$, the solution $\underline{z}(s)$ of

$$\frac{d\underline{z}}{ds} = \{\underline{z}, J\} \qquad \underline{z}(s=0) = \underline{z}_0 \tag{31}$$

is the canonical transformation generated by J.

If the transformation generated by J is a symmetry of H then

$$\begin{aligned} \frac{dH}{ds} = 0 &= \sum_i \frac{\partial H}{\partial z^i} \frac{dz^i}{ds} \\ &= \sum_i \frac{\partial H}{\partial z^i} \{z^i, J\} \\ &= \{H, J\} \\ &= -\{J, H\} \\ &= -\dot{J}. \end{aligned} \tag{32}$$

So J *is a conserved quantity.*

We now consider the case in which the solutions of

$$\frac{d\underline{z}}{ds} = \{\underline{z}, J\} \tag{33}$$

are all closed curves (every orbit is periodic). We will call these closed orbits *loops.* The symmetry transformation is then said to be a *circle action* on phase space.

For example we might consider rotation by θ in J, θ space. In this case phase space looks like a cylinder. The Poisson bracket is

$$\{f, g\} = \frac{\partial f}{\partial \theta}\frac{\partial g}{\partial J} - \frac{\partial f}{\partial J}\frac{\partial g}{\partial \theta}. \tag{34}$$

J generates the dynamics:

$$\frac{d\theta}{ds} = \{\theta, J\} = 1,$$

$$\frac{dJ}{ds} = 0, \tag{35}$$

which just rotates the cylinder.

In studying the dynamics of a Hamiltonian H symmetric under a circle action generated by J, we may make two simplifications which together comprise the *process of reduction*. This procedure was defined by Marsden and Weinstein [32] in a more general setting that we will describe shortly. The process unifies many previously known techniques for simplifying specific examples of Hamiltonian systems.

1. Because J is a constant of the motion, the surface J = constant in phase space is left invariant by the dynamics and so we may restrict attention to it.

2. The symmetry property of H implies that if we take a solution curve $\underline{z}(t)$ of the equation $\dot{\underline{z}} = \{z, H\}$ and let it evolve for a "time" s under the dynamics $\dot{\underline{z}} = \{\underline{z}, J\}$ then we obtain another solution curve of $\dot{\underline{z}} = \{\underline{z}, H\}$. In fact, the dynamics of H takes an entire loop into other entire loops.

The dynamics around loops is easy to solve for:

$$\dot{\theta} = \frac{\partial H}{\partial J}. \tag{36}$$

Notice that θ is not uniquely defined but $\dot{\theta}$ is. We are interested in the problem of finding the dynamics from loop to loop. We want to project the original dynamics on phase space P down to a space P/S^1 whose points represent whole loops in P. Let us call P/S^1, the *space of loops* and $\pi : P \rightarrow P/S^1$, the projection mapping loops in P to points in P/S^1. For example, when $P = J, \theta$ space the projection mapping takes J, θ to J. Thus the second simplification is to consider dynamics on the space of loops P/S^1.

Performing both of these operations—restricting to J = constant and considering the space of loops—leaves us with a space

$$R \equiv P/S^1|_{J=constant} \tag{37}$$

with two dimensions less than P called the *reduced space*.

We have seen that the dynamics on P naturally determines dynamics on R. The key importance of R is that R's *dynamics is itself Hamiltonian*. For this statement to make sense we need to find a Hamiltonian and a Poisson bracket on R. These are the so called *reduced Hamiltonian* and *reduced Poisson bracket*.

The original Hamiltonian H on P is constant on loops by the symmetry condition. We may take the value of the reduced Hamiltonian at a point of R to be the value of H on the corresponding loop in P.

To take the reduced Poisson bracket of two functions f and g on R, we consider any two functions $\hat{f}$ and $\hat{g}$ on P which are constant on loops and agree with f and g when restricted to J = constant and projected by π to R. The Poisson bracket on P of $\hat{f}$ and $\hat{g}$ will be constant on loops and its value of J = constant will be independent of how $\hat{f}$ and $\hat{g}$ were extended as functions on J (because they are constant on loops: $\{\hat{f}, J\} = 0$ and $\{\hat{g}, J\} = 0$ so $\{\hat{f}, \hat{g}\}$ is independent of $\partial\hat{f}/\partial J$ and $\partial\hat{g}/\partial J$). Thus the value of the reduced Poisson bracket on R of f and g is the value on the corresponding loop in P of the Poisson bracket of any two extensions $\hat{f}$, $\hat{g}$ that are constant on loops.

In examples we often introduce a coordinate θ describing the position on a loop. We may then treat P/S^1 as the set $\theta = 0$ (at least locally). In this case R is the subset θ = constant, J = constant of P. The reduced Hamiltonian on R is just the value of H on this subset of P. To calculate the value of the Poisson bracket of two functions on P on this surface, we need only their first derivatives there.

If the functions are constant on loops (i.e., independent of θ), then the derivative $\partial/\partial\theta$ is zero. The dependence on J is irrelevant, so we may take the derivative $\partial/\partial J$ to be zero. Plugging these two expressions into the Poisson bracket on P gives us the expression for the reduced Poisson bracket on R.

6. EXAMPLE: CENTRIFUGAL FORCE

We consider a particle on a two-dimensional plane moving in a rotationally symmetric potential. The phase space is then T^*R^2 with coordinates x, y, p_x, p_y. The Poisson bracket is the canonical one:

$$\{f, g\} = \frac{\partial f}{\partial x}\frac{\partial g}{\partial p_x} - \frac{\partial f}{\partial p_x}\frac{\partial g}{\partial x} + \frac{\partial f}{\partial y}\frac{\partial g}{\partial p_y} - \frac{\partial f}{\partial p_y}\frac{\partial g}{\partial y}. \tag{38}$$

The Hamiltonian is taken to be

$$H = \frac{1}{2m}(p_x^2 + p_y^2) + V\left(\sqrt{x^2 + y^2}\right). \tag{39}$$

The symmetry on phase space is given by the evolution of the equations

$$\frac{dx}{ds} = -y, \quad \frac{dy}{ds} = x,$$

$$\frac{dp_x}{ds} = -p_y, \quad \frac{dp_y}{ds} = p_x. \tag{40}$$

We may think of a point in phase space as a point in the plane (x, y) with a vector (p_x, p_y) attached. The action of the symmetry is to rotate the plane about the origin, vector and all.

The Hamiltonian depends only on the radial distance and the magnitude of the momentum vector and so is clearly left invariant by this rotation. The rotation is a canonical transformation with generator J satisfying

$$\frac{df}{ds} = \{f, J\} = x\frac{\partial f}{\partial y} - y\frac{\partial f}{\partial x} + p_x\frac{\partial f}{\partial p_y} - p_y\frac{\partial f}{\partial p_x} \tag{41}$$

for any f. Taking $f = x, y, p_x, p_y$ gives:

$$\frac{\partial J}{\partial p_x} = -y \quad \frac{\partial J}{\partial p_y} = x \quad \frac{\partial J}{\partial x} = p_y \quad \frac{\partial J}{\partial y} = -p_x. \tag{42}$$

Thus we see that the generator is $J = xp_y - yp_x$, i.e., the angular momentum. We may label a loop by the value of x, p_x, and p_y when $y = 0$ and $x \geqslant 0$. J on this subset is just xp_y. These then form coordinates on the space of loops P/S^1.

To get the reduced space we set J to the constant value μ. Thus we may take the coordinates on $P/S^1|_{J=\mu}$ to be x and p_x when $y = 0$ and $p_y = \mu/x$. On R we have:

$$\frac{\partial}{\partial \theta} = x\frac{\partial}{\partial y} + p_x\frac{\partial}{\partial p_y} - \frac{\mu}{x}\frac{\partial}{\partial p_x}. \tag{43}$$

Setting this to zero gives the reduced bracket by plugging

$$\frac{\partial}{\partial y} = \frac{\mu}{x^2}\frac{\partial}{\partial p_x} \quad \text{and} \quad \frac{\partial}{\partial p_y} = 0 \tag{44}$$

into the expression for $\{f, g\}$:

$$\{f, g\} = \frac{\partial f}{\partial x}\frac{\partial g}{\partial p_x} - \frac{\partial f}{\partial p_x}\frac{\partial g}{\partial x}. \tag{45}$$

The reduced bracket in this case is just the canonical bracket on x, p_x space.

The reduced Hamiltonian is obtained by restricting the original Hamiltonian to our subset and is given by

$$H = \frac{p_x^2}{2m} + \left[\frac{\mu^2}{2m}\frac{1}{x^2} + V(x)\right]. \tag{46}$$

Note the effective potential due to reduction, that represents the *centrifugal force.*

7. HIGHER DIMENSIONAL SYMMETRIES [32]

Quite often physical systems are blessed with more than one dimension of symmetry. In keeping with the philosophy of not making unphysical choices it is natural to consider the process of reduction in the presence of an arbitrary Lie group of symmetry. A Lie group is a group which is also a manifold, such that the group operations respect the smoothness structure. A Hamiltonian system with symmetry consists of a Poisson manifold M, a Hamiltonian H, and a group G that acts on M so as to preserve both H and the Poisson bracket $\{,\}$. The tangent space of G at its identity may be identified with the Lie algebra g of the group and represents group elements infinitesimally close to the identity. The action of an infinitesimal element of G on M perturbs each point of M by an infinitesimal amount. Thus each element v of the Lie Algebra of G naturally determines a vector field on M. The action of the one-dimensional subgroup to which v is tangent on M is given by the flow of this vector field. That the group action preserves the Poisson bracket, implies that this vector field is actually Hamiltonian. Thus we may associate to v a Hamiltonian function which generates this vector field (at least locally). If G is n-dimensional, and we pick a basis for g, then the group action gives us n corresponding Hamiltonian functions on M. So as not to prefer one basis over another, we collect these n numbers at each point of M into a vector. This vector pairs naturally with an element of g (to give the value of the function which generates the action of that element) and so the collection of n Hamiltonians is a vector in the dual of the Lie algebra g^* at each point of M. Thus with every Hamiltonian group action of G on M, there is a natural map called the *momentum map* from M to g^* which collects together the generators of the infinitesimal action of G on M. For a mechanical system in R^3 which is translation invariant, the momentum map associates with each point in phase space the total linear momentum of the system in that state. If the Hamiltonian is rotationally symmetric, then the momentum map gives the total angular momentum in each state (thus we see that angular momentum is not naturally a vector in R^3, rather it takes its values in the dual of the Lie algebra of the rotation group, $so(3)^*$). When we talked about reduction in the one dimensional case above, the generator of the action J was the momentum map.

Does reduction work for higher dimensional symmetries? If the group is commutative, then we may apply the one-dimensional procedure repeatedly to eliminate two degrees of freedom for each dimension of symmetry. If we are able to eliminate all dimensions of phase space in this way, then the system is *integrable.* If orbits are bounded then the group is a torus in this case. Locally we may define angle variables on the toroidal group orbits and the corresponding action variables form the momentum map. Recall that there were two steps in the reduction of systems with one dimension of symmetry, each of which eliminated one dimension of the phase space and that either could be performed first. One was to restrict to a level set of the generating function and the other was to drop down to the orbit space (space of loops). For noncommutative groups, we may again perform either of these two operations, but each gets in the way of subsequently performing the other. The main issue here is that while the Hamiltonian is invariant under the group action, the momentum map is not. Consider the example of a mechanical system in a spherically symmetric potential so that the rotation group acts on phase space as a symmetry and the momentum map is the total angular momentum. While the energy is left unchanged as we rotate the state, the angular momentum is rotated just like a vector in R^3. This action of $SO(3)$ on the dual of its Lie algebra is known as the *coadjoint action.*

Let us digress a bit on the structure of Lie groups to make this point clearer. As shown in the diagram in Fig. 1, every Lie group has three natural actions on itself. If h is an element of G, then we may multiply on the left by h to get the action $L_h\cdot$, we may multiply on the right by h^{-1} (inverse so that $R_{fh} = R_f R_h$) to get R_h, and conjugate by h (i.e., $c \mapsto hch^{-1}$) to get the action $AD_h\cdot$. Conjugation captures the noncommutativity of the group that is at issue here. $AD_h\cdot$ leaves the identity invariant (since $h \cdot e \cdot h^{-1} = e$). We may therefore take the derivative of $AD_h\cdot$ at the identity to get a linear map from the Lie algebra to itself denoted $Ad_h\cdot$. Ad is actually a representation of G on its Lie algebra sometimes called the fundamental representation. If we take the derivative of $Ad_h\cdot$ in the h variable we get an action ad of the Lie algebra on itself. The action of an element $u \in g$ is none other than the Lie bracket with u, so $d_u \cdot v = [u, v]$. We have seen that the dual of the Lie algebra g^* plays an important role in Hamiltonian symmetries. Any time you have a linear transformation L acting on a vector space V, you can define its adjoint L^* acting on V^* by requiring that $< L^*a, v > = < a, Lv >$. The adjoint of $Ad_h\cdot$ is called the coadjoint action of G on the dual of its Lie algebra g^* and is written Ad_h^*. The action of the rotation group on angular momenta that we discovered above is an example of this. One usually requires that a momentum mapping be *equivariant* as in this example. This means that the value of the momentum map varies as the group acts on the phase space according to the coadjoint action: $J(g \cdot x) = Ad_g^* \cdot J(x)$.

Let us now try to mimic the reduction procedure in this noncommutative case. First we restrict attention to the subset of phase space $J = \mu$, a constant. The dynamics restricts to this subset because J is a constant of the motion. The whole group G does not act on this subset, however, because a general element of G will change the value of J. The subgroup of G which leaves μ invariant under the coadjoint action (known as the isotropy subgroup G_μ) will act on this subset and we may drop the dynamics down to its orbit space. The resulting space $M|_{J=\mu}/G_\mu$ has a natural symplectic structure and the Hamiltonian restricted to it generates the projected dynamics. For the rotation group example, we restrict to states with a given total angular momentum (eliminating 3 dimensions) and then forget about the angle of rotation about the axis defined by that angular momentum (eliminating one more). The result is a phase space of four dimensions lower than we started with.

We may obtain the same result in another way. Consider the orbit of a particular element μ of the dual of the Lie algebra under the coadjoint action. This *coadjoint orbit* O_μ has a natural symplectic structure we will discuss in a moment. For the rotation group, the coadjoint orbits are spheres of constant total angular momentum (and the origin). The orbit space of M modulo G has a natural Poisson structure (the bracket of G invariant functions is G invariant) which is not typically symplectic. The symplectic leaves project onto the coadjoint orbits under the momentum map. The inverse image of a whole coadjoint orbit under the momentum map is left invariant under the group action on M. The orbit space $M|_{J^{-1}O_\mu}/G$ is the same reduced space we constructed above. For the rotation group this consists of restricting to states with a given total magnitude of angular momentum and then modding out by the whole rotation group.

An important example of reduction applies to mechanical systems whose configuration space is the symmetry group itself. We will see that the free rigid body and the perfect fluid are examples of this type, a fact first realized by Arnol'd [33]. The phase space M is the T^*G and the G action is the canonical lift to T^*G of left or right multiplication. The G orbits have one point in each fiber and so we may identify the orbit space with the cotangent space at the identity, i.e., the dual of the Lie algebra. The momentum map is then the identity and the coadjoint orbits receive a natural symplectic structure, being the reduced spaces. These symplectic structures are known as Kirillov-Kostant-Souriau (KKS) symplectic structures. If we just consider the orbit space T^*G/G then we obtain a natural Poisson bracket on g^* already known to Sophus Lie (and so called the Lie-Poisson bracket). Explicitly it is:

$$\{f, g\}(\alpha) = \langle \alpha, \left[\frac{\delta f}{\delta \alpha}, \frac{\delta g}{\delta \alpha}\right] \rangle \tag{47}$$

where $\alpha \in g^*$, f and g are functions on g^*, $[,]$ is the Lie algebra bracket, and $\langle , \rangle$ is the natural pairing of g and g^*. This bracket is behind many of the nontrivial Poisson structures recently discovered in various areas of physics.

To specify the configuration of a free rigid body, we give some reference configuration and every other configuration is uniquely specified by giving the element of the rotation group that acts on the reference to give the desired one. Thus the configuration space is identifiable with the group $SO(3)$ itself. The state including angular velocity is naturally a point in $TSO(3)$ and the state including angular momentum is a point in $T^*SO(3)$. A priori, there is no way of comparing the angular velocity or momentum in one configuration with that in another. Using the group action, however, we may push all velocities to velocities at the identity (i.e., velocities on the reference configuration) which may be identified as elements of the Lie algebra. Both left and right multiplication can bring us to the identity since they act on the group transitively. Consider a path at the identity (for example a rotation about the z axis) to which a given element of g is tangent. Left multiplication by $h \in G$ means move along the path and then rotate by h. Thus the path is associated with the body and we get the angular velocity in the body-fixed frame. Multiplying on the right means rotate first by h, then follow the path. The path applied is independent of the configuration of the body (described by h) and so its tangent represents angular velocity in the space-fixed frame. Similarly, left multiplication gives angular momentum in the body-fixed frame and right multiplication gives it in the space-fixed frame. At a configuration represented by $h \in G$, the map from g to g that takes spatial angular velocity to body angular velocity is the adjoint action of h. Similarly, the map from g^* to g^* that takes spatial angular momentum to body angular momentum is the coadjoint action. The energy only depends on the angular momentum in the body (the orientation in space in irrelevant for a free rigid body) and so the Hamiltonian on $T^*SO(3)$ is invariant under the cotangent lift of left multiplication and we are indeed in the situation described above. If we drop down to the orbit space of this left multiplication, we get a Poisson bracket and Hamiltonian on the three dimensional space of angular momenta in the body. The dynamics on this space is exactly Euler's equations. The Poisson bracket is explicitly given by

$$\{J_x, J_y\} = J_z \tag{48}$$

plus cyclic permutations. The total angular momentum $J_x^2 + J_y^2 + J_z^2$ is a Casimir function and so is automatically conserved. The coadjoint orbits (and so the symplectic leaves and bones) are the spheres of constant total angular momentum and the origin. The area element on the spheres is the two-form which is the KKS symplectic structure.

In an exactly analogous way, we may consider the Hamiltonian structure of a perfect fluid. If we choose a reference configuration, then to get any other configuration we apply a unique diffeomorphism (volume preserving if the fluid is incompressible). Thus the configuration space may be identified with the group of diffeormophisms of the region in which the fluid resides. The state of the fluid plus its velocity field is represented by a point in the tangent bundle of the group. The phase space gives the state of the fluid plus the momentum density and so is the cotangent bundle of the group. Again we may identify velocities and momenta with elements of the Lie algebra and its dual by left or right multiplication. Right multiplication gives the Eulerian velocity or momentum field in space. Left multiplication gives them for material points in the reference configuration. Here, in contrast to the rigid body case, the energy depends only on the spatial momentum (which fluid particle is where is irrelevant) and so the Hamiltonian is right invariant. Dropping to the orbit space gives us dynamics for the spatial momentum density, i.e., Euler's fluid equations, in Hamiltonian form.

For gases and plasmas, the state of the system is represented by the particle distribution function on single-particle phase space. This distribution function evolves by the action of symplectomorphisms (i.e., canonical transformations) of this phase space. The group of symplectomorphisms has the Hamiltonian vector fields as its Lie algebra. We may identify this with the space of functions on the phase space with the Lie bracket being the Poisson bracket of functions. The dual of the Lie algebra is then densities on phase space, which we may use to describe the kinetic state of plasmas and gases. The coadjoint action just pushes the density around by the symplectomorphism. One coadjoint orbit comes from considering a delta distribution on phase space. The symplectomorphisms push it all over phase space to give a coadjoint orbit that is identifiable with the original phase space. In fact, the KKS symplectic structure is exactly the original symplectic structure. This shows that every symplectic manifold is a coadjoint orbit. A delta distribution whose support is a loop or torus in phase space shows that the space loops with a given action (or tori with given actions around their fundamental loops) form a symplectic manifold.

8. GEOMETRIC HAMILTONIAN PERTURBATION THEORY

Let us now relate this geometric Hamiltonian mechanics to the geometric perturbation theory we discussed earlier. We will see that the Jth order perturbed dynamics has a natural Hamiltonian structure if the exact dynamics does.

The first thing to note is that the path space dynamics is Hamiltonian. This is not surprising if we think of the path space as a kind of direct integral of the phase spaces at each ϵ. The dynamics at different ϵ's are completely independent (except for the fact that the paths are smooth). If we had the product of only two Hamiltonian systems (instead of a continuum of them) then we would get the correct dynamics from a symplectic structure which is the sum of the pullback to the product of the individual symplectic structures and a Hamiltonian which is the sum of the pulled back Hamiltonians. Extending this construction to a continuum of multiplicands leads to the symplectic structure:

$$\tilde{\omega}_p(\tilde{V}_1, \tilde{V}_2) \equiv \int_0^1 \omega_{p(\epsilon)}(V_1(\epsilon, p(\epsilon)), V_2(\epsilon, p(\epsilon)))\, d\epsilon. \tag{49}$$

The analog of the sum of Hamiltonians is

$$\tilde{H}(p) \equiv \int_0^1 H(\epsilon, p(\epsilon))\, d\epsilon. \tag{50}$$

The dynamics these two generate is indeed the correct path space dynamics. In the case of a product of a finite number of Hamiltonian systems, we are actually allowed to take any linear combination of the symplectic structures (instead of a straight sum) as long as no coefficient vanishes and we take the same linear combination of Hamiltonians. If a coefficient vanishes, that factor has no dynamics. For our perturbation dynamics then, we want to ignore the region in the interval that is away from $\epsilon = 0$.

In fact if we substitute the Jth derivative of a delta function into the integrals in (49) and (50) we get the correct perturbation dynamics on JM. If the Poisson bracket on M is $\{x^a, x^b\} = J_{ab}$ then the bracket on JM is

$$\{x_k^a, x_m^b\} = J^{ab}\, \frac{k!m!}{J!}\, \delta_{k,J-m} \tag{51}$$

and the Hamiltonian is

$$\overline{H}(x_0, \ldots, x_J) \equiv \left.\frac{d^J}{d\epsilon^J}\right|_{\epsilon=0} H(\epsilon, x_0 + \epsilon x_1 + \ldots + \frac{\epsilon}{J!}\, x_J). \tag{52}$$

Together these give the correct perturbation dynamics. Notice that the 0th order variables are paired with Jth order variables, 1st order with $(J-1)$th order, etc.

From the above coordinate description it is not clear that this bracket is in fact intrinsic. We may show this by considering the *iterated tangent bundle to M*. The tangent bundle to a symplectic manifold has a natural symplectic structure. If ω is the structure on M, then we may use it to identify TM and T^*M. T^*M has a natural symplectic structure, which we defined in (29). The structure on TM is obtained by pulling T^*M's back using the identification supplied by ω. This operation may be iterated to give symplectic structures on the iterated tangent bundles TTM, $TTTM$, $TTTTM$, etc. The Jth order jets naturally embed into the Jth iterated tangent bundle. If the symplectic structure on T^JM is pulled back to JM we obtain the jet Poisson bracket in Eq. (51).

The symplectic structure on TM may be thought of as the first derivative of the original symplectic structure [34]. The jet bracket may be thought of as the Jth derivative. Choose J sheets spaced evenly in $I \times M$. The path dynamics projects down to the product of these sheets. We may map this structure to JM with arbitrary coefficients. If these coefficients are chosen to give a nonsingular result as the sheet spacing goes to zero, we again obtain the jet symplectic structure and Hamiltonian. This shows that the perturbation bracket and Hamiltonian are in essence Jth derivatives of the path structures.

We have seen that when the Poisson bracket is degenerate, nondegenerate symplectic leaves and bones are *injected* into the Poisson manifold as submanifolds. If a closed two-form is degenerate, then we *project* out the degenerate directions to obtain a symplectic manifold. The fact that the two-form is closed implies that the annihilated directions satisfy the conditions of Frobenius's theorem and so lie tangent to smooth submanifolds which we may then project along (at least locally). We have used an example of this construction above. If we insert the Jth derivative of a delta function into the path symplectic integral (49), we obtain a degenerate, closed two-form on the path space P_1M. The projection e eliminating the degenerate directions is exactly the projection from path space down to the jet space JM. The resulting symplectic structure is the jet perturbation structure. If we have a Hamiltonian system with an invariant submanifold, we may attempt to obtain the restricted dynamics in Hamiltonian form by pulling back the symplectic structure. The resulting two-form will be closed but may not be nondegenerate. If things are nice globally, we may apply the above projection. A special case of this demonstrates that the jet construction contains as a special case the linearized dynamics of a Hamiltonian system around a fixed point. We consider the 2-jet space $2M$. The submanifold of jets with base point equal to the fixed point is an invariant submanifold. Because the zero order base directions are paired with the second order directions in (51), restricting to a given basepoint makes the second-order directions degenerate. Projecting these out leaves us with only the first-order jets at the fixed point (i.e., the tangent space there). These are paired with themselves by the second-order bracket according to the original symplectic structure at the fixed point. The second-order Hamiltonian (52) gives the quadratic piece of the Taylor expansion in the x_1 variables. Together these give the linearized flow in the tangent space of the fixed point as a Hamiltonian system. The situation in Poisson manifolds is more complex [29]. If the fixed point is a symplectic leaf, we take the Poisson bracket at the point, the quadratic part of the Hamiltonian in the leaf direction, and the linear part of the Hamiltonian across leaves. The bones are more difficult.

We have seen how important symmetry and its related concepts are in Hamiltonian mechanics. How do the symmetry operations intermix with the perturbation operations? A Hamiltonian G action on M lifts to both the path space PM (just push the whole path around by the group action) and the jets space JM (just push the jet around). The corresponding momentum map is just the integral along a path of the M momentum map and the same integral with the Jth derivative of a delta function thrown in. Both are equivariant.

When considering reduction we quickly see that these groups are not of high enough dimension. A four-dimensional phase space with a one-dimensional symmetry drops down to two dimensions. The first order perturbation space has eight dimensions. In the presence of symmetry we expect to be able to drop this down to the first-order perturbation space of the two-dimensional reduced space. The above group action can only eliminate two dimensions instead of the needed four and so we expect a larger group to act. This is indeed the case. It makes sense to multiply two paths in a group by multiplying pointwise. Thus PG is an infinite dimensional "Lie" group and its "Lie" algebra is the path space of G's Lie algebra g. PG has a Hamiltonian action on the path space PM by multiplying the point $p(\epsilon)$ by the group element $g(\epsilon)$. The momentum map sends a path in M to a path in g^* gotten by applying M's momentum map to each ϵ. In an exactly analogous way, we may define the group JG of J-jets of paths in G with Lie algebra being J-jets of paths in g. This acts in a Hamiltonian and equivariant way on the perturbation space JM. The momentum map is obtained by extending a jet to any consistent path, taking the path momentum map to Pg^* and dropping down to Jg^*.

The process of reduction commutes with taking the path space or jet space. The jet or path space of the reduced space is the reduced space of the jet or path space by the jet or path group.

We have seen the central importance of the dual of the Lie algebra and the coadjoint orbits with their KKS symplectic structure for physics. We have seen that any symplectic manifold may be thought of as a coadjoint orbit in the dual of the Lie algebra of some group. It turns out that if M is a coadjoint orbit in the dual of G's Lie algebra then the perturbation space JM with the jet symplectic structure is naturally a coadjoint orbit in the dual of the Lie algebra of the jet group JG and the jet bracket (51) is the natural KKS symplectic structure.

These relations are at the heart of a new framework for singular Lie transform perturbation theory about which we will report elsewhere. Here we discuss only the first order method of averaging.

9. THE METHOD OF AVERAGING FOR HAMILTONIAN SYSTEMS

Many of the interesting physical regularities we find in diverse systems are caused by the presence of processes that operate on widely separated time scales. The basic simplification this entails is that the fast degrees of freedom act almost as if the slow variables are constant and the slow degrees of freedom are affected only by the average behavior of the fast variables. Bogoliubov in particular has used this separation of scales with great success in many examples. For example, he obtains the Boltzmann equation from the BBGKY hierarchy of evolution equations for correlation functions by holding the one-particle distribution functions fixed while determining the fast evolution of the higher correlations, and then substituting the result in as the collision term driving the one-particle evolution. One makes a similar separation in calculating fluid quantities like viscosity, thermal conductivity, diffusion or electrical conductivity from an underlying kinetic description. In studying complex situations with slow heavy nuclei and fast light electrons in molecular and solid state physics, one often holds the nuclei fixed, calculates the electron ground state and energy as a function of the nuclei positions and then uses them to define an effective potential in which the nuclei move. We have seen that in the presence of an exact symmetry, the symmetry directions may be completely eliminated by the process of reduction. One often finds that the effect of "forgetting" these degrees of freedom is to introduce an amended potential into the Hamiltonian and a "magnetic" piece to the Poisson bracket of the reduced system. We have the centrifugal force coming out as an effective potential earlier. We will report on a version of this reduction procedure which begins by including the "angle of the earth" as a dynamical variable and reduces by the earth's rotation and the rotation of the system together. The resulting reduced space gives the centrifugal force as an amended potential in the reduced Hamiltonian and the Coriolus force as a new term in the Poisson bracket.

When we introduce a perturbation which breaks a symmetry we no longer have exactly conserved quantities. It is easy to prove an "approximate Noether's theorem," however, which says that the momentum map for a slightly broken symmetry evolves slowly:

$$X_J \cdot H = \{H, J\} = \epsilon \quad \text{implies} \quad \dot{J} = \{J, H\} = -\epsilon. \tag{53}$$

In the special case where the unperturbed dynamics is entirely composed of periodic orbits, the action of the orbit through each point is the momentum map of a circle symmetry of the unperturbed Hamiltonian. As we turn on a perturbation which breaks this symmetry, the motion will still be primarily around the loops, but it will slowly drift from loop to loop. Because the symmetry is broken, different points on a loop will move toward different loops. As the perturbation is made smaller, though, phase points orbit many times near a given loop before drifting away. This suggests (correctly) that the perturbation a point feels will asymptotically be the same as the average around an unperturbed loop. Because this average is the same for all points on a loop, for small perturbations entire loops drift onto other entire loops. We may therefore drop the dynamics down to the loop space. In fact one can prove that for a general (even dissipative) system where the unperturbed dynamics X_0 is entirely composed of periodic orbits, the motion of a point under the flow of $X_0 + \epsilon X_1$ projected down to the loop space remains within ϵ for a time $1/\epsilon$ of the orbit of a corresponding point on the loop space under the flow of the average of X_1 around each loop projected down [35]. In the Hamiltonian case we break the circle symmetry of H_0 to get the perturbed system $H_0 + H_1$. We average H_1 around the loops to get $\overline{H}_1$. $H_0 + \overline{H}_1$ is again invariant under the circle action and so we may perform reduction. The reduced dynamics is the slow dynamics on the reduced space and the fact that we may restrict to a constant value of the momentum map, shows that it is actually conserved to within order ϵ for time $1/\epsilon$. This is because loops are taken to loops to this order and the action of a loop (i.e., the integral of the symplectic form ω over a sheet whose boundary is the loop) is left invariant under a canonical transformation (like the flow of the perturbed system) since ω is. Kruskal has shown that there is actually a quantity which is conserved to all orders in ϵ for time $1/\epsilon$ [36] (we will report on a geometric formulation of this result in a future paper). Getting results valid for times longer than $1/\epsilon$ is extremely important physically, but so far no general theory exists.

Let us relate this procedure to the perturbation structures we developed in previous sections. We have an action of the circle group S^1 on M. This lifts to an action of PS^1 on PM and JS^1 on JM. The unperturbed Hamiltonian is invariant under the S^1 action on M, but the path and perturbation Hamiltonians are not invariant under PS^1 and JS^1. We would like to change the action of PS^1 on PM so as to leave the path Hamiltonian invariant and so allow reduction. Since the resulting action should still be Hamiltonian, we look for an ϵ-dependent canonical transformation of $I \times M$ which is the identity at $\epsilon = 0$ and which pushes the PS^1 action into a symmetry. The method of Lie transforms attempts to do this at the perturbation level, letting the canonical transformation be the flow of an ϵ-dependent Hamiltonian, which is then obtained order by order. Here we need only consider the first order action of $1S^1 \sim TS^1$ on $1M \sim TM$. We know that the action will be perturbed so that the value of the reduced Hamiltonian is the average of the perturbed Hamiltonian around the untransformed circles. TM has twice the dimension of M. Reducing by TS^1 eliminates four dimensions. The resulting dynamical vector field has no unperturbed component. One may think of this as the reason for getting results good for time $1/\epsilon$ (it is the action of the unperturbed flow on the perturbation which causes this level of secularity). In this situation it makes intrinsic sense to project the 1st order vector field down to M, where it represents the slow dynamics.

A loop in a two-dimensional phase space (like an orbit of a simple harmonic oscillator) may be thought of in three ways. It is one-dimensional, one dimension less than two, and half of two. Each has an important generalization to higher dimensional Hamiltonian systems. In the presence of a slowly varying Hamiltonian, we have already seen that the action of a one-dimensional loop is conserved. There is an analogous result for half dimensional Lagrangian tori. Kubo [37] has shown that for a system ergodic on an energy surface (which has one dimension less than phase space), the volume enclosed is adiabatically invariant under slow variation or parameters. Roughly, since the motion is ergodic, every orbit changes according to the average of the perturbation over the energy surface; thus the entire energy surface changes by the same energy and so is taken to another energy surface; but the volume enclosed by a surface is preserved under a canonical transformation by Liouville's theorem. For a large number of degrees of freedom this leads to the adiabatic invariance of the entropy in statistical mechanics.

The funny potentials and Poisson brackets that result from reduction contain the average effect of the fast on the slow degrees of freedom. Capturing this effect is the content of many physically useful theories. It is interesting to note that in the late eighteenth century, the idea that all potential energies were really kinetic energies of hidden or forgotten degrees of freedom was one the main motivations for the development of kinetic theory. We may use averaging to see how this comes about. If we slowly move a ping pong paddle up and down from a table with a ping pong ball bouncing very rapidly between the paddle and the table, then we will feel a varying force due to the average momenta imparted due to the impacts of the ball. In phase space the ball describes a rectangle and so the action is given by $J = 4LmV$ where L is the distance from the paddle to the table and V is the speed of the ball. Because this is invariant under slow paddle movements, the ball velocity goes as $1/L$. The momentum transferred on each impact is $2mV$ and there are $V/2L$ impacts per unit of time, so the average force felt goes like $V^2 \sim 1/L^2$. Thus, starting with no potential energy at all, we end up with a $1/L$ effective potential for the paddle!

For a harmonic oscillator, the energy is the product of the action and the frequency: $H = \omega J$. If we have a weight hanging on a string undergoing small amplitude oscillations as we slowly pull the string, the change in pendulum energy is the change in $J\omega$. J remains constant and $\omega \sim \sqrt{g/L}$ so we feel a $1/\sqrt{L}$ potential. We get other potentials if we ask for the force we feel if we tune a guitar string as someone plays it or the acoustic pressure on the water if we fill up a shower as someone sings in it. The effective force due to the fast degrees of freedom may sometimes stabilize an unstable fixed point of the slow system. Ordinarily an inverted pendulum is unstable and falls to the position with the weight hanging downward. If we shake the support of the pendulum periodically hard enough and fast enough, the inverted position is stabilized! An even more spectacular version of this effect occurs if you shake an inverted cup of fluid and stabilize the Rayleigh-Taylor instability which ordinarily causes the fluid to spill out (it is easiest to actually do the experiment with a high viscosity fluid like motor oil). The idea of RF stabilization is to stabilize unstable modes of a plasma (say in a tokamak) by bathing it in a high frequency radio wave. Some of the modern airplanes with wings in a forward facing delta are actually operated in an aerodynamically unstable regime that is stabilized by the fast dynamics of a computer controlled feedback loop. This allows for great maneuverability (since the plane would like to turn anyway!).

Quite often it is very useful to split out the main dynamics of a system and linearize the rest, treating them as fast oscillations. Thus one takes a fluid, elastic, or plasma medium and treats its evolution as slow overall development of the background medium with fast oscillations occurring on top of it. The effect of the oscillations is to change or *renormalize* the dynamics of the background. N.G. van Kampen [40] has called into question the usual treatments of constrained mechanical systems. One usually just writes down the Lagrangian for such a system in generalized coordinates which respect the constraints. Physically, though, one supposes that there is some large potential normal to the constraint surface. The system will execute rapid oscillation in the normal direction and slow evolution along it. If the width of the constraining potential well varies with the mechanical coordinates, then as we have seen the adiabatic invariance will give rise to a new pseudopotential which affects the mechanical motion. In a plasma we treat the slowly varying background as a dielectric medium in which waves propagate according to WKB theory. The waves affect the background (introducing a radiation pressure in the dynamics) via pondermotive forces. If we have a charged particle in the presence of a wave with a slowly varying amplitude, the particle will oscillate back and forth with the wave. It feels more of a push in going down an amplitude gradient than in going up one, leading to an overall average force described by the *pondermotive potential*. This kind of separation is the basis of plasma quasilinear theory. We have extended the geometric perturbation theory to some of these singular perturbation problems. We will report elsewhere on a Hamiltonian treatment of an eikonal theory for linear or nonlinear waves (which is related to the averaged Lagrangian treatment of Whitham [38]). Here let us demonstrate the efficacy of a global geometric approach only with the simple example of $\mathbf{E} \times \mathbf{B}$ drift. A particle in a constant magnetic field executes perfect circles. If there is, in addition, an electric field then the radius of the circles is greater in low potential regions and smaller in high potential regions. Thus the circular orbits do not close and the particle drifts *perpendicularly* to the electric field. A Hamiltonian treatment of more complicated versions of this so-called guiding center motion has been previously given [39]. This work required great cleverness in the choice of physically relevant coordinates. We would like to demonstrate, in this simple version, how a coordinate free approach would lead us to the correct answer, with no previous knowledge.

10. EXAMPLE: E × B DRIFT

In the simplest situation we have a charged particle in the x, y plane moving in the presence of a constant magnetic field **B** which points in the $\hat{z}$ direction and a small constant electric field **E** which points in the $\hat{x}$ direction. We introduce the phase space $P \sim T^*R^2$ with coordinates (x, y, p_x, p_y) (we use mechanical momenta $\mathbf{p} = m\mathbf{v}$ here). The correct dynamics in the presence of a magnetic field may be described in a Hamiltonian formulation in two ways. The standard approach is to introduce the unphysical vector potential **A** and to work with canonical momenta $\mathbf{p} = m\mathbf{v} - (e/c)\mathbf{A}$. Here we use the physical momenta and magnetic field, but a noncanonical Poisson bracket

$$\{f, g\} = f_x g_{px} - f_{px} g_x + f_y g_{py} - f_{py} g_y + \frac{eB}{c} (f_{px} g_{py} - f_{py} g_{px}). \tag{54}$$

We obtain the correct dynamics in this case with the Hamiltonian

$$H = H_0 + \epsilon H_1 = \frac{1}{2m} (p_x^2 + p_y^2) - \epsilon eEx. \tag{55}$$

The dynamics is then

$$\dot{x} = \frac{p_x}{m} \ , \ \dot{y} = \frac{p_y}{m},$$

$$\dot{p}_x = \frac{eB}{mc} p_y + \epsilon eE \ , \ \dot{p}_y = - \frac{eB}{mc} p_x. \tag{56}$$

The unperturbed situation here is just a charged particle on a plane in a constant magnetic field. Every orbit in this situation is a closed loop. Thus the unperturbed system has a circle symmetry:

$$\dot{x} = \frac{p_x}{m} \ , \ \dot{y} = \frac{p_y}{m},$$

$$\dot{p}_x = \frac{eB}{mc} p_y \ , \ \dot{p}_y = - \frac{eB}{mc} p_x. \tag{57}$$

The generator of this symmetry (i.e., the momentum map) is none other than the unperturbed Hamiltonian itself:

$$H_0 = \frac{1}{2m} (p_x^2 + p_y^2). \tag{58}$$

Let us obtain the reduced phase space and Poisson bracket for this symmetry action. First we look at the space of loops P/S^1. Each circular particle orbit has exactly one point where $p_y = 0$ and $p_x \geqslant 0$. We may label a loop by the values of x, y, p_x at this point. Next we restrict to the set where the momentum map is a constant: $H_0 = \alpha$. The reduced space is

$$R = P/S^1|_{H_0 = \alpha}$$

and may be coordinatized by the values of x and y when $p_x = \sqrt{2m\alpha}$ and $p_y = 0$. The reduced Poisson bracket $\{,\}_\alpha$ of two functions $f(x, y)$ and $g(x, y)$ is obtained by extending them to P in such a way that

$$\left.\frac{\partial \hat{f}}{\partial p_x}\right|_{p_x = \sqrt{2m\alpha}, p_y = 0} = 0 \tag{59}$$

and

$$\{\hat{f}, H_0\} = 0 = \frac{\sqrt{2m\alpha}}{m} \frac{\partial \hat{f}}{\partial x} - \frac{eB}{mc} \sqrt{2m\alpha} \frac{\partial \hat{f}}{\partial p_y}. \tag{60}$$

Thus we replace $\partial/\partial p_x$ by 0 and $\partial/\partial p_y$ by $(c/eB)\partial/\partial x$ to get:

$$\{f, g\}_\alpha = \frac{c}{eB} (f_y g_x - f_x g_y). \tag{61}$$

Thus we see that the original spatial coordinates x and y now play the role of canonically conjugate variables in the reduced space. The factor of $1/B$ in the bracket appeared in Littlejohn's work [39]. The full system is not invariant under our circle action. If we average the perturbation Hamiltonian H_1 around the circles, we do obtain a circle symmetric system. The average of the potential ϵeEx around a loop is just the value when $p_y = 0$. Thus the reduced averaged Hamiltonian is

$$\overline{H}_\alpha(x, y) = \alpha - \epsilon eEx. \tag{62}$$

The reduced averaged dynamics is then

$$\dot{x} = \{x, \overline{H}_\alpha\}_\alpha = 0$$

$$\dot{y} = \{y, \overline{H}_\alpha\}_\alpha = \frac{c}{eB} (-\epsilon eE) = -\epsilon \frac{cE}{B}. \tag{63}$$

This is indeed the $\mathbf{E} \times \mathbf{B}$ drift dynamics.

ACKNOWLEDGMENTS

I would like to thank Ted Courant, Allan Kaufman, Robert Littlejohn, Jerry Marsden, Rich Montgomery, and Alan Weinstein for many suggestions and discussions regarding these ideas.

REFERENCES

[1] J. L. Lagrange, "Mémoire sur la théorie des variations de éléments des planètes," Mém. Cl. Sci. Math. Phys. Inst. France, 1-72 (1808).

[2] A. Weinstein, "Symplectic Geometry," Bull. Am. Math. Soc. **5** 1-13 (1981).

[3] H. Poincaré, *Les méthodes nouvelles de la mécanique céleste*, Vols. 1, 2, 3 (Gauthier-Villars, Paris, 1892; Dover, New York, 1957).

[4] E. Noether, Nachr. Gesell. Wiss. Göttingen **2**, 136 (1918).

[5] P. A. M. Dirac, *The Principles of Quantum Mechanics*, (Oxford University Press, 1958).

[6] As in L. D. Landau and E. M. Lifshitz, *Course of Theoretical Physics*, Vols. 1-10 (Peramon, New York, 1960-1981).

[7] R. Abraham and J. E. Marsden, *Foundations of Mechanics*, 2nd edition (Benjamin/Cummings, Reading, Mass., 1978).

[8] V. I. Arnol'd, *Mathematical Methods of Classical Mechanics* (Springer, Berlin, 1978).

[9] J. R. Cary, "Lie transform perturbation theory of Hamiltonian systems," Phys. Rep. **79**, 131 (1981).

[10] P. R. Chernoff and J. E. Marsden, *Properties of Infinite Dimensional Hamiltonian Systems*, Lecture Notes in Mathematics, **425** (Springer, New York, 1974).

[11] P. J. Morrison and J. M. Greene, "Noncanonical Hamiltonian density formulation of hydrodynamics and ideal magnetohydrodynamics," Phys. Rev. Lett. **45**, 790-794 (1980).

[12] J. E. Marsden and A. Weinstein, "Coadjoint orbits, vortices, and Clebsch variables for incompressible fluids," Physica **7D**, 305-323 (1983).

[13] J. E. Marsden, T. Ratiu, and A. Weinstein, "Semi-direct products and reduction in mechanics," Trans. Am. Math. Soc. **281**, pp. 147-177 (1984).

[14] P. J. Morrison, "The Maxwell-Vlasov equations as a continuous Hamiltonian systems," Phys. Lett. **80A**, 383-396 (1980).

[15] J. Marsden and A. Weinstein, "The Hamiltonian structure of the Maxwell-Vlasov equations," Physica **4D**, 394-406 (1982).

[16] W. Pauli, *Die Allgemeinen Prinzipien der Wellenmeckanik*, Hand. der Phys., Vol 24, Part 1 (Springer, Berlin, 1933); *General Principles of Quantum Mechanics* (Springer, Berlin, 1981).

[17] D. Holm and B. Kupershmidt, "Poisson brackets and Clebsch representations for magnetohydrodynamics, multifluid plasmas, and elasticity," Physica **D**.

[18] J. E. Marsden and T. J. R. Hughes, *Mathematical Foundations of Elasticity* (Prentice-Hall, Englewood Cliffs, New Jersey, 1983).

[19] R. G. Spencer and A. N. Kaufman, "Hamiltonian structure of two-fluid plasma dynamics," Phys. Rev. **A25**, 2437-2439 (1982).

[20] J. Gibbons, D. D. Holm, and B. Kupershmidt, "Gauge-invariant Poisson brackets for chromohydrodynamics," Phys. Lett. **90A**, 281-283 (1982).

[21] D. D. Holm and B. A. Kupershmidt, "Poisson structures of superfluids," submitted to Phys. Lett. **A**.

[22] L. Faddeev and V. E. Zakharov, "Korteweg-de Vries as a completely integrable Hamiltonian system," Functional Anal. Appl. **5**, 280 (1971).

[23] D. D. Holm, J. E. Marsden, T. Ratiu, and A. Weinstein, "A priori estimates for nonlinear stability of fluids and plasmas," Preprint, 1984.

[24] V. Guillemin and S. Sternberg, *Symplectic Techniques in Physics*, (Cambridge University Press, Cambridge, 1984).

[25] S. Omohundro, "Hamiltonian structures in perturbation theory," submitted to J. of Math. Phys. (1984).

[26] S. Omohundro, "A Hamiltonian approach to wave modulation," Poster presented at Sherwood Plasma Theory Meeting, Lake Tahoe, April 1984 (in preparation for publication).

[27] A. H. Nayfeh, *Perturbation Methods* (John Wiley, New York, 1973).

[28] J. Kevorkian and J. D. Cole, *Perturbation Methods in Applied Mathematics* (Springer, New York, 1981).

[29] A. Weinstein, "The local structure of Poisson manifolds," J. Diff. Geom. **18**, 523-557 (1983).

[30] J. E. Marsden, *Lectures on Geometric Methods in Mathematical Physics*, CBMS-NSF Reg. Conf. Series **37** (PA:SIAM, Philadelphia, 1981).

[31] P. Olver, "Hamiltonian perturbation theory and water waves," *Fluids and Plasmas: Geometry and Dynamics*, J. Marsden, Ed., Contemporary Mathematics, **28**, 231 (1984).

[32] J. Marsden and A. Weinstein, "Reduction of symplectic manifolds with symmetry," Rep. Math. Phys. **5**, 121-130 (1974).

[33] V. Arnol'd, "Sur la géometrie differentielle des groupes de Lie de dimension infinie et ses applications a hydrodynamique des fluids parfaits," *Ann. Inst. Fourier Grenoble* **16(1)** 319-361 (1966).

[34] J. Kijowski and W. Tulczyjew, *A symplectic framework for field theories*, Lecture Notes in Physics **107** (Springer, New York, 1979).

[35] V. Arnol'd, *Geometrical Methods in the Theory of Ordinary Differential Equations* (Springer, New York, 1983).

[36] M. Kruskal, "Asymptotic theory of Hamiltonian and other systems with all solutions nearly periodic," J. Math. Phys. **3**, 806-828 (1962).

[37] R. Kubo, H. Ichimura, T. Usui, and N. Hashitsume, *Statistical Mechanics* (North-Holland, Amsterdam, 1965).

[38] G. Whitham, *Linear and Nonlinear Waves* (John Wiley, New York, 1974).

[39] R. Littlejohn, J. Plasma Physics **29**, 111 (1983).

[40] N. van Kampen, "Constraints," Preprint, March 1983.

LIE POINT TRANSFORMATION GROUP SOLUTIONS OF THE NONLINEAR VLASOV-MAXWELL EQUATIONS

B. Abraham-Shrauner
Department of Electrical Engineering
Washington University in St. Louis
St. Louis, Missouri 63130

I. INTRODUCTION

The method of solution of nonlinear ordinary and partial differential equations by Lie group invariance under infinitesimal point transformations [1-22] is presented here for the exact nonlinear Vlasov-Maxwell equations of plasmas [23-30]. The intent is the systematic determination of exact solutions that exhibit time dependence in addition to the space dependence of the electromagnetic fields and the space and momentum (velocity) dependence of the one-particle distribution functions found for equilibrium Bernstein-Greene-Kruskal (BGK) solutions [31-39]. The development is limited to the Vlasov-Maxwell equations partly in order to develop one topic fairly completely, partly because these are the fundamental equations for collisionless, classical plasmas and partly because until recently [40-42] the sole exact time and phase space-dependent solutions were the uniformly moving BGK equilibria [36]. The method of Lie group invariance has been applied in other disciplines: hydrodynamics, general relativity, heat transfer, nuclear transport, etc. Some overlap of disciplines has led to the occasional use of Lie point group invariance for the solution of fluid equations for plasmas but, usually, plasma physicists have eschewed the Lie point group method for solving plasma differential equations. Some exceptions are magnetohydrodynamic equations for a θ-pinch [43], magnetohydrodynamic blast waves [44], and three-dimensional flows with reconnection [45].

Sophus Lie [1-5] developed this method for solving nonlinear ordinary (as well as linear ordinary) and linear partial differential equations in the last century. His aim was to construct a general and systematic method of integration of differential equations analogous to Abel's method for algebraic equations in which the invariance under transformation could be used to simplify the equations. Several of his original papers have been translated into English [19], however, the long three-volume treatise he wrote with F. Engel is in German [5]. Early in this century simplified versions of his analysis appeared [6-12]. After the 1920s the Lie group method of solution of differential equations fell into disuse. The method survived in an attenuated form in similarity transformations in fluid theory [46-49] and in selected tricks for the solution of first and second order ordinary differential equations [50-51]. When (in the 1960s) nonlinear scientific problems started again to be attacked analytically and directly rather than by a piece-wise linearization approximation or by some perturbation scheme, Lie group methods for solving differential equations reappeared. The Soviets were the most active early on in reintroducing these methods and pursuing research in this area [15, 21] but others also participated [17, 22]. Three thrusts are evident in the research. One activity is the translation of Lie group theory into modern mathematical terminology [16, 19, 52-55] where frequently the solution of differential equations is not the main interest. The second activity is the application of known Lie group point transformation methods to modern scientific problems [17,18,21,22,25-30, 43-45]. The third project is the generalization of the Lie group point transformation method to systems of nonlinear second-order partial differential equations [18,21], and to integral-differential equations [22,28,29] as well as the generalization of point transformations [56-67]. The generalizations are usually contact (tangent) and Lie-Bäcklund transformations that originated in the last century. The discussion here is on the point transformations because the effort in solving the Vlasov-Maxwell equations has been focused on these transformations. Contact transformations may not give any new results since for more than one dependent variable, contact transformations reduce to point transformations. The application of Lie-Bäcklund transformations is much less systematic; consequently the possibility of additional solutions is less assured.

A simple classical model of a plasma is assumed here. The plasma is a collistionless, fully ionized gas which may be overall neutral or nonneutral: partly neutralized or a beam of one particle species. Quantum effects are ignored and the Vlasov equation is nonrelativistic although some unpublished calculations exist for the relativistic Vlasov equation (Roberts, private communication). The Vlasov-Maxwell equations are partially self-consistent in that the particle distribution functions from the Vlasov equation appear in the usual velocity (momentum) integrals for the charge and current density but external electromagnetic fields whose sources are outside the plasma and a neutralizing background plasma with the background charge and current densities a function of position and time may be included.

The Vlasov-Maxwell equations have been solved most frequently by a perturbation scheme that gives the linearized equations in lowest order and with suitable averaging the ponderomotive force or quasilinear terms in the next order, etc. Nevertheless, some processes are inherently nonlinear for which perturbation solutions give misleading answers. The first exact, nonlinear Vlasov-Maxwell solutions were equilibria. The definitive paper was by Bernstein, Greene, and Kruskal [31]; their names are now attached to exact, nonlinear Vlasov-Maxwell equilibria. They showed for a one-dimensional plasma that if the scalar potential and three of the four distribution functions: untrapped (trapped) electron (ion) are specified, the remaining distribution function is determined from an integral equation. Their formalism was enlarged to include magnetic BGK equilibria of various types [32-38].

The BGK equilibria viewed from a moving inertial reference frame are stationary waves. The time dependence for Galilean transformations occurs as $\vec{r} - \vec{v}_p t$ where $\vec{v}_p$, the apparent phase velocity, is the relative velocity of the plasma in a moving inertial reference frame with respect to the observer in the laboratory reference frame. The transformation from a BGK equilibrium to a stationary wave is similar to a Galilean boost in soliton theory. Baranov [25] discovered the first different time dependence for a one-species plasma in a fixed immobile neutralizing background. For the one-dimensional Vlasov-Maxwell equations he reported similarity and oscillating BGK solutions found by Lie point transformations of the moment equations. A closed Lie algebra for the group generators was given. These results seem to have been largely ignored outside the U.S.S.R. Subsequently, Lewis and Symon [40] derived exact, time-dependent solutions of the Vlasov-Poisson equations by using Liouville invariants. Their approach was feasible since time-dependent invariants of the Liouville equation (the Vlasov equation for our case) quadratic in the momenta had already been developed [68-71]. To their solution was appended the conservation of current density and additional freedom in the time dependence [41]. The time-dependent Vlasov-Maxwell solutions with an invariant quadratic in momenta can be calculated from BGK equilibria in some reference frame by a suitable coordinate transformation; this was demonstrated explicitly for the one-dimensional plasma [41].

Next, invariance of differential equations by Lie point transformations was applied to expand the class of time-dependent Vlasov-Maxwell solutions [26-30]. A rather general analysis of the Lie point group for the one-dimensional Vlasov-Maxwell plasma exists now; the analysis of the Lie point group for the Vlasov-Maxwell equations with electromagnetic fields is incompletely known. Four different approaches to finding self-consistent Vlasov-Maxwell solutions by the Lie point group invariance are discussed here. In three cases the Lie group invariance of the Vlasov equation alone is calculated and the resulting distribution function is substituted into the charge and current density integrals in Maxwell's equations. This approach follows the traditional method of solution of the Vlasov-Maxwell equations. In the fourth method the Lie Group invariance of the complete Vlasov-Maxwell set is calculated.

A short review of the invariance of ordinary and partial differential equations under Lie point transformations is first presented. Then the results now in hand for the Vlasov-Maxwell equations are discussed. The time-dependent solutions for the prototype plasma in Cartesian coordinates with one nonignorable coordinate and a longitudinal electric field is treated in detail. Other examples include: a drift plasma with $\vec{B}_o \perp \vec{E}(x,t)$ (longitudinal $\vec{E}$) and a plasma with transverse $\vec{E}(x,t)$ and $\vec{H}(x,t)$ which are mutually perpendicular.

2. INVARIANCE OF DIFFERENTIAL EQUATIONS UNDER LIE POINT TRANSFORMATIONS

Nonlinear ordinary and partial differential equations may be simplified or reduced to quadratures when invariance under a Lie group point transformation occurs. Ordinary differential equations invariant under a one-parameter group can have the number of independent variables reduced by one. Frequently, these reductions in order and in the number of independent variables are sufficient to reduce the differential equation to quadratures or to a general functional form.

We start with the definition of invariance under a continuous point transformation for a function F(x,y) in a plane. F(x,y) is invariant under a transformation if its form remains unchanged. Assume a finite continuous point transformation in the plane:

$$x_1 = \phi(x,y;a), \qquad y_1 = \psi(x,y;a), \tag{1}$$

where the parameter a of the transformation can be any complex number but is usually real. The function F under this transformation is invariant,

$$F(x_1,y_1) = F(x,y) = C = \text{constant} . \tag{2}$$

The finite point transformation (1) constitutes a one-parameter group in that two successive transformations produce a transformation in the group (closure), for three successive transformations the product of the first two or last two transformations may be considered as one transformation with no difference in results (associativity), and the identity and inverse transformations exist. The group may also be viewed as a binary operation under closure and associativity on the set of transformation parameters A where unique identity (a_o) and inverse (a^{-1}) elements exist. For example,

$$F = \exp[(x-y)^2] = \exp[(x_1-y_1)^2] \text{ if } x_1 = x + a,\ y_1 = y + a \tag{3}$$

for a spatial transformation along the line y = x. New variables $\bar{x} = \frac{x-y}{\sqrt{2}}$, $\bar{y} = \frac{x+y}{\sqrt{2}}$; can be defined so that F is a function of $\bar{x}$ alone. In general, if (1) and (2) hold, F depends on a combination of x and y such that F is unchanged under the transformation. Reducing the number of variables is the major advantage gained from applying the Lie point group invariance to functions or differential equations. For an arbitrary function F, guessing the finite transformation (1) may not be easy. On the other hand, finding a function $\bar{x}$ from (1) such that F is some function of $\bar{x}$ is usually easier.

An aid in obtaining the finite point transformation is to look at the infinitesimal point transformation. An expansion of the point transformation about the identity element a_o, $a = a_o + \delta a$, leads to the infinitesimal transformation:

$$x_1 - x = \delta x = \xi(x,y)\delta a, \qquad y_1 - y = \delta y = \eta(x,y)\delta a, \tag{4}$$

where higher order powers of δa can be neglected if one of (ξ,η) does not vanish for all x and y and neither is infinite. Then,

$$\xi(x,y) = \frac{\delta x}{\delta a} = \left(\frac{\partial \phi}{\partial a}\right)_{a_o}, \quad \eta(x,y) = \frac{\delta y}{\delta a} = \left(\frac{\partial \psi}{\partial a}\right)_{a_o} \tag{5}$$

are the characteristic functions and $\frac{\delta}{\delta a}$ is called the Lie derivative [18]. If $F(x_1,y_1)$ is expanded in a Taylor series about (x,y), a necessary and sufficient condition that F be invariant under the infinitesimal transformation is that

$$UF = \xi\frac{\partial F}{\partial x} + \eta\,\frac{\partial F}{\partial y} = 0, \tag{6}$$

where UF represents the infinitesimal transformation and U is the group generator. The arguments about invariance can be repeated for more variables and more parameters (r -parameter group for r parameters) and for sets of nonlinear ordinary and partial differential equations.

The fundamental question is, how do we find the group of transformations? We consider only two variables first. The group generator U is found from (6) for a known F where only the ratio of the coordinate functions is determined. The characteristic equations

$$\frac{dx}{\xi(x,y)} = \frac{dy}{\eta(x,y)} = \frac{dF}{0} \tag{7}$$

are found from (6) and the total differential of F [9],

$$dF = \frac{\partial F}{\partial x}dx + \frac{\partial F}{\partial y}dy = 0, \tag{8}$$

where the proportionality in (7) follows immediately. The third term in (7) is not a mistake by a hapless student, but is standard notation meaning that dF is zero times the other ratios. F depends on the constant found by integrating the first two terms. Since we have the ratio of the coordinate functions, the first two terms give a first order nonlinear (it may be linear at times) ordinary differential equation. Under certain general conditions a unique solution exists for that equation [11,50]; finding it may be difficult, nevertheless. The solution when found is written

$$u(x_1,y_1) = u(x,y) = c_1 = \text{constant} . \tag{9}$$

The finite point transformation can be found if one sets the first two ratios in (7) equal to $\frac{da}{1}$ and integrates both the second ratio and the new one where for $y(x,c_1)$ substitutes the solution from (9). The result is

$$v(x_1,y_1) - a + a_o = v(x,y) = c_2 = \text{constant} . \tag{10}$$

Equations (9) and (10) constitute the finite transformation; to find (1) solve (9) and (10) for x_1 and y_1. The finite Lie point transformations may also be found from a Lie series

$$x_1 = \exp[(a-a_0)U]x, \quad y_1 = \exp[(a-a_0)U]y \ . \tag{11}$$

If F is invariant under a one-parameter Lie transformation group, then in principle, one can find a group generator in the canonical form $U = \frac{\partial}{\partial \bar{y}}$ where canonical coordinates $(\bar{x},\bar{y})$ can be found such that F is invariant under spatial translation in $\bar{y}$. For that case the ratios in (7) equal $\frac{d\bar{y}}{1}$ and the results in (9) and (10) follow if a is replaced by $\bar{y}$. Consequently, $\bar{x} = u(x,y)$ and $\bar{y} = v(x,y)$ are suitable canonical coordinates but are not unique. These results can be extended to a larger number of variables.

The reduction in the number of variables can be extended to an ordinary differential equation

$$F(x,y,y',y'',\ldots) = 0 \ , \tag{12}$$

where the prime denotes differentiation with respect to x, $y' = \frac{dy}{dx}$ etc. Equation (12) is invariant under a continuous Lie group of one-parameter transformations. The derivatives are treated, in a sense, as additional variables. The condition for invariance under a one-parameter Lie transformation group for first order ordinary differential equations is

$$U'F(x,y,y') = 0 \tag{13}$$

with U' the once-extended group generator,

$$U' = \xi(x,y)\frac{\partial}{\partial x} + \eta(x,y)\frac{\partial}{\partial y} + \eta'(x,y,y')\frac{\partial}{\partial y'} \ ,$$

and the first extension of the coordinate functions

$$\eta'(x,y,y') = \frac{\delta}{\delta a}\left(\frac{dy}{dx}\right) = \frac{\partial}{\partial a}\left(\frac{dy_1}{dx_1}\right)_{a_0} = \frac{\partial\eta}{\partial x} + \left(\frac{\partial\eta}{\partial y} - \frac{\partial\xi}{\partial x}\right)y' - \frac{\partial\xi}{\partial y}y'^2 \ .$$

If one knows the characteristic functions ξ and η, the general form of the differential equation (13) can be derived by integrating the characteristic equations of (13). Yet many scientific problems start from differential equations. For the first order differential equation

$$y' - f(x,y) = \frac{dy}{dx} - f(x,y) = 0 \ , \tag{14}$$

as already stated, a solution exists for certain conditions on f and then (14) is invariant under an infinite number of infinitesimal Lie point groups. In practice, the determination of the Lie group (group generator usually) is by guesswork or by comparison with tables [8,16] of general forms of differential equations for known groups. A Riccati equation has been given [8,18,22] that does not seem to have solutions unless the group is known. Nonetheless, many equations turn out to be integrable. If (14) is invariant under a group, it can be rewritten in canonical coordinates:

$$\frac{d\bar{y}}{d\bar{x}} - f(\bar{x}) = 0 , \tag{15}$$

and reduces to quadratures by separation of variables where f changes form in general.

Higher order ordinary, nonlinear differential equations may also be invariant under one or more Lie one-parameter point transformation groups but they also may not be invariant under any group. The nth extension of the group generator can be derived in a similar manner to that for the once-extended group. For a second order differential equation

$$F(x,y,y',y'') = 0 , \tag{16}$$

the twice-extended group generator

$$U'' = \xi(x,y) \frac{\partial}{\partial x} + \eta(x,y) \frac{\partial}{\partial y} + \eta'(x,y,y') \frac{\partial}{\partial y'} + \eta''(x,y,y',y'') \frac{\partial}{\partial y''} \tag{17}$$

for

$$\eta'' = \frac{\partial^2\eta}{\partial x^2} + \left(2 \frac{\partial^2\eta}{\partial x\partial y} - \frac{\partial^2\xi}{\partial x^2}\right)y' + \left(\frac{\partial^2\eta}{\partial y^2} - 2 \frac{\partial^2\eta}{\partial x\partial y}\right)y'^2 - \frac{\partial^2\xi}{\partial y^2} y'^3$$
$$+ \left(\frac{\partial\eta}{\partial y} - 2 \frac{\partial\xi}{\partial x} - 3 \frac{\partial\xi}{\partial y} y'\right)y''$$

acts on F to give the invariance condition

$$U''F(x,y,y',y'') = 0 . \tag{18}$$

A large class of the second order differential equations encountered in scientific problems are of the simpler form

$$y'' - f(x,y,y') = 0 . \tag{19}$$

Then (19) substituted into (18) yields a partial differential equation

$$H(x,y,y') = 0 , \tag{20}$$

where y'' is replaced in η'' by (19), and (20) holds for all values of (x,y,y'). Hence, as (19) must remain unchanged under Lie group invariance, (20) is an identity in (x,y,y'). Usually, y' is the easiest variable on which to impose the identity condition.

For example, if f does not depend on y', H becomes a cubic equation in y' for which the coefficient of each power of y' vanishes. In many cases the only solution to (20) is $\xi = \eta = 0$ or no Lie point transformation group exists for the second order differential equation. In fact, a second order differential equation is invariant under at most eight linearly independent Lie point transformation groups, in contrast to the case for first order ordinary differential equations which are usually invariant under an infinite number of Lie point transformation groups.

Partial differential equations may be invariant under Lie infinitesimal transformation groups. The early work dealt mostly with linear partial differential equations equivalent to a set of characteristic equations of the form already treated. A partial differential equation which is pertinent because the Vlasov equation can be expressed in that form is

$$Af = P(x,y,z)\frac{\partial f}{\partial x} + Q(x,y,z)\frac{\partial f}{\partial y} + R(x,y,z)\frac{\partial f}{\partial z} = 0 \tag{21}$$

with $f(x,y,z)$ the unknown function. If f is also invariant under a Lie infinitesimal point group,

$$Uf = \xi(x,y,z)\frac{\partial f}{\partial x} + \eta(x,y,z)\frac{\partial f}{\partial y} + \sigma(x,y,z)\frac{\partial f}{\partial z} = 0, \tag{22}$$

where (22) is a generalization of (6). A common solution exists [8] if (21) and (22) hold together with

$$[U,A]\, f = \lambda(x,y,z)Af\ . \tag{23}$$

Either partial differential equation has two independent solutions since the set of characteristic equations yields two independent solutions $u(x,y,z)$ and $v(x,y,z)$. If u and v are determined from (22), (21) can be rewritten as

$$Au\frac{\partial f}{\partial u} + Av\frac{\partial f}{\partial v} = 0\ . \tag{24}$$

The group generator U may be found by several means to be discussed later. Once U is determined, the canonical variables u and v may be calculated where the partial differential equation now has two rather than three independent variables.

For nonlinear partial differential equations the total number of variables, dependent and independent, may be reduced by one if the set of equations is invariant under a one-parameter Lie group. The method here leads to particular solutions rather than the general solution, so that boundary conditions may need to be considered in some examples. The set of nonlinear partial differential equations may be tested for invariance under a Lie point transformation group by a generalized extended group generator. For the highest order derivatives of second order the extended group generator U'' is

$$U'' = \xi_j(x_j,u^i)\frac{\partial}{\partial x_j} + \eta^i(x_j,u^i)\frac{\partial}{\partial u^i} + \eta^i_j\frac{\partial}{\partial u^i_{,j}} + \eta^i_{jk}\frac{\partial}{\partial u^i_{,jk}}, \tag{25}$$

with x_j, $j = 1,\dots,n$, the independent variables, u^i, $i = 1,\dots,m$, the dependent variables, the derivatives of u^i with respect to x_j are denoted by

$$u^i_{,j} = \frac{\partial u^i}{\partial x_j},\quad u^i_{,jk} = \frac{\partial^2 u^i}{\partial x_j \partial x_k},\quad k = 1,\dots,n\ ,$$

and the first and second extensions η^i_j and η^i_{jk} respectively [18] are

$$\eta^i_j = \frac{\partial\eta^i}{\partial x_j} + \frac{\partial\eta^i}{\partial u^\mu} u^\mu{}_{,j} - \frac{\partial\xi_\nu}{\partial x_j} u^i{}_{,\nu} - \frac{\partial\xi_\nu}{\partial u^\mu} u^\mu{}_{,j} u^i{}_{,\nu} ,$$

$$\eta^i_{jk} = \frac{\partial^2\eta^i}{\partial x_j\partial x_k} + \frac{\partial^2\eta^i}{\partial x_j\partial u^\mu} u^\mu{}_{,k} + \frac{\partial^2\eta^i}{\partial x_k\partial u^\mu} u^\mu{}_{,j} - \frac{\partial^2\xi_\nu}{\partial x_j\partial x_k} u^i{}_{,\nu} + \frac{\partial\eta^i}{\partial u^\mu} u^\mu{}_{,jk}$$

$$- \frac{\partial\xi_\nu}{\partial x_j} u^i{}_{,k\nu} - \frac{\partial\xi_\nu}{\partial x_k} u^i{}_{,j\nu} + \frac{\partial^2\eta^i}{\partial u^\lambda\partial u^\mu} u^\lambda{}_{,j}u^\mu{}_{,k} - \frac{\partial^2\xi_\nu}{\partial x_j\partial u^\mu} u^\mu{}_{,k} u^i{}_{,\nu}$$

$$- \frac{\partial^2\xi_\nu}{\partial x_k\partial u^\mu} u^\mu{}_{,j} u^i{}_{,\nu} - \frac{\partial\xi_\nu}{\partial u^\mu} [u^i{}_{,\nu}u^\mu{}_{,jk} + u^\mu{}_{,j} u^i{}_{,\nu k} + u^\mu{}_{,k} u^i{}_{,j\nu}]$$

$$- \frac{\partial^2\xi_\nu}{\partial u^\lambda\partial u^\mu} u^\lambda{}_{,k} u^\mu{}_{,j}u^i{}_{,\nu} ,$$

and higher order extensions $\eta^i_{j\ldots n}$ may be defined if higher order derivatives are present. The Einstein convention of summation over repeated indices is used.

For a set of partial differential equations

$$F^\beta(x_j,u^i,u^i{}_{,j},u^i{}_{,jk}) = 0 \tag{26}$$

with second order derivatives, the highest derivative is invariant under a Lie point transformation if

$$U''F^\beta(x_j,u^i,u^i{}_{,j},u^i{}_{,jk}) = 0 \tag{27}$$

for all β together with (26) [18]. We saw for the second order ordinary differential equation that in η'' the y'' was replaced by f. That operation is equivalent to requiring that not only does the group generator which acts on the differential equation vanish, but that the differential equation holds. For a first order ordinary differential equation, y' replaced by f in the invariance condition (13) leads to a partial differential equation in the coordinate functions [22]. Here, the constraint that the set of second order partial differential equations holds must also be applied. One could solve each partial differential equation for a different highest order derivative that appears in that equation and substitute for these derivatives in η^i_{jk} and possibly η^i_j. Another procedure [28] which has proven useful is really a generalization of the invariance of a path curve $F(x,y) = 0$ under a group (6) [8]. Since both UF and F vanish, then

$$UF = \lambda(x,y)F \tag{28}$$

holds or F is a factor of UF where it is assumed that F has no repeated factors. For a set of partial differential equations the condition has been given as

$$U''F^{\beta} = \sum_{\gamma=1}^{N} \lambda_{\beta\gamma}(x_j, u^i)F^{\gamma} \tag{29}$$

for $F^{\beta}(x_j, u^i, u^i{}_{,j}, u^i{}_{,jk}) = 0$. Actually, $\lambda_{\beta\gamma}$ may also depend on the derivatives [8]. In addition, one may remark that no formal proof has been given that (29) with $\lambda_{\beta\gamma}$ dependent on derivatives is equivalent to (27) with the constraints (26).

3. LIE POINT TRANSFORMATION GROUP SOLUTIONS OF THE NONLINEAR VLASOV-MAXWELL EQUATIONS

The Lie point group generator, canonical variables and solutions for the one-particle distribution function f_{β}, the electric field $\vec{E}$ and the magnetic flux density $\vec{B}$ are discussed next for the exact, nonlinear Vlasov-Maxwell equations. The Vlasov equation is

$$\frac{\partial f_{\beta}}{\partial t} + \vec{v} \cdot \frac{\partial f_{\beta}}{\partial \vec{r}} + \frac{q_{\beta}}{m_{\beta}} [\vec{E} + \vec{E}_a + \vec{v} \times (\vec{B} + \vec{B}_a)] \cdot \frac{\partial f_{\beta}}{\partial \vec{v}} = 0, \tag{30}$$

with $f_{\beta}(t,\vec{r},\vec{v})$, $\vec{E}(t,\vec{r})$, $\vec{B}(t,\vec{r})$ depending on the time t, spatial displacement $\vec{r}$, velocity $\vec{v}$, particle charge q_{β}, particle mass m_{β} and the subscript a denotes the applied fields whose sources are outside the plasma. In some applications the Liouville form is preferable,

$$\frac{\partial \bar{f}_{\beta}}{\partial t} + \{\bar{f}_{\beta}, H_{\beta}\} = 0 \tag{31}$$

where H_{β} denotes single-particle Hamiltonian

$$H_{\beta} = \frac{(\vec{p}-q_{\beta}\vec{A}) \cdot (\vec{p}-q_{\beta}\vec{A})}{2m_{\beta}} + q_{\beta}\phi , \tag{32}$$

$\vec{p}$ the momentum, and the curly brackets are Poisson brackets. The fields are constructed in terms of the vector potential $\vec{A}(t,\vec{r})$ and the scalar potential $\phi(t,\vec{r})$: $\vec{E} = -\nabla\phi - \frac{\partial \vec{A}}{\partial t}$, $\vec{B} = \nabla\times\vec{A}$. The Vlasov equation in the Liouville form is expressed in canonical variables with $\vec{v} = \vec{p} - q_{\beta}\vec{A}$. Maxwell's equations are:

$$\nabla\cdot\vec{B} = 0 \ , \quad \nabla\times\vec{E} + \frac{\partial \vec{B}}{\partial t} = 0 , \tag{33),(34}$$

$$\nabla\cdot\vec{D} - \rho_{\varepsilon} = 0, \ \rho_{\varepsilon} = \sum_{\beta} q_{\beta} \int d\vec{v} f_{\beta} + q_b n_b = \sum_{\beta} q_{\beta} \int d\vec{p} \bar{f}_{\beta} + q_b n_b , \tag{35}$$

$$\nabla\times\vec{H} - \frac{\partial\vec{D}}{\partial t} - \vec{J} = 0,\ \vec{J} = \sum_\beta q_\beta \int d\vec{v}\vec{v}f_\beta + q_b n_b u_b =$$

$$\sum_\beta \frac{q_\beta}{m_\beta} \int d\vec{p}(\vec{p}-q_\beta\vec{A})\bar{f}_\beta + q_b n_b \vec{u}_b\ , \tag{36}$$

where the subscript b denotes the background plasma and SI units are used.

Equilibrium solutions of (30) or (31), and (33) - (36) exist if the time derivatives are set equal to zero. The vanishing of the Poisson brackets of $\bar{f}_\beta$ and H_β shows that the distribution function depends on various constants of the motion: conjugate momenta of cyclic coordinates and the particle energy. Uniformly moving plasma equilibria are stationary waves in a laboratory reference frame where a Galilean coordinate transformation $\vec{r}' = \vec{r} - \vec{v}_p t$, $\vec{v}' = \vec{v} - v_p$ relates the primed plasma coordinates to the unprimed laboratory coordinates with $\vec{v}_p$ the uniform relative velocity. The time dependence of stationary waves is trivial here.

The search for nontrivial time-dependent solutions of the Vlasov-Maxwell equations began with an unmagnetized, one-dimensional (Cartesian) plasma with a longitudinal electric field. Four methods are presented and these are designated as: the Vlasov-Maxwell characteristic equations method, the Vlasov-Maxwell invariant paths method, the Vlasov invariant paths method and the Vlasov Green's function method. These are discussed in order.

The Vlasov characteristic equations method [26] calculates the Lie point group invariance of the characteristic equations of the Vlasov equation alone under infinitesimal transformations. The one-dimensional Vlasov equation reduces to

$$\frac{\partial f_\beta}{\partial t} + v\,\frac{\partial f_\beta}{\partial x} + \frac{q_\beta}{m_\beta}\,E(t,x)\,\frac{\partial f_\beta}{\partial v} = 0 \tag{37}$$

and its characteristic equations are

$$\frac{dt}{1} = \frac{dx}{v} = \frac{dv}{\frac{q_\beta}{m_\beta}E(t,x)}\ . \tag{38}$$

The solution of (38) is equivalent to the solution of (37). Our approach was motivated by the Lie invariance of linear partial differential equations and the integration of equations of motion for Vlasov invariants. In this approach the Vlasov equation is a linear partial differential equation when the Lie group generator is calculated. Once the group is determined, the constraints imposed by Maxwell's equations make the electric field not just a given function of (t,x) but also a functional of the distribution function so that the Vlasov equation becomes nonlinear. Eliminating v from the set of two ordinary differential equations (38), we have the equation of motion of a single particle in an electric field

$$\frac{d^2x}{dt^2} - \frac{q_\beta}{m_\beta}\,E(t,x) = 0,\ \frac{dx}{dt} = v,\ \ddot{x} = \frac{d^2x}{dt^2},\ \dot{x} = \frac{dx}{dt}\ . \tag{39}$$

Equation (39) has the form of (19) with the Lorentz force velocity independent. Next apply U" to (39) and substitute for $\ddot{x}$ from (39) its value in η''. The general result for (19) in (18) appeared in an Appendix by Cohen [8]; for this electric field the condition for invariance under a Lie group transformation simplifies to

$$\left(\frac{\partial \eta}{\partial x} - 2\frac{\partial \xi}{\partial t}\right)\frac{q_\beta}{m_\beta}E - \xi\frac{q_\beta}{m_\beta}\frac{\partial E}{\partial t} - \eta\frac{q_\beta}{m_\beta}\frac{\partial E}{\partial x} + \frac{\partial^2 \eta}{\partial t^2}$$
$$+\left(2\frac{\partial^2 \eta}{\partial t \partial x} - \frac{\partial^2 \xi}{\partial t^2} - 3\frac{\partial \xi}{\partial x}\frac{q_\beta}{m_\beta}E\right)\dot{x} + \left(\frac{\partial^2 \eta}{\partial x^2} - 2\frac{\partial^2 \xi}{\partial t \partial x}\right)\dot{x}^2 - \frac{\partial^2 \xi}{\partial x^2}\dot{x}^3 = 0 \qquad (40)$$

where we find a cubic equation in $\dot{x}$ since (ξ,η,E) do not depend on $\dot{x}$. This relation is an identity in $(t,x,\dot{x})$ because the invariance under an infinitesimal group transformation requires that the form of the differential equation be unchanged. Consequently, if one focuses on the $\dot{x}$ dependence, one concludes that the coefficient of each power of x must be set equal to zero. Four partial differential equations result that despite their complicated aspect can be solved. Their solution was facilitated by the recent discovery of nontrivial time-dependent Vlasov solutions. The Lewis-Leach variables $(\bar{t},\bar{x})$

$$\bar{x} = \frac{x-\alpha}{\rho}, \quad \bar{t} = \int \frac{dt}{\rho^2} \qquad (41)$$

for $\alpha(t)$ and $\rho(t)$ given by subsidiary differential equations are the variables for the one-dimensional Vlasov-Maxwell equations reduced to the BGK form [40,41]. In these variables the Vlasov equation is translationally invariant in $\bar{t}$. Since

$$U\bar{x} = 0 = \bar{\eta}, \quad U\bar{t} = 1 = \bar{\xi} \qquad (42)$$

hold, (ξ,η) can be determined from the two algebraic equations (42) and (6). Supplied with a group generator, one is assured that some solution of the four partial differential equations exists.

Simplicity dictates that the equation from the coefficient of $\dot{x}^3$ be integrated first. The linear dependence in x of ξ leads to E at most linear in x unless ξ is restricted to dependence on t only. The group for the one-dimensional harmonic oscillator was worked out by Leach; we do not consider it further here. The equations from $\dot{x}^2$ and $\dot{x}$ are integrated to give a form for η. The equation from $\dot{x}^0$ is tricky, but guided by the Lewis-Symon solution [40] we can assume a form for E that splits the equation by x-dependence into three equations that can be integrated by comparison with the previous solutions. The coordinate functions in the group generator U are:

$$\xi = \rho^2, \quad \eta = (\rho\dot{\rho} + \frac{N}{2}) + \rho^2\dot{\alpha} - \rho\dot{\rho}\alpha\,, \qquad (43)$$

with the subsidiary differential equations

$$m_\beta\ddot{\rho} + q_\beta\Omega^2(t)\rho = \frac{q_\beta k_\beta}{\rho^3}, \qquad (44a)$$

$$m_\beta \ddot{\alpha} + q_\beta \Omega^2(t)\alpha = q_\beta\left(F(t) - \frac{N}{2\rho^3}\int dt\rho F(t)\right) \quad , \tag{44b}$$

where N and k_β are integration constants. The subscript β has been omitted on p, α and N but, in general, these differ for different particle species. In the equation for the electric field, F(t) and $\Omega^2(t)$ are arbitrary functions of t:

$$E = F(t) - \Omega^2(t)x - \frac{e^{N\bar{t}/2}}{\rho^3}\frac{dU_e(\bar{x})}{d\bar{x}} \quad , \tag{45}$$

with U_e an arbitrary function of

$$\bar{x} = \left(\frac{x-\alpha}{\rho}\right) e^{-\frac{N}{2}\bar{t}} - \frac{N}{2}\int \frac{dt\alpha}{\rho^3} e^{-\frac{N}{2}\bar{t}} \tag{46}$$

and $\bar{t}$ is given in (41). The relations (43) - (46) reduce to those for a Liouville invariant, of which the distribution function is a function, quadratic in the momenta for N = 0.

Now the functional dependence of the distribution function may be determined from properties of the complete system of partial differential equations for f_β: (23), the Vlasov equation (37) $V_\beta f_\beta = 0$, and the Lie group invariance condition (22) ,

$$U' f_\beta = \xi(t)\frac{\partial f_\beta}{\partial t} + \eta(t,x)\frac{\partial f_\beta}{\partial x} + \eta'(t,x,v)\frac{\partial f_\beta}{\partial v} = 0 \, , \tag{47}$$

for η' from (13) if $v = \dot{x}$ and we exclude x-dependence of ξ. Here (23) takes the appropriate form for the group generator just calculated:

$$[U', V_\beta]f_\beta = -\frac{\partial \xi}{\partial t} V_\beta f_\beta = 0 \; . \tag{48}$$

Since (37), (47) and (48) constitute a complete system, a common solution f_β exists. In addition, both (37) and (47) have two independent solutions as can be seen by the fact that linear partial differential equations in three independent variables have two characteristic ordinary differential equations. From (43), (47) together with (13) we have the characteristic equations of (47),

$$\frac{dt}{\xi(t)} = \frac{dx}{\eta(t,x)} = \frac{d\dot{x}}{\eta'(t,x,\dot{x})} \; . \tag{49}$$

The first two terms integrate to give $\bar{x}$ (46). The first and third terms integrated with x replaced by its value from (46) in η' give ν :

$$\nu = e^{-\frac{N}{2}\bar{t}}\left[\rho(\dot{x}-\dot{\alpha}) - \dot{\rho}(x-\alpha) - \frac{N\alpha}{2\rho} - \frac{N^2}{4} e^{\frac{N}{2}\bar{t}} \int \frac{dt\alpha}{\rho^3} e^{-\frac{N}{2}\bar{t}}\right] , \tag{50}$$

where $(\bar{x},\nu)$ are the canonical variables. The Vlasov equation (37) can be written in these variables by applying the chain rule and taking derivatives with respect to the canonical variables where $\bar{x}, \nu' = \nu - \frac{N}{2}\bar{x}$ are more useful since the coefficient of the x derivative is now the new velocity variable v'. The equation just mentioned can be written in these new variables; but a more general equation occurs if a coordinate transformation from (t,x,v) to $(\bar{t},\bar{x},v')$ is performed:

$$\frac{\partial f_\beta}{\partial \bar{t}} + \nu' \frac{\partial f_\beta}{\partial \bar{x}} - (N\nu' + \frac{q_\beta}{m_\beta}\frac{dU_e'}{d\bar{x}})\frac{\partial f_\beta}{\partial \nu'} = 0 \ , \ U_e' = U_e(\bar{x}) + \frac{k_\beta \bar{x}^2}{2} + \frac{m_\beta}{q_\beta}\frac{N^2\bar{x}^2}{8} \ . \tag{51}$$

Either $f_\beta = g(\bar{x},\nu')e^{-M\bar{t}}$, M a constant, or a stationary solution $\frac{\partial f_\beta}{\partial \bar{t}} = 0$ is a feasible alternative. The stationary solution is the generalized BGK solution in that unlike the result for the Liouville invariant quadratic in momenta, it has a velocity dependent Lorentz force. The characteristic equation

$$\frac{d\nu'}{d\bar{x}} = - \frac{\left(N\nu' + \frac{q_\beta}{m_\beta}\frac{dU_e'}{d\bar{x}}\right)}{\nu'} \tag{52}$$

integrates to the BGK invariant in Lewis-Leach variables for $N = 0$ and invariants for a "potential" U_e' quadratic in $\bar{x}$. The f_β is a function of the invariant which is the energy in the Lewis-Leach variables for $N = 0$.

The Vlasov characteristic equations method is frequently convenient for systems with one nonignorable coordinate. If in (31) a new Hamiltonian $H_\beta(t,Q_1,Q_2,Q_3,P_1)$ is substituted where generalized coordinates replace the old phase space coordinates and Q_1 and P_1 are conjugate variables, the Vlasov equation again has three independent variables (t,Q_1,P_1) with Q_2,Q_3 as parameters. In many cases an equation of motion for an effective particle can be found from the characteristic equations. Once that is known one may be able to write down the solution by inspection or by extending the result (40) to velocity dependent forces. If on the other hand the Vlasov equation is written in velocity variables (30), more than three variables may appear in the derivatives. Identification of cyclic coordinates can be used to reduce the equation just described to a complete system constructed for more independent variables [8].

The constraints imposed by Maxwell's equations further limit the group generator. A complete analysis has not been performed for the Vlasov characteristic equations method. The earlier time-dependent solutions found from Liouville invariants [40, 41] are shown to be a special case for the Vlasov equation and the substitution into Maxwell's equations follows as before. For a single-species plasma in an immobile uniform background, Baranov [25] found the solution by Lie group invariance of moment equations for a BGK equilibrium oscillating at a plasma frequency. This was a special case of a more recent solution [41]. For multi-species plasmas the Liouville invariants quadratic in momenta lead to nontrivial time dependence only if the charge to mass ratios are the same or if a is quadratic in the time t. In an effort to find less restricted time dependences, a multi-species plasma with no background was treated [27]. Maxwell's equations reduce to

$$\frac{\partial E}{\partial x} - \frac{1}{\varepsilon}\sum_{\beta} q_\beta \int dvf_\beta - \frac{q_b u_b}{\varepsilon} = 0 , \tag{53}$$

$$\frac{\partial E}{\partial t} + \frac{1}{\varepsilon}\sum_{\beta} q_\beta \int dvvf_\beta + \frac{q_b n_b u_b}{\varepsilon} = 0 . \tag{54}$$

The form of E in (45) can be chosen, but $\Omega^2(t) = 0$ for the multi-species case as we need x, and hence (p,α), to be independent of particle species. The E chosen is close to the O'Neil model for Landau damping of plasma oscillations. Let

$$E(t,x) = \frac{E_0}{\rho^3} e^{\frac{N}{2}t} \sin(\beta\bar{x}+\Gamma) = E_0 \exp[-\int_0^t \gamma(t)dt]\sin(\beta\bar{x}+\Gamma) \tag{55}$$

where β and Γ are the wave number and phase angle respectively. Since the electric field must be independent of particle species, arguments can be given to demonstrate that ρ, N, and α are the same for all species; consequently (44a) and (44b) simplify. Maxwell's equations (53) and (54) reduce to

$$\beta E_o\cos(\beta\bar{x}+\Gamma) - \sum_{\beta} \frac{q_\beta}{\varepsilon} \int d\nu' f_\beta(\nu',\bar{x}) = 0 , \tag{56}$$

$$NE_o\sin(\beta\bar{x}+\Gamma) + \sum_{\beta} \frac{q_\beta}{\varepsilon} \int d\nu'\nu' f_\beta(\nu',\bar{x}) = 0 . \tag{57}$$

The reduction involves changing the integration variable v to ν' and putting all the time dependence in Gauss' law on one side of the equation and setting it equal to a constant. That gives $\exp(-\frac{N}{2}t)/\rho^3$ = constant. This condition together with (44a) leads to the time dependence of ρ:

$$\rho = (-\frac{Nt}{3} + 1)^{1/2}, \quad N = -6(-\frac{q_\beta k_\beta}{m_\beta})^{1/2} . \tag{58}$$

The electric field is

$$E(t,x) = E_o e^{N\bar{t}}\sin(\beta\bar{x}-\beta\alpha_o) = \frac{E_o}{\rho^6}\sin[\beta\rho^2(x - \frac{\omega_o t}{\beta} - \frac{\omega_1 t}{\beta\rho^2})] \tag{59}$$

with α_0, ω_0, ω_1 constants and the damping decrement $\gamma(t) = \frac{-N}{2}$.

Next, the Vlasov-Maxwell invariant-paths method is presented where the Lie point group invariance of the entire Vlasov-Maxwell set is performed. This method was developed by Roberts [28] after looking at the Vlasov characteristic equations method, the work of Baranov, and the calculations of Lie group invariance by Axford that showed that the group generator acting on a partial differential equation produced a constant multiple of that equation. The condition for Lie point-group invariance of the Vlasov-Maxwell set of equations is that the once-extended group generator ((25) with all derivatives with respect to $u^i{}_{,jk}$ equal to zero) acting on each member of the Vlasov-Maxwell set produces a linear combination of the original equations (29). The Vlasov equation is denoted by $V_\beta f_\beta = 0$, Gauss' law by G, and the current density equation by J. Then

$$U'V_\beta f_\beta = \sum_{\gamma=1}^{N} \lambda_{\beta\gamma} V_\gamma f_\gamma \ , \tag{60}$$

$$U'G = \lambda_{N+1,N+1} G + \lambda_{N+1,N+2} J \ , \tag{61}$$

$$U'J = \lambda_{N+2,N+1} G + \lambda_{N+2,N+2} J \ , \tag{62}$$

where $\lambda_{\alpha\beta}$ is a function of the dependent and independent variables denoted by $x_1 = t$, $x_2 = x$, $x_3 = v$, $u^i = f_\beta$, $\beta = 1,\ldots,N$, $u^{N+1} = E$, $u^{N+2} = n_b$, $u^{N+3} = u_b$. The charge and current density integral are also subject to infinitesimal transformations [22-25] but one must be careful of the notation. If one uses U' from(25) on Maxwell's equations (53) and (54), the velocity derivatives are zero since that variable is in the integration. Hence, we define for a definite integral I, $U'I = (\frac{\partial I}{\partial a})_{a_0}$.

Consider $I = \int \prod_{j=M}^{N} dx_j x_M f(x_1,x_2,\ldots,x_N)$, $2 < M < N$.

Then

$$U'I = \int \prod_{j=M}^{N} dx_j \ \{ U'[x_M f(x_1,x_2,\ldots,x_N)] + x_m f \sum_{j=M}^{N} \frac{\partial \xi_i}{\partial x_i} \} \ . \tag{63}$$

Now, the relations (60) to (62) are again identities in the variables and derivatives. Hence, the coefficients of the different products of derivatives are equated to zero which produces a set of partial differential equations. The solution of these equations is more intricate than the comparable set found from (40). Certain simplifications occur as the integral terms are grouped separately from other terms. Products of derivatives are not cancelled such that ξ_i are independent of u^i. The coordinate functions for a single species with background plasma are

$$\xi_t = (b_4-b_5)t + a_1, \quad \xi_x = (2b_4-b_5)x + b_3 t + b_2 + g, \quad \xi_v = b_4 v + b_3 + \dot{g} \ ,$$

$$\eta^E = b_5 E + \frac{m}{q}\ddot{g}, \quad \eta^f = (2b_5 - 3b_4)f, \quad \eta^n = (2b_5 - 2b_4)n_b,$$

$$\eta^u = b_4 u_b + b_3 + \dot{g} - \frac{m\varepsilon}{q q_b n_b}\dddot{g} \ , \tag{65}$$

for g(t) arbitrary and for the multi-species plasma, $g = 0$ and charge and mass assume subscripts. The b_j, $j = 1,\ldots,5$ are integration constants.

The canonical coordinates can be determined by integration of the characteristic equations

$$\sum_j \frac{dx_i}{\xi_j} = \frac{du^i}{\eta^i} \ , \tag{66}$$

where the sum over j is only for the independent variables, upon which u^i depends. Equation (66) follows from the invariant surface condition [18,29] that relates the group generator in independent and dependent variables on the same footing to the group generator with dependent variables as explicit functions of the independent variables.

The canonical variables determined by integration of (66) are:

$$\zeta_1 = xR^{(b_5-2b_4)} - \int dtR^{(b_5-2b_4)} (b_3t+b_2+g)S ,$$

$$\zeta_2 = vR^{-b_4} - \int dtR^{-b_4} (b_3+\dot{g})S ,$$

$$u_b = \tilde{u}_b(\zeta_1)R^{b_4} + R^{b_4} \int dtR^{-b_4}(b_3+\dot{g} - \frac{m\varepsilon}{qq_bn_b})S, \quad n_b = \tilde{n}_b(\zeta_1)R^{(2b_5-2b_4)} ,$$

$$E = \tilde{E}(\zeta_1)R^{b_5} + R^{b_5} \int dtR^{-b_5}(\frac{m}{q} \ddot{g})S, \quad f = \tilde{f}(\zeta_1,\zeta_2)R^{(2b_5-3b_4)} , \tag{67}$$

with $R = \exp \{\int \frac{dt}{(b_4-b_5)t+b_1}\}$, $S = \frac{1}{R}\frac{dR}{dt}$. The three independent variables have been reduced to two canonical variables (ζ_1,ζ_2) which are quite similar to $\bar{x}$ and v. Substituted into the Vlasov-Maxwell equations, these relations give

$$(2b_5-3b_4) \tilde{f} + (b_5-2b_4)\zeta_1 \frac{\partial \tilde{f}}{\partial \zeta_1} + \zeta_2 \frac{\partial \tilde{f}}{\partial \zeta_q} - b_4\zeta_2 \frac{\partial \tilde{f}}{\partial \zeta_2} + \frac{q}{m} \tilde{E} \frac{\partial \tilde{f}}{\partial \zeta_2} = 0 , \tag{68}$$

$$b_5\tilde{E} + \frac{q_b}{\varepsilon} \tilde{n}_b\tilde{u}_b + (b_5-2b_4)\zeta_1 \frac{d\tilde{E}}{d\zeta_1} + \frac{q}{\varepsilon} \int d\zeta_2\zeta_2\tilde{f} = 0 , \tag{69}$$

$$\frac{d\tilde{E}}{d\zeta_1} - \frac{q_b\tilde{n}_b}{\varepsilon} - \frac{q}{\varepsilon} \int d\zeta_2\tilde{f} = 0. \tag{70}$$

The characteristic equations from (68) relate $\tilde{f}$ and $\tilde{E}$ in a complicated manner unless $3b_4 = 2b_5$, in which case a simpler result found for the complete system in the Vlasov characteristic equations method results.

A Lie algebra can be defined for the above. The group operators are U_1 through U_6, where the appropriate parameter b_1 to b_5 or the function g is kept nonzero. The operators U_1 to U_5 form a Lie algebra in that the following relation holds

$$[U_i,U_j] = \sum_k C_{ijk}U_k \tag{71}$$

with C_{ijk} constants. Baranov found a finite Lie algebra for a single-species plasma in an immobile, uniform background in which U_6 splits into two generators and U_4 and U_5 coalesce. For the general case the Lie Algebra may be infinite.

The next approach is the Vlasov Green's function method. For the one-dimensional Vlasov-Maxwell equations with a longitudinal field, the feasibility has been checked but a detailed investigation has not been made. The electric field in the Vlasov equation is given in terms of a free space Green's function. This form of the Vlasov equation was suggested by a referee [26] in that he recommended considering the Vlasov equation used by Montgomery and Tidman [72]. They had the Coulomb potential of a charged particle in the integral for the electric field which is just the free space Green's function. The Green's function G here is for charged sheets because the spatial variation is one-dimensional. The Vlasov equation (37) for $\frac{\partial^2 G}{\partial x^2} = -\delta(x-x')$ becomes

$$\frac{\partial f_\beta}{\partial t} + v\frac{\partial f_\beta}{\partial x} + \frac{q_\beta}{m_\beta}\sum_\gamma \frac{q_\gamma}{\varepsilon}\int dx' \int dv' \frac{\partial G}{\partial x}(x-x')f_\gamma(t,x,v')\frac{\partial f_\beta}{\partial v} = 0 . \tag{72}$$

In unpublished calculations (B. Abraham-Shrauner) the group generator was calculated and found to be the same for (72) as found in (65), but the background plasma was omitted. The electric field contribution due to the background plasma must be included separately in (72). Equation (60) was used for calculation of the group generator. The Green's function method has not been developed, but holds promise for bounded plasmas where the free space Green's function is replaced by the appropriate Green's function for the plasma geometry.

The fourth method is the Vlasov invariant paths method [29]. Here the Lie point-group invariance is considered for E and f_β on the same footing by (60). The group generator is of the same form as found by the Vlasov characteristic equations method except for η^E and η^β, which can be found from the invariant surface condition. In addition, ξ_t which is ξ in (43), cannot depend on x. This is the first significantly different result found when the invariant paths and characteristic equations methods are compared. The electric field was reduced to quadratures by this new approach, whereas in (45) the arbitrary functions appear in the subsidiary equations (44). However, $\Omega^2(t)$ can be eliminated directly in terms of ρ and derivatives from (44), and F(t) obeys a linear first order differential equation so that E in (45) can also be reduced to quadratures with ρ and α and their derivatives appearing.

The four methods for determining the Lie point-group generator and from that the functional dependences of the distribution functions, background plasma and electric field have been discussed for the one-dimensional Vlasov-Maxwell equations with a longitudinal electric field. The results are the same except for the case of an electric field linear in x for which a more general group generator was found by the characteristic equations method. The characteristic equations method substitutes for the Vlasov equation the equivalent set of characteristic equations. The Lie group invariance of the second order ordinary different equation, the equation of motion of a charged particle, determines the group generator. The equation for the distribution function is found by forming a complete system, or by coordinate transformation to canonical variables. The method is rapid for certain systems with one nonignorable coordinate if the results are known, but it does have subsidiary functions. The Vlasov-Maxwell invariant paths method is direct in application, the the lengthy part of the calculation being the determination of the group generator. It appears to be the easiest for the construction of a Lie algebra. The Vlasov Green's function method appears to be the most efficient for calculating the group generator from scratch. None of the last three methods has been worked out completely for $\lambda_{\beta\gamma}$ a function of derivatives. A simple argument predicts no dependence of $\lambda_{\beta\gamma}$ on derivatives of the electric field. The possibility of other derivatives has not yet been excluded.

4. LIE POINT-GROUP TRANSFORMATION OF THE VLASOV-MAXWELL EQUATIONS FOR LONGITUDINAL $\vec{E} \perp \vec{B}_o$ AND $\vec{E} \,||\, \vec{B}_o$.

The addition of a magnetic field to a plasma complicates the determination of the Lie point-group generator that characterizes the Lie infinitesimal transformation under which the Vlasov-Maxwell equations are invariant. The simplest case physically is of a plasma with a longitudinal electric field parallel to a uniform magnetic field. A coordinate transformation to a reference frame rotating with the cyclotron frequency reduces, the Vlasov equation for $\vec{E}\,||\,\vec{B}_o$ to the one-dimensional Vlasov equation (37) in the rotating frame.

The drift plasma with a longitudinal electric field perpendicular to a uniform magnetic field can give rise to drift waves. The presence of the crossed fields breaks some of the symmetry; one expects the Lie point transformation group to be more circumscribed. That is found [30]. The determination of the group generator for this plasma is quite involved but insight into the character of these Lie group solutions is gained. The Vlasov equations are

$$\frac{\partial f_\beta}{\partial t} + v_x \frac{\partial f_\beta}{\partial x} + \frac{q_\beta}{m_\beta} E(t,x) \frac{\partial f_\beta}{\partial v_x} + \frac{q_\beta}{m_\beta} B_o(v_y \frac{\partial f_\beta}{\partial v_x} - v_x \frac{\partial f_\beta}{\partial v_y}) = 0 \tag{73}$$

in velocity variables and

$$\frac{\partial \bar{f}_\beta}{\partial t} + \frac{p_x}{m_\beta} \frac{\partial \bar{f}_\beta}{\partial x} + q_\beta \left[E(t,x) - \frac{B_o}{m_\beta} (p_y + q_\beta B_o x)\right] \frac{\partial \bar{f}_\beta}{\partial p_x} = 0 \tag{74}$$

in momentum variables. In the calculations, (73) has been used for the invariant paths method and (74) for the Vlasov characteristic equations method. Calculations of the Lie group generator have been performed for three plasmas: (a) a single-species plasma with neutralizing background, (b) a single-species plasma and (c) a multi-species plasma for the first two cases mentioned above.

For the Vlasov Green's function method the group generator

$$U = \xi_t \frac{\partial}{\partial t} + \xi_x \frac{\partial}{\partial x} + \xi_{v_x} \frac{\partial}{\partial v_x} + \xi_{v_y} \frac{\partial}{\partial v_y} + \eta^f \frac{\partial}{\partial f} \tag{75}$$

has the coordinate functions

$$\xi_t = b_1, \qquad \xi_x = b_2 \cos(\omega_{UH} t + b_3),$$

$$\xi_{v_x} = \frac{d\xi_x}{dt}, \qquad \xi_{v_y} = \omega_c \xi_x, \qquad \eta^f = 0, \tag{76}$$

for the single-species plasma with a constant-density neutralizing background. The cyclotron frequency is $\omega_c = \frac{q}{m} B_o$, where the subscript β has been dropped, and the upper hybrid frequency is

$$\omega_{UH} = (\omega_p^2+\omega_c^2)^{1/2} \quad , \quad \omega_p^2 = \frac{-qq_b n_b}{\varepsilon} , \tag{77}$$

with ω_p the plasma frequency. The same group generator is found for the single-species plasma, but in the absence of a background plasma the upper hybrid frequency collapses to the cyclotron frequency. Only a trivial time dependence of stationary waves occurs for a multi-species plasma. The electric field has been replaced by $E_g + E_b(t,x)$, where

$$\frac{\partial E_b}{\partial x} = \frac{q_b n_b}{\varepsilon} , \tag{78}$$

and E_g is the Green's function electric field in (72). The lambdas are assumed to be functions of the independent and dependent variables.

The characteristic equations together with (76) can be easily integrated as they are all separable and the canonical coordinates are found from the four separate differential equations. In addition, the conservation of p_y is imposed which considerably constrains the form of the coordinate functions so that the simple form in (76) occurs. With (74) the conservation of p_y is automatic. The canonical coordinates are

$$\zeta_1 = \omega_c x - v_y, \qquad \zeta_2 = x - \int \zeta_x dt ,$$

$$\zeta_3 = v_x - \zeta_x, \qquad \zeta_4 = v_y - \omega_c \int \zeta_x dt. \tag{79}$$

Only three of these are independent coordinates and in addition, ξ_1 is proportional to p_y and a constant of the motion since y is cyclic. Consequently, only two variables appear as derivatives in the transformed Vlasov equation, which is found by the chain rule and the Vlasov operator acting on the first three variables in (79). Then (73) becomes

$$\zeta_3 \frac{\partial f_\beta}{\partial \zeta_2} + (\omega_c \zeta_1 - \omega_{UH}^2 \zeta_2 + \frac{q}{m} E_g (\zeta_2) + C_D) \frac{\partial f}{\partial \zeta_3} = 0 , \tag{80}$$

where the subscript β is dropped for a single species and C_D is a constant. The distribution function is an arbitrary function of p_y (which is proportional to ζ_1) and p_z, and a function of the invariant found from (80). The characteristic equation integrated from (80) is the BGK energy in the coordinates (ζ_3, ζ_4) in a single-species plasma, but is a more complicated function if a constant density background is present.

The constraints on Maxwell's equations show that for the single-species plasma only a trivial time dependence can occur, but for a single-species plasma with a constant density background the oscillatory behavior at the upper hybrid frequency is found. The reason is that a y-component of the particle current density is balanced by the background current density or vanishes. The motion of the background plasma at the upper hybrid frequency may be unphysical in certain cases.

Only a trivial time dependence is determined by the Vlasov-Maxwell invariant paths method for background charges and current densities which are functions of (t,x). If the background current density depends on the spatial derivative of the electric field for a constant density background and the electric field is of the form above (78) where E_g is now the dependent variable, then (76) is found.

The Vlasov characteristic equations method applied to (74) gives a trivial time dependence for a multi-species plasma. The Lorentz force in (74) defines an effective electric field such that the one-dimensional results discussed in III apply. Of course, the actual electric field E is not a function of p_y. The single-species case is incomplete.

5. LIE POINT-GROUP FOR TRANSVERSE $\bar{E} \perp \bar{B}$

Exact solutions of the nonlinear Vlasov-Maxwell equations with full electromagnetic fields are of wider interest than the examples discussed so far. Yet, discovering such solutions has been harder as the length and intricacy of the calculation increases, whereas the constraints appear to reduce the variable dependence of the group generator. The transformation groups for the Maxwell equations were given early in the development of relativity [23,24]. The sources were electrons or ponderable bodies as functions of space and time. The groups are Lorentz, dilation, and conformal. Nevertheless, Bateman indicated that transformations may be applied to a particular problem that do not work, in general. One expects the Vlasov equation to restrict further the group of transformations allowed. The invariant quadratic in the momenta for the Vlasov equation with a three-dimensional electromagnetic field has also been derived [42], but the constraints imposed by Maxwell's equations have not been deduced. One stretching group has been found for the relativistic Vlasov-Maxwell equations by the Vlasov-Maxwell invariant paths method (Dana Roberts, private communication).

In order to check that a simple plasma configuration with the electromagnetic fields can be invariant under Lie point-groups, a multi-species plasma with $\vec{E} = E_y(t,x)\hat{y}$ and $\vec{B} = B_z(t,x)\hat{z}$ has been chosen. First, the group generator was calculated for (30). Only stretching transformations are found. The stretching transformation is $x_1 = ay$, $y_1 = by$ where special names are: affine a or b = 0, similitude or perspective (a=b). An alternative approach employs the Liouville form of the Vlasov equation (31). The equations are

$$\text{Vlasov:}\quad \frac{\partial \bar{f}_\beta}{\partial t} + \frac{p_x}{m_\beta}\frac{\partial \bar{f}_\beta}{\partial x} + \frac{q_\beta}{m_\beta}\frac{\partial A_y}{\partial x}(p_y - q_\beta A_y)\frac{\partial \bar{f}_\beta}{\partial p_x} = 0\ , \tag{81}$$

Generalized Ampere´s law:

$$\frac{\partial^2 A_y}{\partial x^2} - \mu\varepsilon \frac{\partial^2 A_y}{\partial t^2} = - \mu \sum_B \frac{q_\beta}{m_\beta} \int d\vec{p}(p_y - q_\beta A_y)\bar{f}_\beta , \tag{82}$$

Gauss´ law, and the current density for vanishing charge and current density. Faraday´s law is an identity and the vanishing of the z-component of the current density is ignored since the p_z-dependence of the distribution function is arbitrary. The inhomogeneous wave equation for the vector potential has second order derivatives so that the full U" (25) applies. The alternate approach seems the more economical in terms of calculation of the two.

The group generator is found to be

$$U = \xi_t \frac{\partial}{\partial t} + \xi_x \frac{\partial}{\partial x} + \xi_{p_x} \frac{\partial}{\partial p_x} + \xi_{p_y} \frac{\partial}{\partial p_y} + \eta^\beta \frac{\partial}{\partial \bar{f}_\beta} + \eta^A \frac{\partial}{\partial A_y} . \tag{83}$$

The coordinate functions

$$\xi_t = (b_1 - b_2)t + b_3, \qquad \xi_x = (b_1 - b_2 + b_6)x + b_4, \ \xi_{p_x} = b_6 p_x , \ \xi_{p_x} = 0$$

$$\eta^\beta = b_1 f_\beta + b_5, \qquad \eta^A = b_6 A_y \tag{84}$$

were determined by the Vlasov invariant paths method. When the constraints of Maxwell´s equations were added, $b_5 = b_6 = 0$ and $b_2 = -\frac{3}{2} b_1$ for the integration constants b_j follow.

ACKNOWLEDGEMENTS

Dana Roberts is thanked for many discussions on the material in this article. This research was supported by the National Science Foundation and the National Aeronautics and Space Administration.

REFERENCES

[1]. S. Lie, "Begrundung einer Invariantentheorie der Berührungs transformationen," Math. Ann. 8,215 (1874).

[2]. S. Lie, "Zur Analytischen Theorie der Beruhrungstransformationen," Kristiania Forh. Aeret., 237 (1874).

[3]. S. Lie, "Theorie der Transformationsgruppen," Math. Ann. 16, 441 (1880).

[4]. S. Lie, "Über die Integration durch bestimme Integrale von einer Klasse lineare partieller Differentialgleichungen," Arch. for Math. Naturvidensk. 6, 328 (1881).

[5]. S. Lie and F. Engel, Theorie der Transformationsgruppen, Vol. I (1888), Vol. II (1890), Vol. III (1893) (B. G. Teubner, Leipzig; reprinted by Chelsea, New York, 1970).

[6]. J. M. Page, Ordinary Differential Equations with an Introduction to Lie's Theory of the Group of One Parameter, (MacMillan, London, 1897).
[7]. J. E. Wright, Invariants of Quadratic Differential Forms (Hafner, New York, 1908).
[8]. A. Cohen, An Introduction to the Lie Theory of One-Parameter Groups with Applications to the Solution of Differential Equations (Heath, Boston, 1911; reprinted by Stechert, 1931).
[9]. E. Goursat, Differential Equations, Vol. II, Pt. II, trans. E. R. Hedrick and O. Dunkel (Ginn, New York, 1917).
[10]. L. E. Dickson, "Differential equations from the group standpoint," Ann. Math. 25, 287 (1924).
[11]. E. L. Ince, Ordinary Differential Equations (Dover, New York, 1950).
[12]. J. E. Campbell, Introductory Treatise on Lie's Theory of Finite Continuous Transformations (Chelsea, New York, 1966; original ed., 1903).
[13]. L. P. Eisenhart, Continuous Groups of Transformations (Princeton Univ. Press, Princeton, 1933; also Dover, New York, 1961).
[14]. G. Kowalwski, Integrationsmethoden der Lieschen Theorie (Akad. Verlagsgesellschaft M. B. H., Leipzig, 1933).
[15]. L. V. Ovsiannikov, Gruppovye Svoystva Differentsialny Uraveri, (Novosibirsk, 1962) (Group Properties of Differential Equations, trans. G. Bluman, 1976).
[16]. L. Markus, (Group Theory and Differential Equations, Tech. Rep. 4, U. S. Dept. of Commerce, Washington, D. C. (1968) (unpublished).
[17]. W. F. Ames, Nonlinear Partial Differential Equations in Engineering, Vol. I (1965) and Vol. II (1972) (Academic, New York).
[18]. G. W. Bluman and J. D. Cole, Similarity Methods for Differential Equations (Springer, New York, 1974).
[19]. R. Hermann, Lie Groups: History, Frontiers and Applications, Vol. I: Sophus Lie's 1880 Transformation Group Paper, trans. M. Ackerman (1975), Vol. III: Sophus Lie's 1884 Differential Invariant Paper (1976) (Math Sci Press, Boston).
[20]. J. M. Hill, Solution of Differential Equations by Means of One-Parameter Groups (Pitman, Boston, 1982).
[21]. L. V. Ovsiannikov, Group Analysis of Differential Equations, trans. edited by W. F. Ames (Academic, New York, 1982).
[22]. R. Axford, Lectures on Lie Groups and Systems of Ordinary and Partial Differential Equations (Los Alamos, 1983).
[23]. H. Bateman, "The transformation of the electrodynamical equations," Proc. London Math. Soc. 223 (1909).
[24]. E. Cunningham, "The principle of relativity in electrodynamics and an extension thereof," Proc. London Math Soc. 77 (1909).
[25]. V. B. Baranov, "Symmetry of one-dimensional high frequency motion of a collisionless plasma," Sov. Phys. Tech. Phys. 21, 720 (1976).
[26]. B. Abraham-Shrauner, "Lie transformation group solutions of the nonlinear one-dimensional Vlasov Equation," Bull. APS 28, 1105 (1983); J. Math Phys. (submitted) (1984).
[27]. B. Abraham-Shrauner, "Exact sinusoidal electric field of nonlinear one-dimensional Vlasov-Maxwell equations," J. Plasma Phys. (to be published) (1984).
[28]. D. Roberts, "The general Lie group and similarity solutions for the one-dimensional Vlasov-Maxwell equations," J. Plasma Phys. (submitted) (1984).
[29]. D. Roberts, "An alternate approach to finding and using the Lie group of the Vlasov equation," IEEE International Conf. on Plasma Science, late paper. J. Math. Phys. (to be submitted for publication) (1984).
[30]. B. Abraham-Shrauner, "Lie group invariance of the Vlasov Equation for $\vec{E} \perp \vec{B}_0$," IEEE International Conf. on Plasma Science, 119 (1984).
[31]. I. B. Bernstein, J. M. Greene, and M. D. Kruskal, "Exact, nonlinear plasma oscillations," Phys, Rev. 108, 546 (1957).
[32]. R. F. Lutomirski and R. N. Sudan, "Exact nonlinear electromagnetic whistler modes," Phys. Rev. 147, 156 (1966).
[33]. M. J. Laird and F. B. Knox, "Exact solution for charged particle trajectories in an electromagnetic field," Phys. Fluids 8, 755 (1965).

[34]. T. F. Bell, "Nonlinear Alfvén waves in a Vlasov Plasma," Phys. Fluids 8, 1829 (1965).
[35]. B. U. O. Sonnerup and S. Y. Su, "Large amplitude whistler waves in a hot collision-free plasma," Phys. Fluids 10, 462 (1967).
[36]. B. Abraham-Shrauner, "Exact stationery wave solutions of the nonlinear Vlasov equations," Phys. Fluids 11, 1162 (1968).
[37]. R. C. Davidson, Methods in Nonlinear Plasma Theory (Academic, New York, 1972).
[38]. B. Abraham-Shrauner and W. C. Feldman, "Nonlinear Alfvén waves in high speed solar wind streams," J. Geophys. Res. 82, 618 (1977).
[39]. J. L. Schwarzmeier, H. R. Lewis, B. Abraham-Shrauner, and K. R. Symon, "Stability of Bernstein-Greene-Kruskal equilibria," Phys. Fluids 22, 1747 (1979); K. M. King and B. Abraham-Shrauner, "Modified Poisson eigenfunctions for electrostatic Bernstein-Greene-Kruskal equilibria," Phys. Fluids 24, 629 (1981). (Additional references are found in both of these).
[40]. H. R. Lewis and K. R. Symon, "Exact time-dependent solutions of the Vlasov-Poisson equations," Phys. Fluids 27, 192 (1984).
[41]. B. Abraham-Shrauner, "Exact time-dependent solutions of the one-dimensional Vlasov-Maxwell equations," Phys. Fluids 27, 197 (1984).
[42]. H. Ralph Lewis, "Exact invariants quadratic in the momentum for a particle in a three-dimensional electromagnetic field," J. Math. Phys. 25, 1139 (1984).
[43]. B. Abraham-Shrauner, "Nonlinear theta-pinch equilibria," J. Plasma Phys. 26, 419 (1981).
[44]. P. Rosenau, "Magnetohydrodynamic blast waves in a gravitational field," Phys. Fluids 21, 1455 (1978).
[45]. P. Rosenau, "Three-dimensional flow with neutral points," Phys. Fluids 22, 849 (1979)/
[46]. L. I. Sedov, Similarity and Dimensional Methods in Mechanics, trans. M. Friedman and edited by M. Holt (Academic, New York, 1959).
[47]. P. Mora and R. Pellat, "Self-similar expansion of a plasma into a vacuum," Phys. Fluids 22, 2300 (1979).
[48]. F. S. Felber, "Self-similar oscillations of a Z-pinch," Phys. Fluids 25, 643 (1982).
[49]. J. R. Burgan, M. R. Feix, E. Fijalkow, and A. Munier, "Self-similar and asymptotic solutions for a one-dimensional Vlasov Beam," J. Plasma Phys. 29, 139 (1983).
[50]. H. T. Davis, Introduction to Nonlinear Differential and Integral Equations (Dover, New York, 1962).
[51]. C. M. Bender and S. A. Orszag, Advanced Mathematical Methods for Scientists and Engineers (McGraw-Hill, New York, 1978).
[52]. J. D. Talman, Special Functions, A Group Theoretic Approach (Benjamin, New York, 1968).
[53]. B. K. Harrison and F. B. Estabrook, "Geometric approach to invariance groups and solutions of partial differential equations," J. Math. Phys. 12, 653 (1971).
[54]. Y. Choquet-Bruhat, C. Dewitt-Morette, and M. Dillard-Bleick, Analysis, Manifolds, and Physics (North-Holland, Amsterdam, 1978).
[55]. B. Schutz, Geometric Methods of Mathematical physics (Cambridge University Press, Cambridge, 1980).
[56]. R. Anderson, S. Kumei, and C. Wulfman, "Generalization of the concept of invariance of differential equations," Phys. Rev. Lett. 28, 988 (1972).
[57]. R. L Anderson, S. Kumei, and C. E. Wulfman, "Invariants of the equations of wave mechanics; rigid rotator and symmetric top," J. Math. Phys. 14, 1527 (1973).
[58]. S. Kumei, "Invariance transformations, invariance group transformations and invariance groups of the sine-gordon equations," J. Math. Phys. 16, 2461 (1975).
[59]. F. Gonzáles-Gascón, "Notes on the symmetries of systems of differential equations," J. Math. Phys. 18, 1763 (1977).
[60]. R. L. Anderson and N. H. Ibragimov, Lie-Bäcklund Transformations in Applications (SIAM, Philadelphia, 1979).
[61]. C. E. Wulfman, "Systematic methods for determining the continuous transformation groups admitted by differential equations," Symmetries in Science, B. Gruber and R. Millman, Eds. (Plenum, New York, 1980).

[62]. C. W. Bluman and S. Kumei, "On the remarkable diffusion equation $\frac{\partial}{\partial x}[a(a+b)^{-2}\frac{\partial a}{\partial x}] - \frac{\partial a}{\partial t} = 0$," J. Math. Phys. 21, 1019 (1980).

[63]. S. Kumei and G. W. Bluman, "When nonlinear differential equations are equivalent to linear differential equations," SIAM J. Appl. Math. 42, 1157 (1982).

[64]. J. J. Cullen and J. L. Reid, "Lie-Bäcklund groups and the linearization of differential equations," J. Phys. A: Math. Gen. 16, 1889 (1983).

[65]. M. Lakshmanan and K. M. Tamizhmani, Comment on "Method for the exact solution of a nonlinear diffusion-convection equation," Phys. Rev. Lett. 51, 1497 (1983).

[66]. L. Hlavatý, K. B. Wolf, and S. Steinberg, "Integral and Bäcklund transforms within symmetry groups of certain families of nonlinear differential equations," J. Phys. A: Math. Gen. 16, 2917 (1983).

[67]. W. H. Steeb and W. Oevel, "Bäcklund transformation groups of nonlinear evolution equations and the Painlevé property," Z. Naturforsch 38a, 86, (1983).

[68]. H. R. Lewis, "Class of exact invariants for classical and quantum time-dependent harmonic oscillators," J. Math. Phys. 9, 1976 (1968).

[69]. P. G. L. Leach, "An exact invariant for a class of time-dependent anharmonic oscillations with cubic anharmonicity," J. Math. Phys. 22, 465 (1981).

[70]. H. R. Lewis and P. G. L. Leach, "Exact invariants for a class of time-dependent nonlinear Hamiltonian systems," J. Math. Phys. 23, 165 (1982).

[71]. H. R. Lewis and P. G. L. Leach, "A direct approach to finding exact invariants for one-dimensional time-dependent classical Hamiltonians," J. Math. Phys. 23, 2371 (1982).

[72]. D. C. Montgomery and D. A. Tidman, Plasma Kinetic Theory (McGraw-Hill, New York, 1964).

A CONSTRUCTIVE SOLUTION TO THE HAMILTON-JACOBI EQUATION

F. H. Molzahn and T. A. Osborn
Department of Physics and Astronomy
University of Maryland
College Park, Maryland 20742

and

Department of Physics
University of Manitoba
Winnipeg, Manitoba R3T 2N2

1. INTRODUCTION

In this paper we construct an explicit solution to the Hamilton-Jacobi equation. Consider a nonrelativistic classical system composed of N point particles each having mass m and interacting via smooth bounded pair potentials. The Hamiltonian for such a system has the general form

$$H(x,p) = \frac{1}{2m} p^2 + v(x) \quad , \tag{1.1}$$

where $x,p \in \mathbb{R}^d$ denote the vectors giving the position and momenta of all N particles. For particles moving in three dimensions, d=3N. The function v(x) is the total potential energy of the system associated with configuration x. Given H(x,p) the related Hamilton-Jacobi equation is

$$\frac{\partial}{\partial t} S(x,t,y) + \frac{1}{2m} |\nabla_x S(x,t,y)|^2 + v(x) = 0 \quad , \tag{1.2}$$

where $t \in \mathbb{R}$ is the time variable and y represents d independent free parameters.

The Hamilton-Jacobi equation is a nonlinear partial differential equation in d+1 dimensions, so obtaining solutions is a nontrivial task. This is particularly true in the general problem considered here where one cannot expect that v(x) will have symmetry properties that would permit the study of (1.2) by the method of separation of variables. We investigate (1.2) by considering solutions that take the form

$$S(x,t,y) = \frac{m}{2} \frac{|x-y|^2}{t} - \Phi(x,t,y) \quad , \tag{1.3}$$

where Φ is real-valued, an odd function of t, and has the series representation

$$\Phi(x,t,y) = \sum_{n=1}^{\infty} A_{2n-1}(x,y)\, t^{2n-1} \tag{1.4}$$

for some coefficients A_{2n-1}. Time reversal invariance for system (1.1) requires that Hamilton's principal function be odd in the time variable. For this reason we restrict our attention to solutions of (1.2) that are odd in t. If v=0, then $\Phi=0$ and (1.3) becomes the solution of the Hamilton-Jacobi equation that describes free particle motion (with constant velocity) from initial configuration y to final configuration x taking elapsed time t. It will turn out that A_{2n-1} assumes the form of a parametric integral in n dimensions whose integrand has a structure determined by the sum of all labelled tree graphs that can be formed on the vertex set $\{1,2,\ldots,n\}$. It will be shown that the series (1.4) is uniformly and absolutely convergent for all x, y and for t restricted to some finite time interval containing the origin. Given the construction of S, a just described, we will establish that S is a complete integral in the sense of Jacobi's theorem. We supplement Jacobi's theorem by showing that several aspects of its local character can be made uniform. By repeated application of this strengthened Jacobi theorem we analyze the behavior of the classical paths admitted by the Hamiltonian (1.1) and study the fixed-end-point variational problem.

One merit of our constructive representation of the complete integral of the Hamilton-Jacobi equation is that many detailed dynamical properties of the Hamiltonian system (1.1) can be established with elementary analytical methods. That such graphical expansions of the Hamilton-Jacobi equation solutions exist has been previously noted by Marinov [1]. In Sec. II, explicit expressions for A_{2n-1} are determined from a recurrence relation and the convergence properties of series (1.4) are determined. The fact that (1.3) and (1.4) constitute a complete integral of the Hamilton-Jacobi equation is proved in Sec. III. The final section characterizes the dynamical behavior of the system and analyzes the behavior of the fixed-endpoint variational problem. The Appendix contains the tree-graph proof of the explicit form the coefficients A_{2n-1} take.

2. RECURRENCE RELATIONS

This section investigates the natural recurrence relations associated with the time power series expansion of a solution to the Hamilton-Jacobi equation. It is established that the coefficient functions A_{2n-1} have a multiple integral representation with an integrand determined by tree graphs. Finally we conjecture an explicit solution of the form (1.3)-(1.4) and determine its convergence properties.

We first characterize the class of potentials v(x) that will be employed throughout the remainder of the paper. Basically these allowed potentials are real-valued C^∞ functions of x with bounded derivatives of controlled growth. The symbol ∇^α denotes a partial derivative in $\mathbb{R}^d$ with multi-index α.

Definition 1. Let $v:\mathbb{R}^d \to \mathbb{R}$. Then we say $v \in \mathcal{D}$ if

(i) $v \in C^\infty(\mathbb{R}^d)$,

(ii) v is bounded on $\mathbb{R}^d$,

$$\|v\| \equiv \sup_{x\in\mathbb{R}^d} |v(x)| < \infty ,$$

(iii) For every multi-index $\alpha \neq 0$, there is an α independent constant $K \in [0,\infty)$ such that

$$\|\nabla^\alpha v\| \leq \left(\frac{K}{\sqrt{d}}\right)^{|\alpha|} .$$

Throughout the remainder of the paper it is assumed that $v \in \mathcal{D}$, but this hypothesis will not always be written out. This class of potentials is suitable for the N-body problem since no decay or limiting behavior is specified as $|x|\to\infty$. In addition oscillatory potentials like sin kx are within the class $\mathcal{D}$. Note that the assumption of a common mass m for all particles is made purely for notational convenience. The general kinetic energy operator for an N-body system composed of particles with distinct masses m_i $(i=1,\ldots,N)$ can always be brought into the Hamiltonian form (1.1) by a scale transformation of the particle coordinates.

Suppose a solution S(x,t,y) of (1.2) admits the ascending series expansion in t given by (1.3)-(1.4). Now if we formally substitute (1.3)-(1.4) into (1.2) we find that the coefficients of powers of t must satisfy the recurrence relation

$$[(2n-1)+(x-y)\cdot\nabla_x]a_n(x,y) = \delta_{n,1}\, v(x) + \frac{1}{2}\sum_{k=1}^{n-1} \binom{n}{k}(\nabla_1 a_k)(x,y)\cdot(\nabla_1 a_{n-k})(x,y) , \qquad (2.1)$$

where

$$A_{2n-1}(x,y) \equiv \frac{1}{m^{n-1}n!}\, a_n(x,y) , \qquad n \in \mathbb{N} . \qquad (2.2)$$

Here ∇_i is the gradient with respect to the ith vector argument of a_k. The sum on the right of (2.1) is absent if n=1. The factor $\binom{n}{k}$ is the binomial coefficient, and $\mathbb{N}$ is the set of natural numbers. In the following we let the symbol $\tilde{\xi}_i$ represent the linear path in $\mathbb{R}^d$ from y to x:

$$\tilde{\xi}_i = y + \xi_i\,(x-y), \quad \xi_i \in I \equiv [0,1] \ .$$

Our first observation is that the recurrence relation has an integral equivalent.

Lemma 1. Suppose $n \in \mathbb{N}$. Any sequence of continuously differentiable functions $\{a_k\}_{k=1}^n$, $a_k: \mathbb{R}^{2d} \to \mathbb{R}$, satisfies (2.1) if and only if the $\{a_k\}_{k=1}^n$ obey the recursive integral identity:

$$[n=1] \quad a_1(x,y) = \int_I d\xi\, v(\tilde{\xi}),$$

$$[n\geq 2] \quad a_n(x,y) = \frac{1}{2}\sum_{k=1}^{n-1} \binom{n}{k} \int_I d\xi\, \xi^{2n-2} (\nabla_1 a_k)(\tilde{\xi},y)\cdot(\nabla_1 a_{n-k})(\tilde{\xi},y) \ . \qquad (2.3)$$

Proof: Suppose (2.1) is satisfied and $n \geq 2$. Set $x = y+\xi z$ where $\xi \in I$ and $z \in \mathbb{R}^d$ are arbitrary. Notice that

$$\xi \frac{d}{d\xi} a_n(y+\xi z,y) = \xi z\cdot(\nabla_1 a_n)(y+\xi z,y) \ .$$

After multiplying (2.1) by ξ^{2n-2} one finds

$$\frac{d}{d\xi}\,[\xi^{2n-1} a_n(y+\xi z,y)] = \xi^{2n-2}\,\frac{1}{2}\sum_{k=1}^{n-1} \binom{n}{k}(\nabla_1 a_k)(y+\xi z,y)\cdot(\nabla_1 a_{n-k})(y+\xi z,y).$$

Integrating over $\xi \in I$ and setting $x = y+z$, which like z, is arbitrary, we obtain (2.3). A similar argument applies if n=1.

Conversely, let a_n be defined by (2.3). Set $x = y+\lambda z$, $(\lambda>0, z\in \mathbb{R}^d)$ and multiply (2.3) by λ^{2k-1}. Change the integration variable to $\gamma = \lambda\xi$, differentiate with respect to λ and then set $\lambda=1$. Choosing $z = x-y$ results in (2.1). □

Our next step is to guess a formula for $a_n(x,y)$. This guess derives from a study of the WKB approximation [2,3] for the unitary time evolution operator in quantum mechanics, with a generator given by

$$H_{(x)} = -\frac{\hbar^2}{2m}\Delta_x + v(x),$$

where Δ_x is the Laplacian in $\mathbb{R}^d$ and $\hbar$ is Planck's constant.

In order to describe the conjectured formula for $a_n(x,y)$ it is helpful to review the notation used to describe tree graphs. A labelled tree T on n vertices is the ordered pair $T = (V(T),E(T))$. V(T) is the vertex set of T, and consists of n distinct natural numbers, the vertex labels. E(T) is the edge set over V(T) of T, and consists of n-1 unordered pairs of distinct elements of V(T), unless n=1 where $E(T) = \emptyset$. Each element $v \in V(T)$ must appear in at least one pair in E(T). The elements of E(T) are called links.

If we fix a vertex set V, then the symbol $\mathcal{T}V$ will denote the set of all labelled trees on V. According to Cayley's theorem, if V has n elements there are n^{n-2} trees in $\mathcal{T}V$ [4]. If $f: \mathcal{T}V \to B$, where B is a vector space, then the notation

$$\sum_{T \in \mathcal{T}V} f(T)$$

means to sum over all trees in $\mathcal{T}V$.

If $\beta \in E(T)$, then the notation $\beta = \{i_\beta, j_\beta\}$ $(i_\beta, j_\beta \in V(T))$ will be understood consistently to imply $i_\beta < j_\beta$. For each $\beta \in E(T)$ we define the quadratic differential operator b_β

$$b_\beta \equiv \phi(\xi_{i_\beta}, \xi_{j_\beta})\, D_{i_\beta} \cdot D_{j_\beta},$$

where ϕ is the one dimension Green's function on the unit interval I,

$$\phi(\xi,\xi') = \min(\xi,\xi')\,[1-\max(\xi,\xi')] \quad .$$

Here D_i denotes a gradient operator in $\mathbb{R}^d$ which acts only on the potential function v whose argument contains the index i, as follows:

$$D_i\, v(\tilde{\xi}_1)\cdots v(\tilde{\xi}_n) \equiv v(\tilde{\xi}_1)\cdots(\nabla v)\,(\tilde{\xi}_i)\cdots v(\tilde{\xi}_n), \quad i=1,2,\ldots,n.$$

Further, we abbreviate the multiple integral over the n-dimensional unit cube I^n by

$$\int_0^1 d\xi_1 \int_0^1 d\xi_1 \cdots \int_0^1 d\xi_n = \int_{I^n} d^n\xi \quad .$$

We often encounter vertex sets which consist of the first n natural numbers, so it is convenient to introduce the notation

$$V_n \equiv 1\sim n \equiv \{1,2,\dots,n\} .$$

Definition 2. For $n \in \mathbb{N}$, define the functions $a_n : \mathbb{R}^{2d} \to \mathbb{R}$ by the following "tree sums":

$$[n=1] \qquad a_1(x,y) \equiv \int_I d\xi \; v(\tilde{\xi}) \tag{2.4}$$

$$[n \geq 2] \qquad a_n(x,y) \equiv \int_{I^n} d^n\xi \sum_{T \in \mathcal{T} V_n} \Big(\prod_{\beta \in E(T)} b_\beta \Big) \prod_{p=1}^{n} v(\tilde{\xi}_p) . \tag{2.5}$$

The main result of this section is the following:

Proposition 1. The functions $a_n : \mathbb{R}^{2d} \to \mathbb{R}$, defined by (2.4)-(2.5) satisfy the recurrence relation (2.1) for all $n \geq 1$.

Proof: Because of its substantial length we place the proof in the Appendix. There it is shown that if the a_n are given by (2.4)-(2.5), then the recursive integral identity (2.3) is obeyed. By Lemma 1 it follows that recurrence relation (2.1) is satisfied. □

Given the determination of the coefficients a_n in Definition 2, we define a function Φ, which will turn out to be proportional to the part of Hamilton's principal function S, due to the interaction v (see 1.3).

Definition 3. Define $\Phi : \mathbb{R}^{2d+1} \to \mathbb{R}$ by the following series when it converges:

$$\Phi(x,t,y) \equiv \sum_{n=1}^{\infty} \frac{t^{2n-1}}{m^{n-1} n!} a_n(x,y) . \tag{2.6}$$

It is useful to summarize the convergence properties of (2.6). We have

Lemma 2. Let $v \in \mathfrak{D}$ and

$$T = 2K^{-1}(m/e)^{1/2} . \tag{2.7}$$

(i) If $t \in (-T,T)$ is fixed, series (2.6) converges absolutely, and uniformly for $(x,y) \in \mathbb{R}^{2d}$. Moreover if t is restricted to a compact subset of $(-T,T)$ the convergence is also uniform in t.

(ii) For each $(x,y) \in \mathbb{R}^{2d}$, $\Phi(x,t,y)$ is an analytic function of t in the complex disc $|t| < T$.

(iii) If $t \in (-T,T)$ is fixed, Φ is a C^∞ function of $(x,y) \in \mathbb{R}^{2d}$. In fact, if $\nabla_x^{\alpha_1} \nabla_y^{\alpha_2}$ is an arbitrary partial derivative in these variables (multi-indexed by α_1, α_2) then

$$\nabla_x^{\alpha_1} \nabla_y^{\alpha_2} \Phi(x,t,y) = \sum_{n=1}^{\infty} \frac{t^{2n-1}}{m^{n-1}\, n!} \nabla_x^{\alpha_1} \nabla_y^{\alpha_2} a_n(x,y), \tag{2.8}$$

where the series on the right-hand side has the same properties (i), (ii) above as does series (2.6).

Proof: These results are evident consequences of the tree sum definition (2.4)-(2.5) of a_n and the requirement $v \in \mathcal{D}$. For example, using Cayley's theorem, the fact that 2(n-1) gradients occur in $\prod_{\beta \in E(T)} b_\beta$ for a tree T on n vertices, and $0 < \phi(\xi_i, \xi_j) < \frac{1}{4}$, we obtain the bound

$$|a_n(x,y)| < n^{n-2} \left(\frac{K^2}{4}\right)^{n-1}, \qquad n > 2,\ (x,y) \in \mathbb{R}^{2d}.$$

This implies that (2.7) is a lower bound for the radius of convergence of (2.6).

Similarly the estimate (for $n \in \mathbb{N}$, $\alpha_1 + \alpha_2 \neq 0$)

$$|\nabla_x^{\alpha_1} \nabla_y^{\alpha_2} a_n(x,y)| < n^{n-2} \left(\frac{K^2}{4}\right)^{n-1} \left(\frac{nK}{\sqrt{d}}\right)^{|\alpha_1+\alpha_2|}, \tag{2.9}$$

and the smoothness of v lead to (2.8). □

3. THE COMPLETE INTEGRAL

The detailed behavior of the function S(x,t,y) and its relation to the Hamiltonian is considered in this section. In particular it is proved that S(x,t,y) is a complete integral of the Hamilton-Jacobi equation. By the introduction of an appropriate fixed point theorem we indicate how one can make the behavior of the Jacobi theorem uniform in x and y.

Let the Hamiltonian H: $\mathbb{R}^{2d} \to \mathbb{R}$ be given by (1.1). The associated Hamilton-Jacobi equation is

$$\frac{\partial}{\partial\tau} \mathcal{S}(q,\tau) + H\big(q,\nabla_q \mathcal{S}(q,\tau)\big) = 0 \ , \tag{3.1}$$

where q is the position coordinate in $\mathbb{R}^d$, and $\mathcal{S}:\mathbb{R}^{d+1} \to \mathbb{R}$. A class of solutions to (3.1) of particular importance is the complete integral.

Definition 4: A function $\mathcal{S}$: $G \subset \mathbb{R}^{2d+1} \to \mathbb{R}$ will be called a complete integral of the Hamilton-Jacobi equation if

(i) G is a region (a connected open set);

(ii) $\mathcal{S}(q,\tau;Q)$ depends continuously on d independent parameters $Q = (Q_1,Q_2,\ldots,Q_d)$ lying in some (non-empty) region $G_Q \subset \mathbb{R}^d$;

(iii) $\mathcal{S}(q,\tau;Q)$ is continuously differentiable with respect to q and τ;

(iv) $\mathcal{S}$ is a solution of (3.1).

Consider the function S given by:

Definition 5: Suppose $\Phi(x,t,y)$ is the function given by series (2.6). Define $S: \mathbb{R}^d \times (0,T) \times \mathbb{R}^d \to \mathbb{R}$ by

$$S(x,t,y) = \frac{m}{2t} |x-y|^2 - \Phi(x,t,y) \ . \tag{3.2}$$

There is no loss of generality in restricting the time variable in S to be positive. For negative t one can set

$$S(x,t,y) = -S(x,-t,y) \ .$$

Our first conclusion for S is

Theorem 1: The function S, defined in Eq. (3.2) is a complete integral of the Hamilton-Jacobi equation (3.1).

Proof: It is apparent that conditions (i)-(iii) of Definition 4 are satisfied. Note that the d independent parameters are given by the vector $y \in \mathbb{R}^d$. It remains to show that S(x,t,y) is a solution of the Hamilton-Jacobi equation. Direct calculation yields

$$\frac{\partial}{\partial t} S(x,t,y) + H\big(x,\nabla_x S(x,t,y)\big) = -\frac{\partial}{\partial t}\Phi - \frac{1}{t}(x-y)\cdot\nabla_x\Phi + \frac{1}{2m}|\nabla_x\Phi|^2 + v(x) \ . \tag{3.3}$$

Lemma 2 makes it permissible to substitute series (2.6) for Φ and differentiate term by term. Collecting common powers of t, we find that the coefficient of $-(t^2/m)^{n-1}/n!$ is

$$[(2n-1) + (x-y)\cdot\nabla_x]a_n(x,y) - v(x)\delta_{n,1} - \frac{1}{2}\sum_{k=1}^{n-1}\binom{n}{k}(\nabla_1 a_k)(x,y)\cdot(\nabla_1 a_{n-k})(x,y), \tag{3.4}$$

where the sum is taken to be zero if n=1. But since the a_k are defined by the tree sums (2.4)-(2.5) it follows from Proposition 1 that the coefficient (3.4) vanishes identically. Thus the right-hand side of (3.3) vanishes, proving the theorem. □

In view of the fact that S is a complete integral we can continue the investigation of S and the classical dynamics it implies by appealing to Jacobi's theorem. First recall the definition of the Jacobian matrix associated with S:

$$M_{ij}(x,t,y) \equiv -\frac{\partial}{\partial y_i}\frac{\partial}{\partial x_j} S(x,t,y) = \frac{m}{t}\delta_{ij} + \frac{\partial}{\partial y_i}\frac{\partial}{\partial x_j}\Phi(x,t,y) \ , \qquad i,j=1,\dots,d,$$

and its related determinant $D\colon \mathbb{R}^d\times(0,T)\times\mathbb{R}^d\to\mathbb{R}$

$$D(x,t,y) \equiv \det M(x,t,y) \ .$$

With this terminology we can state the famous result of Jacobi:

<u>Theorem 2</u>: Suppose S(x,t,y) is a complete integral of the Hamilton-Jacobi equation. Suppose further that the second partials,

$$\frac{\partial^2 S}{\partial x_i\,\partial y_j} \ , \ \frac{\partial^2 S}{\partial x_i\,\partial t} \ , \ \frac{\partial^2 S}{\partial y_i\,\partial t} \qquad i,j\in 1\sim d \ .$$

exist and are continuous.

For $p_0, y\in\mathbb{R}^d$ let $q(\tau) \equiv q(\tau;y,p_0)$ be a solution to

$$-(\nabla_2 S)(q,\tau,y) = p_0 \tag{3.5}$$

defined by the implicit function theorem in a region where $D(q,\tau,y) \neq 0$. Let $p(\tau) \equiv p(\tau;y,p_0)$ be given [whenever $q(\tau)$ exists] by

$$p(\tau) \equiv (\nabla_1 S)(q(\tau),\tau,y) \ . \tag{3.6}$$

Then $q(\cdot), p(\cdot)$ satisfy Hamilton's equations

$$\dot q(\tau) = (\nabla_2 H)\big(q(\tau),p(\tau)\big) \ , \qquad \dot p(\tau) = -(\nabla_1 H)\big(q(\tau),p(\tau)\big) \ .$$

Proof: See [5]. □

A major limitation on the usefulness of Jacobi's theorem is the local character of the implicit function theorem solution of (3.5). For example, if the value p_o is not assumed by $-(\nabla_2 S)(q,\tau,y)$ as q varies through $\mathbb{R}^d$ and τ varies through $(0,T)$ for fixed y, then (3.5) is empty and so does not define a classical path $q(\cdot)$. A related problem is that it is often difficult to determine the region where $D(q,\tau,y)$ is non-zero. Thus it is of interest that we can replace the implicit function theorem solution of (3.5) by a more flexible and uniform fixed point solution.

Lemma 3. Given $\sigma \in (0,1)$, let $t_\sigma = (1 + c/\sigma)^{-1/2}\,T$, where $c = 2\sqrt{2/\pi}$. Then

$$\left|\frac{t}{m}\frac{\partial}{\partial x_j}\frac{\partial}{\partial y_i}\Phi(x,t,y)\right| < \frac{\sigma}{d} \tag{3.7}$$

for all $t \in (0,t_\sigma)$; $i,j=1,\ldots,d$; $x,y \in \mathbb{R}^d$.

Proof: This is an immediate consequence of the coefficient bound (2.9). □

Lemma 4. Given any $(t,y,p_o) \in (0,t_\sigma)\times\mathbb{R}^{2d}$, $\sigma \in (0,1)$ the equation

$$-(\nabla_2 S)(x,t,y) = p_o \tag{3.8}$$

has a unique solution $x = q(t;y,p_o) \in \mathbb{R}^d$. For fixed $(y,p_o) \in \mathbb{R}^{2d}$, $q(\cdot;y,p_o):(0,t_\sigma) \to \mathbb{R}^d$ is a C^1 function of t.

Proof: Consider first the existence of a unique solution. Eq. (3.8) suggests we examine a function $F: \mathbb{R}^d \to \mathbb{R}^d$ defined by

$$F(x) = y + \frac{t}{m}p_o - \frac{t}{m}(\nabla_2\Phi)(x,t,y)\ . \tag{3.9}$$

Of course, F depends parametrically on y, p_o, and t. We first show F is a contraction mapping of $\mathbb{R}^d$ into $\mathbb{R}^d$. If $x,x' \in \mathbb{R}^d$ are arbitrary, then (3.9) implies

$$F(x) - F(x') = \frac{t}{m}\left[(\nabla_2\Phi)(x',t,y) - (\nabla_2\Phi)(x,t,y)\right]\ .$$

Apply Taylor's formula to the jth component of F and use the Schwartz inequality. One finds

$$|F_j(x)-F_j(x')| \leq |x-x'|\,|\nabla F_j(x+\lambda_j(x'-x))|$$

for some $\lambda_j \in [0,1]$. Since $t < t_\sigma$ estimate (3.7) is valid and

$$|\nabla F_j(x'')| < \frac{\sigma}{\sqrt{d}}$$

for all $x'' \in \mathbb{R}^d$. Computing the Euclidean norm for the difference of F gives

$$|F(x) - F(x')| < \sigma \, |x-x'| \; .$$

Since $\sigma < 1$, F is a contraction mapping. By the contraction principle [6] it possesses a unique fixed point $q \in \mathbb{R}^d$, viz. $F(q) = q$. Since $t \neq 0$ this can be rearranged to give

$$\frac{m}{t}(q-y) + (\nabla_2\Phi)(q,t,y) = p_o \, , \tag{3.10}$$

so that q is a solution of (3.8).

To establish the continuity of $q(t) \equiv q(t;y,p_o)$ and its t-derivative let $t,t' < t_\sigma$ be neighboring values of time. Eq. (3.10) gives

$$q(t)-q(t') = \frac{t-t'}{m} p_o - \frac{t}{m}(\nabla_2\Phi)(q(t),t,y) + \frac{t'}{m}(\nabla_2\Phi)(q(t'),t',y). \tag{3.11}$$

Extract the q(t)-q(t') dependence on the right hand side of (3.11) by using Taylor's formula to write

$$(\nabla_2\Phi)_j(q(t),t',y)-(\nabla_2\Phi)_j(q(t'),t',y)$$

$$= (q(t)-q(t'))\cdot\nabla_x(\nabla_2\Phi)_j(q(t')+\lambda_j\,[q(t)-q(t')],t',y)$$

for some $\lambda_j \in [0,1]$. Now introduce the $d \times d$ matrix

$$[W(t,t')]_{j,i} \equiv \frac{t'}{m}\frac{\partial}{\partial x_i}\frac{\partial}{\partial y_j}\Phi(q(t')+\lambda_j[q(t)-q(t')],t',y)$$

and observe that (3.7) implies that this matrix has the norm bound

$$\|W(t,t')\| < \sigma < 1, \qquad t,t' < t_\sigma \; .$$

Thus $(1 + W(t,t'))^{-1}$ exists and (3.11) is equivalent to

$$q(t)-q(t') = (1 + W(t,t'))^{-1}\, V(t,t'), \tag{3.12}$$

where V(t,t') is the vector-valued function

$$V(t,t') \equiv \frac{t-t'}{m} p_o - \frac{t}{m}(\nabla_2\Phi)(q(t),t,y) + \frac{t'}{m}(\nabla_2\Phi)(q(t),t',y) \; .$$

Since $|V(t,t')| \to 0$ as $t' \to t$, (3.12) shows that q(t) is continuous at t. If one divides (3.12) by $t - t'$ and uses Lemma 2 then the existence of the derivative of q(t) and its continuity readily follows. □

Note that the first part of Lemma 4 implies that Range $(-\nabla_2 S) \equiv -(\nabla_2 S)(\mathbb{R}^d \times (0,T)\times \mathbb{R}^d) = \mathbb{R}^d$. In particular (3.5) can be solved for every $p_0 \in \mathbb{R}^d$.

4. CLASSICAL PATHS, UNIQUENESS AND THE VARIATIONAL PRINCIPLE

By systematic application of Jacobi's theorem we shall derive many of the dynamical properties of the Hamiltonian system (1.1). This section will establish the uniqueness properties of the classical paths, prove the absence of conjugate points and demonstrate that the fixed-endpoint variational problem has only one extremal if $|t| < T$.

Consider first the behavior of the solutions to (3.5) for $t < t_\sigma$, $\sigma < 1$. We have

Lemma 5. For each $y, p_0 \in \mathbb{R}^d$ the C^1 solution, $q(\tau;y,p_0)$ of Eq. (3.8) satisfies

$$\lim_{\tau\to 0^+} q(\tau;y,p_0) = y \ , \tag{4.1}$$

$$\frac{\partial q}{\partial \tau}(0^+;y,p_0) = p_0/m \ . \tag{4.2}$$

Proof: To verify (4.1) note that (3.8) may be written

$$\frac{m}{\tau}[q(\tau;y,p_0)-y] = -(\nabla_2\Phi)\big(q(\tau;y,p_0),\tau,y\big) + p_0 \ .$$

The boundedness of $-\nabla_2\Phi + p_0$ as $\tau \to 0^+$ implies (4.1). The fact that $-(\nabla_2\Phi)(q,\tau,y)$ vanishes uniformly (in q) as $\tau \to 0^+$ gives (4.2). □

The term classical path will denote a C^2 function $q: E \subset \mathbb{R} \to \mathbb{R}^d$ which satisfies Newton's equation

$$m\frac{d^2}{d\tau^2}q(\tau) = -\nabla v\big(q(\tau)\big) \ .$$

Proposition 2. Let $y, p_0 \in \mathbb{R}^d$ and $q(\tau) \equiv q(\tau;y,p_0)$ be the solution of (3.8) given in Lemma 4. Then $q(\tau)$ is a classical path with the following properties:

(i) Initial value uniqueness: Let $(\tilde{q},\tilde{p}): (0,t_\sigma) \to \mathbb{R}^{2d}$ be any solution to Hamilton's equations with $\tilde{q}(0^+) = y$ and $\tilde{p}(0^+) = p_0$. Then,

$$\tilde{q}(\tau) = q(\tau) \ , \qquad \tau \in (0,t_\sigma) \ .$$

(ii) Two-endpoint uniqueness: Consider any two distinct classical paths $q_i:(0,t_\sigma)\to\mathbb{R}^d$ $(i=1,2;\ q_1\neq q_2)$ emanating from the same initial point

$$q_i(0^+) = y\ , \quad i = 1,2\ .$$

Then these paths do not intersect:

$$q_1(\tau) \neq q_2(\tau)\ , \quad \tau\in(0,t_\sigma)\ .$$

(iii) Completeness: As y,p_o varies throughout $\mathbb{R}^{2d}$ the unique solutions $q(\tau)$ to (3.8) exhaust all possible classical paths.

Proof: Because S is a complete integral, Jacobi's theorem augmented with Lemma 4 provides a solution $q(\tau)$ of Hamilton's equations. Lemma 5 shows that the initial position (at $\tau = 0$) of $q(\tau)$ is y and its initial velocity is p_o/m.

(i) If $\tilde{q}(\tau)$ is a solution of Hamilton's equations with initial position y and velocity p_o/m, then since Hamilton's equations have unique solutions $\tilde{q}(\tau) = q(\tau)$.

(ii) Suppose there exists a $\tau^*\in(0,t_\sigma)$ such that

$$q_1(\tau^*) = q_2(\tau^*)\ . \tag{4.3}$$

By hypothesis q_1 and q_2 are solutions of Hamilton's equations. Part (i) shows that q_1 and q_2 must also be solutions of (3.8). The value of $-\nabla_2 S$ is constant along any solution of (3.5) and this constant is the initial momentum. Thus (4.3) implies

$$-(\nabla_2 S)\big(q_1(\tau^*),\tau^*,y\big) = -(\nabla_2 S)\big(q_2(\tau^*),\tau^*,y\big)$$

or

$$\frac{\partial q_1}{\partial\tau}(0^+,y,p_o) = \frac{\partial q_2}{\partial\tau}(0^+,y,p_o)\ .$$

Thus by (i), $q_1(\tau)$ and $q_2(\tau)$ are equal for all $\tau\in(0,t_\sigma)$. This is in contradiction with the assumption that q_1 and q_2 are distinct.

(iii) This is an immediate consequence of (i) and the fact that Range $(-\nabla_2 S) = \mathbb{R}^d$. □

Notice that the two-endpoint trajectory problem is solved by a special choice of p_o. Set

$$p_o = p_o(x,t,y) \equiv -(\nabla_2 S)(x,t,y) \tag{4.4}$$

where $t\in(0,t_\sigma)$. Then the unique classical path

$$q(\tau;x,t,y) \equiv q\big(\tau;y,p_o(x,t,y)\big) \tag{4.5}$$

provided by the solution of (3.8) with p_o given by (4.4) runs from y to x in elapsed time t. That $q(t) = x$ is verified by checking that this endpoint is indeed a solution of (3.8).

A second observation of interest concerns orbital periods. A closed orbit $(q(\cdot),p(\cdot))$ is a periodic solution of Hamilton's equations, i.e., there is a least positive time T_o such that $(q(\cdot),p(\cdot))$ are translation invariant

$$(q(\tau),p(\tau)) = (q(\tau+T_o),p(\tau+T_o)), \text{ all } \tau .$$

Consider a closed orbit $(q(\cdot),p(\cdot))$. One possibility is that the orbit be trivially 'periodic' in the sense that $p(\tau) = 0$ for all τ. In this case the path $q(\tau) = x_o$, a constant, and x_o corresponds to a point of equilibrium for the system: $\nabla v(x_o) = 0$. (Also a least $T_o > 0$ fails to exist).

If the orbit is not trivial, then there is a time, say $\tau = 0$, where $p_o \equiv p(0) \neq 0$. Let $y = q(0)$ and consider two distinct paths emanating from y; one being the original orbit-path and the other having initial momentum $-p(0) \neq p_o$. Since Hamiltonian (1.1) is time-reversal invariant, the second path follows the trace of q in the reverse direction relative to the first. Because the orbit is closed, the two paths must intersect at time $\tau^* = T_o/2$. Proposition 2(ii) implies $\tau^* \geq t_\sigma$. So we have the conclusion that every (nontrivial) closed orbit has period $T_o \geq 2t_\sigma$.

The Jacobi theorem as used in Proposition 2 provides one with the solutions $q(\cdot),p(\cdot)$ of Hamilton's equations provided that $0 < \tau < t_\sigma$. However one may obtain these solutions for all time displacements $\tau > 0$ by an iterative application of Jacobi's theorem. This is done as follows.

Let $y^{(0)},p^{(0)}$ be the arbitrary initial position and momentum of a classical path $Q:[0,\infty)\to\mathbb{R}^d$. We choose some $t \in (0,t_\sigma)$, and let $n \equiv [\tau/t]$ be the greatest integer less than τ/t. The trajectory Q,P (solution to Hamilton's equation's) corresponding to path Q is then defined recursively for all $\tau > 0$:

$$Q(\tau) \equiv q(\tau-nt;\, y^{(n)},\, p^{(n)})$$

$$P(\tau) \equiv (\nabla_1 S)\,(Q(\tau),\, \tau-nt,\, y^{(n)})$$

$$y^{(\ell)} \equiv q(t;\, y^{(\ell-1)},\, p^{(\ell-1)}) = Q(\ell t)$$

$$p^{(\ell)} \equiv (\nabla_1 S)\,(y^{(\ell)},t,y^{(\ell-1)}) = P(\ell t)$$

where $\ell \in \mathbb{N}$. This procedure works because the solutions $q(\cdot;y^{(\ell)},p^{(\ell)})$ of (3.8) are unique and available for all $(p_o,y) \in \mathbb{R}^d \times \mathbb{R}^d$.

Given that $S(x,t,y)$ is a complete integral of the Hamilton-Jacobi equation it is natural to inquire whether or not S is also the Lagrangian action (Hamilton's principal function). If $q(\tau)$ is the classical path defined in (4.5) the Lagrangian action is defined to be

$$S_L(x,t,y) \equiv \int_0^t d\tau\, L\big(q(\tau),\dot{q}(\tau)\big) \,. \tag{4.6}$$

The integrand here is the Lagrange function associated with the Hamiltonian (1.1),

$$L\big(q(\tau),\dot{q}(\tau)\big) = \frac{1}{2}\, m\, \dot{q}(\tau)^2 - v\big(q(\tau)\big) \,.$$

It is known that S_L is a solution of the Hamilton-Jacobi equation (1.2). We demonstrate that S_L is identical with the function $S(x,t,y)$ defined by (3.2) provided $t < t_\sigma$. Let $\mathcal{R}_\sigma$ denote the region $\mathbb{R}^d \times (0,t_\sigma) \times \mathbb{R}^d$.

<u>Proposition 3</u>. $S_L(x,t,y) = S(x,t,y)$ for all $(x,t,y) \in \mathcal{R}_\sigma$.

<u>Proof</u>: Start with formula (4.6) where $q(\tau)$ is the unique classical path from y to x taking time t, i.e., the path defined by (4.5). Jacobi's theorem gives us

$$\dot{q}(\tau) = \frac{1}{m}\, (\nabla_1 S)\big(q(\tau),\tau,y\big) \,.$$

Insert this identity into the integrand of (4.12) and use the Hamilton-Jacobi equation for S. One finds

$$S_L(x,t,y) = \int_0^t d\tau\, [- \frac{\partial}{\partial\tau} S\big(q(\tau),\tau,y\big) - 2v(q(\tau))] \,,$$

where the τ derivative acts only on the second argument of S. Expressing the first term of the integrand as a total derivative with respect to τ, via

$$\frac{\partial}{\partial\tau} S\big(q(\tau),\tau,y\big) = \frac{d}{d\tau} S\big(q(\tau),\tau,y\big) - m\Big|\frac{d}{d\tau}\, q(\tau)\Big|^2$$

leads to

$$S_L(x,t,y) = -S\big(q(\tau),\tau,y\big)\Big|_{\tau=0+}^{t} + 2 \int_0^t d\tau\, L\big(q(\tau),\dot{q}(\tau)\big) \,,$$

or

$$S_L(x,t,y) = S(x,t,y) - \lim_{\tau\to 0+} S\big(q(\tau),\tau,y\big) \,.$$

From the properties of $q(\tau)$ and Φ it is easily seen that the limit vanishes. □

We conclude our analysis by discussing the implications of our solution of the Hamilton-Jacobi equation for the fixed-endpoint variational problem. Let $D(J)$ be a class of functions appropriate for the domain of a functional. The action functional $J:D(J)\to\mathbb{R}$ is defined by

$$J[Q] = \int_0^t d\tau\, L\big(Q(\tau),\, \dot{Q}(\tau)\big)\,, \tag{4.7}$$

where the domain of J is

$$D(J) = \{Q:[0,t]\to\mathbb{R}^d \mid Q \text{ piecewise smooth, } Q(0) = y,\ Q(t) = x\}\ .$$

Each smooth element $Q^* \in D(J)$ where the first variation [5] of (4.7) vanishes is an <u>extremal</u>. The associated $Q^*(\tau)$ is a solution of the Euler-Lagrange equations. In favorable circumstances a extremal may be a local minimum of the functional. The nature of the minimum is characterized by the norm associated with it.

For piecewise-smooth functions Q on [0,t] define two norms by

$$\|Q\|_o \equiv \sup_{\tau\in[0,t]} |Q(t)|\ ,$$

$$\|Q\|_1 \equiv \|Q\|_o + \sup_\tau |\dot{Q}(\tau)|\ .$$

In the $\|\cdot\|_1$ norm the supremum is taken only over the values of $\tau \in [0,t]$ where the derivative exists. An extremal $Q^* \in D(J)$ is a <u>strong local minimum</u> if there exists an $\varepsilon > 0$ such that

$$J[Q^*] < J[Q]\ , \tag{4.8}$$

for all $Q \in D(J)$ satisfying

$$\|Q-Q^*\|_o < \varepsilon\ .$$

On the other hand, J[Q] is said to have a <u>weak local minimum</u> at Q^* if (4.8) is valid for $Q \in D(J)$ obeying

$$\|Q-Q^*\|_1 < \varepsilon\ .$$

Of course, if Q^* is a strong minimum it is also a weak minimum.

In general terms establishing whether or not an extremal is a minimum requires proving that the second variation of J(Q) at $Q=Q^*$ is positive and this in turn is determined in large part by the absence of conjugate points. More precisely [5] given a classical path $q:[0,t]\to\mathbb{R}^d$, a point $\tau_F \in (0,t]$ is said to be <u>conjugate</u> to 0 if there exists a sequence $\{q_n\}$ of neighboring classical paths satisfying

$$q_n(0) = q(0)\ , \qquad q_n(\tau_n) = q(\tau_n) \tag{4.9}$$

where

$$\|q_n - q\|_o \to 0\ , \qquad \tau_n \to \tau_F\ , \qquad \text{as } n\to\infty\ .$$

Lemma 6. For each $(y,p_0) \in \mathbb{R}^{2d}$, let $q(\cdot;y,p_0)$ be the classical path defined by (3.8). Then $q(\cdot;y,p_0)$ has no conjugate points in the interval $(0,t_\sigma)$.

Proof: Equation (4.9) requires the existence of two distinct classical paths q and q_n that have the same initial position y and the same final position for $\tau = \tau_n$. But by Proposition 2 this is impossible if $\tau_n < t_\sigma$. □

With this technical result we can prove

Theorem 3. For each $(x,t,y) \in \Omega_\sigma$ the associated functional $J[\cdot]$ has only one extremal $Q^* \in D(J)$. This extremal is a strong minimum and is given by $Q^*(\cdot) = q(\cdot;x,t,y)$.

Proof: Each extremal Q^* is a solution of the Euler-Lagrange equations that has initial point y, final point x and takes elapsed time t. Proposition 2 shows that there is only one such classical path and this unique path is $Q^*(\cdot) = q(\cdot;x,t,y)$ where q is defined in (4.5).

Sufficient conditions for a strong local minimum are established in Refs. [5] and [7]. The first three of the four conditions are:

(i) $q(\cdot)$ is a classical path.

(ii) The $d \times d$ matrix, $K(q(\tau),\dot{q}(\tau))$, defined by

$$\frac{\partial}{\partial \dot{q}i}\,\frac{\partial}{\partial \dot{q}j}\,L(q(\tau),\dot{q}(\tau)) , \qquad i,j=1,\ldots,d,$$

is positive definite for $\tau \in [0,t]$.

(iii) The path $q(\cdot)$ has no point conjugate to $\tau = 0$ in $(0,t]$.

Jacobi's theorem assures us that (i) is valid. Lemma 6 guarantees that (iii) is fulfilled. (ii) is an immediate consequence of the definition of the Lagrangian.

Let E be the Weierstrass function defined by the time-independent Langrange function $L(q,\dot{q})$. For $Q,\dot{Q},z \quad R^d$

$$E(Q,\dot{Q},z) \equiv L(Q,z) - L(Q,\dot{Q}) - (z-\dot{Q})(\nabla_2 L)(Q,\dot{Q}) .$$

Upon using the Taylor series remainder formula, this can also be written

$$E(Q,\dot{Q},z) = \frac{1}{2}\,(z-\dot{Q})\cdot K\big(Q,\dot{Q} + \Theta(z-\dot{Q})\big)\,(z-\dot{Q})$$

for some value of $\Theta \in (0,1)$.

The fourth condition needed to establish a strong local minimum is

(iv) The E-function $E(Q,\dot{Q},z) > 0$ in a region of $Q,\dot{Q}$ space containing $q(\tau),\dot{q}(\tau)$ for all z R^d.

Noting that K is m times the unit matrix for all arguments, one has

$$E(Q,\dot{Q},z) = \frac{1}{2}\, m|z-\dot{Q}|^2 > 0 .$$

Thus condition (iv) is obeyed. □

A more extensive study of this tree graph constructive solution $S(x,t,y)$ is found in Ref. [3]. In particular we note that it is possible to show that the Jacobian determinant, $D(x,\tau,y)$, is non-vanishing for all $\tau < T$. Thus all the conclusions of this section remain valid if t_σ is replaced by T, and $\mathcal{R}_\sigma$ by $\mathcal{R} \equiv \mathbb{R}^d \times (0,T) \times \mathbb{R}^d$.

APPENDIX: TREE COMBINATORICS

The purpose of this Appendix is to prove Proposition 1. First we make a few additions to our notations for trees and their combinatorics. For a given $k \in \mathbb{N}$, let J_k be a k-element vertex set chosen from the set $\{1,2,\ldots,n+1\}$. There are $\binom{n+1}{k}$ different possible such choices and the sum over all these possibilities we denote by $\sum_{J_k}$. Also, given J_k the associated complement will be taken as

$$J_k^c = \{1,2,\ldots,n+1\} - J_k .$$

Suppose β is some element of the edge set for a given tree $(V(T),E(T))$. If β is removed from the edge set E(T) the tree is broken into two disjoint subtrees T_1^β, T_2^β satisfying

$$V(T) = V(T_1^\beta) \cup V(T_2^\beta) , \quad E(T) = E(T_1^\beta) \cup E(T_2^\beta) \cup \{\beta\} , \quad i_\beta \in V(T_1^\beta) \text{ and } j_\beta \in V(T_2^\beta) .$$

Finally we let $\mathcal{F}$ indicate the set of all labelled trees formed over finite subsets of $\mathbb{N}$.

<u>Lemma 7</u> (Tree Grafting): Let $n \in \mathbb{N}$, and A be a vector space. Suppose $f: \mathcal{F}^2 \times \mathbb{N}^2 \to A$ is a function symmetric in its first two and last two arguments

$$f(T',T'';\ell,m) = f(T'',T';\ell,m) = f(T',T'';m,\ell)$$

i.e., is a function of an unordered pair of trees and an unordered pair of vertex labels. Then

$$\frac{1}{2}\sum_{k=1}^{n}\sum_{J_k}\sum_{T_1\in\mathcal{T}J_k}\sum_{T_2\in\mathcal{T}J_k^c}\sum_{r\in J_k}\sum_{q\in J_k^c} f(T_1,T_2;r,q) = \sum_{T\in\mathcal{T}V_{n+1}}\sum_{\beta\in E(T)} f(T_1^\beta,T_2^\beta;i_\beta,j_\beta) \tag{A.1}$$

Proof: We show that (i) sums of pairs of terms on the left-hand side occur as a single term on the right-hand side, and conversely, (ii) every term on the right side occurs as a pair of terms on the left-hand side.

(i) On the left-hand side fix an arbitrary term in the sum:

$$k = \ell \in V_n \ , \quad J_\ell \ , \quad T_1 = T_1^* \in \mathcal{T}J_\ell \ , \ T_2 = T_2^* \in \mathcal{T}J_\ell^c$$

$$r = r^* \in J_\ell \ , \quad q = q^* \in J_\ell^c \ .$$

However, when $k = n+1-\ell$, there will exist a partner to this term for which $J_{n+1-\ell} = J_\ell^c$ and

$$T_1 = T_2^* \in \mathcal{T}J_{n+1-\ell} = \mathcal{T}J_\ell^c \ , \quad T_2 = T_1^* \in \mathcal{T}J_{n+1-\ell}^c = \mathcal{T}J_\ell$$

$$r = q^* \in J_{n+1-\ell} \ , \quad q = r^* \in J_{n+1-\ell}^c \ .$$

We note these partners are always different terms in the left-hand sum. If $\ell \neq n+1-\ell$, then they arise from different terms in the $\sum_{k=1}^n$ sum. Whereas if $\ell = (n+1)/2$, then because T_1 and T_2 are trees on disjoint vertex sets (J_ℓ, J_ℓ^c), the partners arise from complementary choices of J_k in the sum $\sum_{J_k}$.

The sum of the two partner terms is:

$$\frac{1}{2} f(T_1^*,T_2^*;r^*,q^*) + \frac{1}{2} f(T_2^*,T_1^*;q^*,r^*) = f(T_1^*,T_2^*;r^*,q^*) \tag{A.2}$$

by the symmetry of f. Now T_1^* and T_2^* are two disjoint trees which can be grafted together by the link $\beta = \{r^*,q^*\}$ because $r^* \in J_\ell = V(T_1^*)$ and $q^* \in J_\ell^c = V(T_2^*)$. That is we can define a tree T by

$$V(T) = V(T_1^*) \cup V(T_2^*) = J_\ell \cup J_\ell^c = V_{n+1} \ , \quad E(T) = E(T_1^*) \cup E(T_2^*) \cup \{r^*,q^*\} \ ,$$

so that $T \in \mathcal{T}V_{n+1}$ and $\beta \in E(T)$. Appealing to the symmetry of f in the event $r^* > q^*$, we find that sum $f(T_1^*,T_2^*;r^*,q^*)$ of the left-hand side partner term will occur on the right-hand side of (A.1).

(ii) Let $T \in \mathcal{T}V_{n+1}$ and $\beta \in E(T)$ be given on the right side of (A.1) with T_1^β and T_2^β the disjoint trees defined by removing link β. Since all possible disjoint vertex sets, trees on them and vertex label pairs between them are summed over on the left-hand side of (A.1) there will exist a term where

$$k = \text{cardinality of } V(T_1^\beta) \in V_n\ , \quad J_k = V(T_1^\beta) \text{ and } J_k^c = V(T_2^\beta)\ ,$$

$$T_1 = T_1^\beta\ , \quad T_2 = T_2^\beta\ , \quad r \in \beta\ , \quad q \in \beta \backslash \{r\}\ ,$$

and as in (i) this term will have a partner. □

We remark that in the case where $A = \mathbb{R}$ and $f = 1$ (the unit constant function) then (A.1) is a known result (c.f. Ref. [4], pg. 53). The basic content of **Proposition** 1 is given by

Lemma 8: The functions $a_n : \mathbb{R}^{2d} \to \mathbb{R}$, defined by (2.4)-(2.5) satisfy the recurrence relation (2.3) for all $n > 1$.

Proof: We use induction to implement the proof. The definitions of a_1 shows the recurrence relation is valid for $n = 1$. Now assume that (2.3) is valid for all a_m, $1 \le m \le n$, where a_m is defined by the tree sums (2.4) and (2.5). Let W be the integral

$$W(x,y) \equiv \sum_{k=1}^{n} \frac{1}{2} \binom{n+1}{k} \int_I d\xi\ \xi^{2n}\ (\nabla_1 a_k)(\tilde{\xi},y) \cdot (\nabla_1 a_{n+1-k})(\tilde{\xi},y)\ . \tag{A.3}$$

We must show that

$$W(x,y) = \int_{I^{n+1}} d^{n+1}\xi \sum_{T \in \mathcal{T}V_{n+1}} \Big(\prod_{\beta \in E(T)} b_\beta \Big) \prod_{p=1}^{n+1} v(\tilde{\xi}_p)\ , \tag{A.4}$$

or equivalently we must show that W is a tree sum over the vertex set V_{n+1}. Using (2.4)-(2.5) for $m \le n$ gives

$$W = \sum_{k=1}^{n} \frac{1}{2} \binom{n+1}{k} \int_I d\xi\ \xi^{2n} \Big[\nabla_{\tilde{\xi}} \int_{I^k} d^k\xi \sum_{T \in \mathcal{T}V_k} \Big(\prod_{\beta \in E(T)} b_\beta \Big) \prod_{p=1}^{k} v\big(y+\xi_p(\tilde{\xi}-y)\big)\Big]$$

$$\cdot \Big[\nabla_{\tilde{\xi}} \int_{I^{n+1-k}} d^{n+1-k}\xi \sum_{T' \in \mathcal{T}V_{n+1-k}} \Big(\prod_{\beta' \in E(T')} b_{\beta'} \Big) \prod_{p'=1}^{n+1-k} v\big(y+\xi_{p'}(\tilde{\xi}-y)\big)\Big]\ .$$

For potentials in the class $\mathcal{D}$ we can differentiate with respect to $\tilde{\xi}$ inside the integrals. The $\nabla_{\tilde{\xi}}$ may then act on any of the factors v. Relabel the integration variables ξ_j in the second [ ] factor above $\xi_1 \to \xi_{k+1}, \ldots, \xi_{n+1-k} \to \xi_{n+1}$ then the two multiple integrals may be written as one giving

$$W = \sum_{k=1}^{n} \frac{1}{2} \binom{n+1}{k} \int_I d\xi\, \xi^{2n} \int_{I^{n+1}} d^{n+1}\xi$$

$$\times \{ \sum_{T \in \mathcal{T}V_k} \sum_{T' \in \mathcal{T}k+1\sim n+1} \prod_{\beta \in E(T)} b_\beta \prod_{\beta' \in E(T')} b_{\beta'} (\sum_{r=1}^{k} \sum_{q=k+1}^{n+1} \xi_r \xi_q\, D_r \cdot D_q) \prod_{p=1}^{n+1} v([\xi\xi_p]^\sim) \}.$$

The notation for the argument of the potential is the usual abbreviation $[\xi\xi_p]^\sim = y + \xi\xi_p(x-y)$, and the symbol $k+1\sim n+1$ denotes the vertex set $\{k+1,\ldots,n+1\}$. The overall factor $\binom{n+1}{k}$ may be absorbed by summing over all distinct n-element vertex sets J_k which may be chosen from V_{n+1}, so

$$W = \sum_{k=1}^{n} \frac{1}{2} \int_I d\xi\, \xi^{2n} \int_{I^{n+1}} d^{n+1}\xi \sum_{J_k} \sum_{T \in \mathcal{T}J_k} \sum_{T' \in \mathcal{T}J_k^c} \prod_{\beta \in E(T)} b_\beta$$

$$\times \prod_{\beta' \in E(T')} b_{\beta'} \sum_{r \in J_k} \sum_{q \in J_k^c} \xi_r \xi_q\, D_r \cdot D_q \prod_{p=1}^{n+1} v([\xi\xi_p]^\sim) .$$

We may remove all the ξ dependence in the potentials by the change of variables $\xi_j \to \xi_j/\xi$ $(j \in V_{n+1})$. Each factor ϕ in a b_β will contribute one multiplicative factor of ξ^{-1} due to this change of variables. The altered integration region becomes

$$\int_0^1 d\xi \int_{[0,\xi]^{n+1}} d^{n+1}\xi = \int_{I^{n+1}} d^{n+1}\xi \int_M^1 d\xi ,$$

where $M = \max\{\xi_1,\ldots,\xi_{n+1}\}$. Thus W may be written

$$W = \sum_{k=1}^{n} \frac{1}{2} \int_{I^{n+1}} d^{n+1}\xi\, \{ \int_M^1 d\xi\, \xi^{-2} \sum_{J_k} \sum_{T \in \mathcal{T}J_k} \sum_{T' \in \mathcal{T}J_k^c} [\prod_{\beta \in E(T)} \xi_\beta^<(1-\xi_\beta^>/\xi) D_{i_\beta} \cdot D_{j_\beta}]$$

$$\times [\prod_{\beta' \in E(T')} \xi_{\beta'}^<(1-\xi_{\beta'}^>/\xi) D_{i_{\beta'}} \cdot D_{j_{\beta'}}] \sum_{r \in J_k} \sum_{q \in J_k^c} \xi_r \xi_q\, D_r \cdot D_q \prod_{p=1}^{n+1} v(\tilde{\xi}_p) \} ,$$

where $\xi_\beta^<$ $(\xi_\beta^>)$ is the smallest (greatest) value of the pair $\xi_{i_\beta}, \xi_{j_\beta}$.

Observe that the integrand $\int_M^1 d\xi$ is symmetric under the pair exchange $\ell \leftrightarrow m$ of any two labels $\ell, m \in V_{n+1}$. To see this notice that M and the product of the v's have this property. Next consider any fixed term J_k in $\sum_{J_k}$. If $\ell,m \in J_k$ (or $\ell,m \in J_k^c$) then $\ell \leftrightarrow m$ at most exchanges the identification of two trees in $\mathcal{T}J_k$ (or $\mathcal{T}J_k^c$). But all trees in $\mathcal{T}J_k$ (or $\mathcal{T}J_k^c$) are summed over, so the sum is invariant under the exchange $\ell \leftrightarrow m$. Alternately if $\ell \in J_k$ and $m \in J_k^c$, then $\ell \leftrightarrow m$ just interchanges two terms in the $\sum_{J_k}$ sum, i.e., $J_k \leftrightarrow (J_k-\{\ell\}) \cup \{m\}$ and $J_k^c \leftrightarrow (J_k^c-\{m\}) \cup \{\ell\}$.

Since the integrand is permutation invariant the integral may be ordered:

$$\int_{I^{n+1}} d^{n+1}\xi = (n+1)! \int_{0<\xi_1<\dots<\xi_{n+1}<1} \int d^{n+1}\xi \equiv (n+1)! \int_<^1 d^{n+1}\xi \, .$$

With this ordering implemented, note that

$$M = \xi_{n+1} \, , \quad \xi_\beta^< = \xi_{i_\beta} \, , \quad \xi_\beta^> = \xi_{j_\beta} \, .$$

Thus we arrive at

$$W = (n+1)! \int_<^1 d^{n+1}\xi \int_{\xi_{n+1}}^1 d\xi \, \xi^{-2} \Big\{ \sum_{k=1}^n \frac{1}{2} \sum_{J_k} \sum_{T \in \mathcal{T}J_k} \sum_{T' \in \mathcal{T}J_k^c} \sum_{r \in J_k} \sum_{q \in J_k^c}$$

$$\times \Big[\prod_{\beta \in E(T)} \xi_{i_\beta} (1-\xi_{j_\beta}/\xi) D_{i_\beta} \cdot D_{j_\beta} \Big] \Big[\prod_{\alpha \in E(T')} \xi_{i_\alpha} (1-\xi_{j_\alpha}/\xi) D_{i_\alpha} \cdot D_{j_\alpha} \Big]$$

$$\times \xi_r \xi_q D_r \cdot D_q \Big\} \prod_{p=1}^{n+1} v(\tilde{\xi}_p) \, .$$

To the term in $\{\cdots\}$ we apply Lemma 7 with A chosen to be the vector space of differential operators and f to be

$$f = f(T,T';r,q) \equiv \Big[\prod_{\beta \in E(T)} \cdots \Big] \Big[\prod_{\alpha \in E(T')} \cdots \Big] \xi_r \xi_q D_r \cdot D_q \, ,$$

where the square brackets are those given immediately above. This f has all the required properties for Lemma 7 (note the D_j's commute). Thus, after some rearrangement W may be written

$$W = (n+1)! \int_<^1 d^{n+1}\xi \sum_{T \in \mathcal{T}V_{n+1}} \Big(\prod_{\alpha \in E(T)} \xi_{i_\alpha} D_{i_\alpha} \cdot D_{j_\alpha} \Big) \prod_{p=1}^{n+1} v(\tilde{\xi}_p)$$

$$\times \sum_{\beta \in E(T)} \xi_{j_\beta} \Big\{ \int_{\xi_{n+1}}^1 d\xi \, \xi^{-2} \prod_{\gamma \in E(T) \setminus \{\beta\}} (1-\xi_{j_\gamma}/\xi) \Big\} \tag{A.5}$$

The ξ integral is now elementary and straightforward calculation shows the curly bracket term becomes

$$\{\cdots\} = \sum_{\mu=0}^{n-1} \frac{(-1)^\mu}{\mu+1} \left(\xi_{n+1}^{-(\mu+1)} - 1 \right) \sum_{\underline{\alpha} \subset E(T) \setminus \{\beta\}}^{(\mu)} \xi_{j_{\alpha(1)}} \cdots \xi_{j_{\alpha(\mu)}} \, . \tag{A.6}$$

The sum $\sum^{(\mu)}$ and notation $\underline{\alpha}$ require explanation. First of all, when $\mu=0$, $\sum^{(0)} \equiv 1$.

Otherwise, $\underline{\alpha} \equiv \{\alpha(1),\ldots,\alpha(\mu)\} \subset E(T)\setminus\{\beta\}$ denotes a set of μ distinct elements $\alpha(i) \in E(T)\setminus\{\beta\}$ and the $\sum^{(\mu)}$ specifies a sum over all possible distinct sets $\underline{\alpha}$ of this type.

Performing the summation and a few simple manipulations gives

$$\sum_{\beta \in E(T)} \xi_{j_\beta} \{\cdots\} = \sum_{\mu=0}^{n-1} \left(\left(\xi_{n+1}^{-(\mu+1)} - 1\right)(-1)^{\mu} \sum_{\underline{\alpha} \subset E(T)}^{(\mu+1)} \xi_{j_{\alpha(1)}} \cdots \xi_{j_{\alpha(\mu+1)}}\right.$$

$$= \prod_{\beta \in E(T)} (1-\xi_{j_\beta}) - \prod_{\beta \in E(T)} (1-\xi_{j_\beta}/\xi_{n+1})$$

As β runs through E(T) there is at least one link with $j_\beta = n+1$, so the last product vanishes. Returning to (A.5) we see that it takes the form

$$W = (n+1)! \int_<^1 d^{n+1}\xi \sum_{T \in \mathcal{J}V_{n+1}} \left[\prod_{\alpha \in E(T)} \xi_{i_\alpha}(1-\xi_{j_\alpha}) D_{i_\alpha} \cdot D_{j_\alpha}\right] \prod_{p=1}^{n+1} v(\tilde{\xi}_p) \;.$$

Note that the differential operator in the square bracket is b_α. Observing that the integrand is invariant under permutation of the labels on $\xi_1, \xi_2 \ldots \xi_{n+1}$, we may remove the order restriction and integrate over the n+1 dimensional unit cube giving (A.4).

ACKNOWLEDGEMENT

The authors would like to thank the members of the Nuclear Theory Group at the University of Maryland for their warm hospitality. This work has been supported in part by grants from the Natural Sciences and Engineering Research Council of Canada and the U. S. Department of Energy. F.H.M. gratefully acknowledges a post-graduate scholarship awarded by NSERC.

REFERENCES

[1] M. S. Marinov, J. Phys. A 12, 31 (1979).

[2] Y. Fujiwara, T. A. Osborn, and S. F. J. Wilk, Phys. Rev. A25, 14 (1982).

[3] F. H. Molzahn and T. A. Osborn, "Tree Graphs and the Solution to the Hamilton-Jacobi Equation", University of Maryland preprint #84-074 (1984).

[4] R. J. Wilson, Introduction to Graph Theory (Academic Press, New York, 1979).

[5] I. M. Gelfand and M. V. Fomin, Calculus of Variations (Blaisdell, New York, 1962).

[6] W. Rudin, Principles of Mathematical Analysis, 3rd ed. (McGraw Hill, New York, 1964).

[7] N. I. Akhiezer, The Calculus of Variations (Blaisdell, New York, 1962).

LOCAL AND GLOBAL ASPECTS OF A GENERALIZED HAMILTONIAN THEORY

Robert Cawley
Naval Surface Weapons Center
White Oak, Silver Spring, MD 20910

1. INTRODUCTION

Following local methods first used by Dirac, I have constructed a generalized Hamiltonian theory applicable to a wide class of dissipative dynamical systems. I will describe essential features of the formalism, indicate its derivation, and draw some of the issues of its realization in global terms. A theoretical circumstance exists that does not seem generally to be recognized in modern dynamics. That circumstance is one of the motivations for this paper, in which I describe, in particular, a novel Hamilton-Jacobi scheme.

It comes as a surprise to many physicists unfamiliar with the subject that time-dependent Hamilton-Jacobi theory has been abandoned in modern global formulations of dynamics; while on the other hand, the matter is regarded generally as unimportant, or of little concern, to many mathematicians. Somewhat better appreciated is the fact that the classical subject was restricted to Hamiltonian systems, a situation less troubling to physicists than one might expect it to be to mathematicians, due largely to the interest in quantum mechanics and relative disinterest in classical mechanics on the part of physicists in this century. Two technical matters attend the foregoing circumstance: (1) time-dependent Hamilton-Jacobi theory is a coordinate transformation theory, and (2) its implications involve the time inherently. In my opinion, these features are also foci of its theoretical interest and its potential concern to physicists.

In classical mechanics, the main solution problem of dynamical systems is understood to be that of effecting quadratures. For systems whose equations can be derived from a Hamiltonian, $H = H(q,p,t)$, the strategy of Hamilton-Jacobi theory[1] is to solve first a partial differential equation,

$$H_0(q,\partial R_0/\partial q,t) + \partial R_0/\partial t = 0, \qquad H_0\text{: arbitrary,} \tag{1.1}$$

for the generator $R_0 = R_0(q,P,t)$ of a transformation that is explicitly time dependent, but canonical in that the new variables, Q, P, constants of integration for eq. (1), obey a new system of Hamilton's equations governed by $H + \partial R_0/\partial t \equiv K = K(Q,P,t)$. A shortcoming of this strategy is that the necessity to perform eliminations to secure the transformation from R_0 requires use of the implicit function theorem, where the ranges of the variables are not generally known[2]. Results, consequently, have only a local character. Analogously, Lie transformations depending on a parameter[3], in which the feature of mixed old and new variables, along with the restriction to Hamiltonian systems[4], can be removed, generate perturbation expansions whose convergence is generally only asymptotic.

In global dynamics the strategy of quadratures is largely abandoned, and instead generic properties of dynamical systems, defined as flows or as differentiable vector fields on manifolds, are sought. Since a flow[5] is a differentiable mapping $\phi: \mathbb{R} \times M \to M$, with $\phi(0,p) = p$, $p \in M$, and $\phi(s,\phi(t,p)) = \phi(s+t,p)$, the corresponding notion is that of an autonomous system represented by a differentiable vector field $X \in \mathcal{X}(M)$, i.e., $\partial\phi(t,p)/\partial t = X(\phi(t,p))$. Here $\mathcal{X}(M)$ is a topological space, normally specified in terms of differentiability properties of the mappings of M. Nonautonomous systems can be treated by the well-known trick of replacing the system $\dot{x} = F(x,t)$ by one of higher dimension, appending an $(n + 1)$-th equation, $\dot{x}^{n+1} = 1$. This can be useful in special cases, but for general properties, such as stability or asymptotic behavior, other kinds of approaches to "embedding the time" may be desirable, as in the notion of a skew-product flow[6]. Using methods of local analysis, I have obtained some results having structural features suggesting the possibility of a global framework in which the role of the time finds another kind of treatment, one giving expression to the notion of a "moving coordinate system in phase space". Indeed, the Hamiltonian transformation theory of classical mechanics, described in the first paragraph, can be viewed as fundamental, and in just this way: the dynamical specification of a system is to be presented by <u>two</u> quantities, the motion of a point on a phase space, with the help of a Hamiltonian K, <u>and</u> the motion of a physical coordinate system for observation (picture), which is fixed by Eq. (1.1). In the Heisenberg picture observation is made from a system at rest and K = H, while in the Schrödinger picture the system co-moves with the phase fluid and K=0. The results I will describe have been constructed[7] for the system,

$$df_{2n}(x,\dot{x},t)/dt - f_{1n}(x,\dot{x},t) = 0, \quad \det \|\partial f_{2n}/\partial \dot{x}^m\| \neq 0, \qquad (1.2)$$

$n,m = 1$ to N, and f_{1n}, f_{2n} arbitrary. Notice that eq. (2) subsumes the traditional classical domain of Hamiltonian mechanics, and includes also time-asymmetric systems. Extension to singular systems involving zero determinant, and phase spaces of odd dimension should not present prohibitive obstacles.

Eq. (2) is produced from a stationary action condition for a universal Lagrangian $\mathcal{L}$, whose specification reduces to giving an equivalence class of functions of 2N additional quantities, $(z_1,\ldots,z_N,\dot{z}_1,\ldots,\dot{z}_N)$, which play the role of universal constraints [7]. The class is specified, in fact, by the 2N-vector components $f_{11},\ldots f_{1N}, f_{21},\ldots,f_{2N}$. The phase space is 4N-dimensional, but the system motion is confined to a universal hypersurface $\mathbb{U}^{2N}$ defined by the canonical constraints, $z^n \approx 0$, $p_{x^n} \approx 0$. $\mathbb{U}^{2N}$ is the usual phase space. There are no gauge functions, and the universal Hamiltonian $\mathcal{H}$ corresponds also to an equivalence class of 2N quantities. The nontrivial canonical equations are

$$dw^{\mu}/dt \approx F^{\mu}(w,t), \ \mu = 1,\ldots,2N, \tag{1.3}$$

where $(w^1,\ldots,w^{2N}) = (x^1,\ldots,x^N, p_{z^1},\ldots,p_{z^N})$ and the F^{μ} are determined from the f's, via the Legendre transformation from $\mathcal{L}$. F^{μ} need not be "Hamiltonian," viz. constructed from a function $H(w,t)$, but it can be, if and only if a gauge choice exists for the f's for which f can be written as a 2N-gradient of some function $L = L\,(x,\dot{x},t)$, which is the case defining the traditional scope of classical mechanics.

If the traditional subject of Hamiltonian mechanics be characterized, roughly speaking, as an avatar of symplectic geometry, the generalized Hamiltonian theory I will describe may be characterized as a generalized symplectic geometry.

Many of the results I present have been derived previously, so I only sketch their derivation, or even just state them. A few new results will be given but the principal purpose is to lay out essential features of the theory to draw some of the issues, and explore options, of globalization. I also give some illustrative examples of application.

There are two main areas where issues of globalization arise. The first centers around methods of constrained Hamiltonian dynamics which I use, having generalized them from local techniques of analysis introduced originally by Dirac[8]; the second has to do with the generalized Hamilton-Jacobi theory and time-dependent coordinate transformations.

In Section 2, I provide background for Dirac's local theory of constraints. In Section 3, I describe essential features of the universal

Lagrangian and Hamiltonian theories needed later. In Section 4, I outline the coordinate covariance theory preserving the basic structures introduced in Section 3. In Section 5, I present a line element (Finslerian) geometry formulation. In Section 6, I give the new Hamilton-Jacobi theory, and in Section 7, I illustrate its application by simple examples. I consider globalization in Section 8 and conclude the paper in Section 9.

2. DIRAC'S METHOD FOR CONSTRAINED SYSTEMS - HISTORICAL SUMMARY

In 1950, Dirac [8] introduced a method of analysis for treating dynamical systems for which the Hessian matrix

$$A = \left\| \frac{\partial^2 L}{\partial \dot{x}^m \partial \dot{x}^n} \right\| \qquad , \; m,n = 1,\ldots, N, \tag{2.1}$$

has rank less than N, where it is assumed that the system is specified by the Lagrangian L. A central feature of Dirac's theory was the variation procedure he employed in his first paper, in which equations defining the momenta,

$$p_n = \partial L/\partial \dot{x}^n, \qquad n = 1,\ldots, N, \tag{2.2}$$

were said to be "weak" in the sense that under independent variations of the $x,\dot{x},p$ by small amounts of order ε, they are violated by amounts of order ε. Equations valid to accuracy ε under the variation were said to be "strong". Dirac went on to derive properties of weak and strong equality, such as $A \approx 0$, $B \approx 0 \Rightarrow A \cdot B \simeq 0$, where $\approx$ denotes weak and $\simeq$ strong equality. In the case that the N quantities $\partial L/\partial \dot{x}^n$ are independent functions of the velocities, called the standard case, the rank of A is N, and Eqs. (2.2) determine each $\dot{x}$ as a function of the x,p by the implicit function theorem. When this fails, elimination of the x from Eqs. (2.2) leads to one or more relations

$$\phi(x,p) \approx 0, \tag{2.3}$$

where Dirac assumed that the functions ϕ were to be expressed in such a fashion that the violation of Eq. (2.3) under the variations is to be of order ε.

With these underpinnings, Dirac developed an algorithm for determination of the Hamiltonian in the presence of constraints. Constraints are functions such as appear in Eq. (2.3), which, because they issue from Eqs. (2.2), are termed primary, together with further functions called secondary constraints which arise from consistency conditions expressing a requirement that all the constraints obey

$$\dot{\Phi}(x,p) \approx 0, \tag{2.4}$$

where the time derivatives are evaluated by the Hamiltonian identified with the algorithm, and which I will denote by H_G. Constraints fall into two types, first class and second class. Second class constraints have the feature that they can be arranged in canonically conjugate pairs (possibly following a canonical transformation), and their role in the Hamilton-Dirac formalism was clearly understood from the beginning (see, e.g., [9]). Their presence is correctly accounted for in the formalism by one of two equivalent procedures: (i) replacing Poisson brackets with Dirac brackets or (ii) "freezing" the affected conjugate pairs of degrees of freedom of the canonical variables, which means omitting their contributions from the sums defining the Poisson bracket, or in other words, "reducing the phase space".

The role of the first class constraints, however, has not been clearly understood at all. A first class constraint is defined as having weakly vanishing Poisson bracket with every other constraint. So its canonically conjugate object, or more succinctly, the generator of its infinitestimal translations, cannot also be a constraint. This was recognized by Bergmann and Goldberg[10], who, going a little bit beyond Dirac's approach, proposed that only those infinitesimal generators should be admitted to the status of "dynamical variable" which satisfy the condition that the constraints -- all of them -- are left invariant. In other words, the degrees of freedom corresponding to all the constraints, together with those corresponding to each of their conjugate partners (viz. the generators of their translations) should be frozen. This "grand reduction of the phase space" was implemented by means of the Bergmann-Goldberg bracket.

This is how the reduced phase space (RPS) was born.* It was later born again in disguised form, or perhaps better, it was (partially)

---

*Note that the defining context for "reduced phase space" here is distinct from that arising in the articles by Martin Kummer and Stephen Omohundro in this volume.

resurrected -- by Dirac himself. Dirac had earlier derived a basic form of the Hamiltonian[11]

$$H_G \approx H' + \sum_g v_g \phi_g, \tag{2.5}$$

where H' is a function fixed by the algorithm, the v_g are arbitrary, so-called gauge functions of the time, and the ϕ_g are a collection of first class constraints generated by the primary first class subset. Dirac subsequently conjectured[12] that the set of ϕ_g should be comprised of the entire collection of primary and secondary first class constraints. In Dirac's analysis, the ϕ_g are to be a minimal collection necessary for closure under the conditions,

$$\{\phi_g,\phi_{g'}\} \approx \sum_{g''} \gamma_{gg'g''}(x,p)\phi_{g''}, \tag{2.6a}$$

$$\{\phi_g,H'\} \approx \sum_{g''} h_{gg''}(x,p)\ \phi_{g''}. \tag{2.6b}$$

I refer to Eqs. (2.6a,b) as Dirac's test. When Dirac's test identifies $\{\phi_g\}$ as the entire collection, the Hamiltonian is called the extended Hamiltonian, H_E, and Dirac's conjecture reads

$$H_G \approx H_E. \tag{2.7}$$

If I denote the generator of translations of ϕ_g by $\tilde{\phi}_g$, eq. (2.7) gives

$$\dot{\tilde{\phi}}_g \approx \{\dot{\tilde{\phi}}_g,H'\} + v_g, \quad \text{all } g, \tag{2.8}$$

and all of the $\tilde{\phi}_g$ enjoy arbitrary time-dependence "governed" by an independent collection of arbitrary gauge functions, v_g. Evidently, something of the basis of the Bergmann-Goldberg prescription is recovered if Dirac's conjecture, Eq. (2.7), is true, and the RPS has returned as a proper candidate for the dynamical arena. Some of the frozen degrees of freedom are understood to be gauge freedoms, and their values are supposed not to represent descriptive features of the "physical state".

The first counter-examples to Dirac's conjecture were found by Allcock[13] and by me[14], and further counter-examples subsequently, of a pathological nature, by Frenkel[15] and again by me[16].

On the other hand, Dirac's conjecture was recovered by Allcock[17] for a restricted class of "linearization-stable" Lagrangian systems, a restriction violated by the Lagrangians defining the examples of Refs. [13]-[16]. So, happily, there was no contradiction.

Another approach to the recovery of the Dirac conjecture was provided by diStefano[18], who abandoned Dirac's definitions of weak and strong equality described above. In her approach, two functions f,g of the canonical variables which are equal on the constraint hypersurface in phase space are weakly equal if there exists a set of functions $\{h_j(x,p) \mid j = 1,\ldots,J\}$, such that

$$[h_j,\ldots [h_1,f],\ldots] \neq [h_j,\ldots, [h_1,g],\ldots]. \quad (2.9)$$

Any equality not a weak equality is then called a strong equality, although this notion now seems to bear little relation to Dirac's idea of strong equality. In diStefano's treatment, Dirac's conjecture is recovered for the examples of Refs. 13 and 14.

Early geometrical treatments of Dirac's theory included matters of translation and articulation, such as classification of submanifolds identified locally by Dirac. Typical of early papers was Sniatycki's formulation in terms of symplectic geometry, in which global existence of Dirac brackets is proved[19]. Gotay, Nester, and Hinds[20] constructed a more detailed algorithm, but the controversy over whether secondary first class constraints ought to generate gauge transformations remained unresolved since this could not be proved. Nevertheless, those authors insisted that a physical interpretation criterion identified the extended Hamiltonian as a uniquely correct choice, and the RPS as the proper dynamical arena once again.

Finally, two more recent efforts to find a satisfactory formulation of constrained Hamiltonian dynamics are the work of Sugano and his collaborators[21], and that of Rusk and Skinner[22]. Sugano's interesting treatment is differential geometric and is not a formulation in Lagrangian terms (tangent bundle TQ to the configuration manifold Q) or Hamiltonian terms (T*Q) alone, but in terms of both together. Recall Dirac's starting point, Eqs.(2.2), where variations of all three of the $x, \dot{x}$ and p are independent. The relationship of this work to the more purely geometric treatment of Skinner and Rusk, on $TQ + T^*Q$ (Whitney Sum) may be close, but has not been examined so far, to my knowledge. In each case, the Dirac conjecture appears to be false.

The counter-example to Dirac's conjecture advanced in [14] is the simplest of a class; in fact, the absence of gauge functions and inappropriateness of the RPS is quite significant, and I discuss this now in detail. A slight modification of that example is defined by the Lagrangian

$$L(x,y,z,\dot{x},\dot{y},\dot{z}) = - kx\cdot z + \dot{x}\cdot\dot{z} + \frac{1}{2}yz^2; \tag{2.10}$$

the Euler-Lagrange equations are

$$\ddot{z} + kz = 0, \; -\frac{1}{2}z^2 = 0, \; \ddot{x} + kx - yz = 0, \tag{2.11}$$

which gives a simple equation for x, the restriction to z = 0, and undetermined y-motion. There is one primary constraint,

$$p_y \approx 0, \tag{2.12}$$

the total Hamiltonian is

$$H_T \approx H + up_y, \tag{2.13}$$

where u is an undetermined multiplier, and

$$H = kx\cdot z + p_z\cdot p_x - \frac{1}{2}yz^2. \tag{2.14}$$

The consistency condition to eq. (2.12) is

$$\dot{p}_y \approx \{p_y, H_T\} \approx \frac{1}{2}z^2 \approx 0, \tag{2.15}$$

which, following Dirac[8], gives the secondary constraint

$$z \approx 0, \tag{2.16}$$

whence

$$\dot{z} \approx \{z, H_T\} \approx p_x \approx 0. \tag{2.17}$$

The algorithm terminates because $\dot{p}_x \approx -kz \approx 0$ is already met owing to Eq. (2.16), so u remains undetermined. The three constraints, z, p_x and p_y are first class, but p_y all by itself constitutes the minimal collection specified by Dirac's test, Eqs. (2.6a,b). The reason is that $z \approx 0$ implies $z^2 \simeq 0$ (recall the exposition in the first paragraph of this section), so that $-\frac{1}{2}yz^2$ should be dropped from H, and we have

$$H \simeq H' \simeq kx\cdot z + p_z\cdot p_x, \tag{2.18}$$

with also

$$H_G \simeq H' + up_y. \tag{2.19}$$

Since the y-motion is arbitrary and p_y vanishes, we can drop the gauge term and simply take Eq. (2.18) as the Hamiltonian, thereby "freezing" the yp_y-sector of the phase space.

There is a certain symmetry to this example; if we could somehow also drop the yz^2-term from L we'd have

$$L \to - kx\cdot z + \dot{x}\cdot\dot{z}. \tag{2.20}$$

In the next section I show how this can be done by an extension of Dirac's methods. This will result in a new kind of object, the universal Lagrangian, together with its universal Hamiltonian. The constraints will take a new universal meaning also, and we shall be led to our first level of extension of Dirac's procedures.

3. UNIVERSAL LAGRANGIANS AND HAMILTONIANS

The Euler-Lagrange equations issuing from the Lagrangian

$$L_0(x,z,\dot{x},\dot{z},t) = \sum_n f_{1n}(x,\dot{x},t)z^n + \sum_n f_{2n}(x,\dot{x},t)\dot{z}^n \tag{3.1}$$

are those promised in Section 1, namely,

$$G_n(x,\dot{x},\ddot{x},t) = \frac{d}{dt}f_{2n}(x,\dot{x},t) - f_{1n}(x,\dot{x},t) = 0, \quad n = 1,\dots,N, \tag{3.2}$$

together with another N equations with which the subsidiary conditions

$$z^n = 0, \qquad n = 1,\dots,N, \tag{3.3}$$

are always consistent. Using the standard Lagrange multiplier trick of appending a new coordinate y, Eqs. (3.3) will ensue from the new choice of Lagrangian,

$$L_1 = L_0 + \sum_n \frac{1}{2} y^n (z^n)^2, \tag{3.4}$$

in a manner exactly similar to that from the Lagrangian of eq. (2.10). The generalized Hamiltonian that results, by a parallel treatment, is a modest modification of Eq. (2.19) containing now N gauge functions, $\{u^n\}$, and having

$$H' \simeq \sum_n F_{1n}(x,p_z,t)z^n + \sum_n F_2^n(x,p_z,t)p_{x^n}, \tag{3.5}$$

where

$$F_{1n}(x,p_z,t) = -f_{1n}(x,\dot{x},t), \tag{3.6}$$

$$F_2^n(x,p_z,t) = \dot{x}^n, \tag{3.7}$$

in which the $\dot{x}$ are to be determined as functions of the canonical variables from the N equations

$$p_{z^n} \approx f_{2n}(x,\dot{x},t), \tag{3.8}$$

which is uniquely possible if and only if

$$|\mathcal{Q}| = \det\|\partial f_{2n}/\partial \dot{x}^m\| \neq 0. \tag{3.9}$$

The additional secondary constraints are like (2.17),

$$\dot{z}^n \approx p_{x^n} \approx 0. \tag{3.10}$$

So far, I have not departed from Dirac's original methods, but I will do so shortly. My starting point is the observation that if two Lagrangians, L_0 and $\bar{L}_0$, differ by a function of the $x,\dot{x},z,\dot{z}$, and t that vanishes at least quadratically in the 2N variables $z,\dot{z}$, the resulting Hamiltonian function $\bar{H}'$ has the property

$$H' \simeq \bar{H}', \tag{3.11}$$

provided only that the local condition (3.9) holds. Furthermore, Eqs. (3.2) still also hold. To express this fact, I write

$$\bar{L}_0 \simeq L_0 \to \mathcal{L}, \tag{3.12}$$

and replace Eq. (3.1) with the generalized strong eguation

$$\mathcal{L} \simeq \sum_n f_{1n}(x,\dot{x},t)z^n + \sum_n f_{2n}(x,\dot{x},t)\dot{z}^n. \tag{3.13}$$

To signify the new sense of the meaning of $\mathcal{L}$, I also rewrite Eq. (3.5) as

$$\mathcal{H} \simeq \sum_n F_{1n}(x,p_z,t)z^n + \sum_n F_2^n(x,p_z,t)p_{x^n}. \tag{3.14}$$

See that the last term of Eq. (3.4) vanishes strongly now, and formally is absent. Moreover, the Euler-Lagrange equations generated by $\mathcal{L}$ from variation of the x^n comprise an infinite equivalence class, in general having only $z^n(t) = 0$ as a consistent solution. This is a procedure which is alternative, therefore, to that of (3.4) to secure Eqs. (3.3). Equation (3.14) does follow again, but $z^n \approx 0$ and its corollary $p_{x^n} \approx 0$, guaranteed by (3.9), have different meanings. In addition, these constraint equations are universal.

I will present a remarkable alternative approach to gaining Eqs. (3.13) and (3.14) in Section 5.

It is important to recognize that with these procedures I have changed the meaning of Lagrangian and Hamiltonian. Thus, the "Lagrangian" is no longer to be understood as the function appearing on the right side of Eq. (3.13), but instead as a "strongly specified" function, or in other words, an equivalence class of functions. Correspondingly, in Eq. (3.14) so is $\mathcal{H}$. I call $\mathcal{L}$ and $\mathcal{H}$ the universal Lagrangian and universal Hamiltonian for the system. The matrix $\mathcal{A}$ generalizes the Hessian in (2.1) and, following Dirac, I refer to the condition (3.9) as that of the standard case.

Notice that here the Bergmann-Goldberg reduction procedure freezes the z and p_z coordinates, along with the p_x and x coordinates. The RPS is zero-dimensional, and all the dynamics is wiped out! It is not a useful procedure for the present universal formalism.

The variable change,

$$\xi^n = x^n + \frac{1}{2} z^n, \qquad \xi'^n = x^n - \frac{1}{2} z^n, \tag{3.15a,b}$$

having as canonical partners,

$$p_n = p_{z^n} + \frac{1}{2} p_{x^n}, \quad -p'_n = p_{z^n} - \frac{1}{2} p_{x^n}, \tag{3.16a,b}$$

gives

$$\mathcal{L}(\xi,\dot{\xi};\xi',\dot{\xi}',t) \simeq L(\xi,\dot{\xi},t) - L(\xi',\dot{\xi}',t) + \delta\mathcal{L}(\xi,\xi';\dot{\xi},\dot{\xi}',t), \tag{3.17}$$

$$\mathcal{H}(\xi,p;\xi',-p',t) \approx H(\xi,p,t)-H(\xi',-p't) + \delta\mathcal{H}(\xi,p;\xi',-p',t), \tag{3.18}$$

where L, H may be chosen arbitrarily. The universal constraints, Eqs. (3.3), and (3.10) are now

$$\xi^n - \xi'^n \approx 0, \; p_n + p_n' \approx 0. \tag{3.19}$$

Equations (3.17) and (3.18) are called principal decompositions of $\mathcal{L}$ and $\mathcal{H}$. The variable sets, $(\xi,\dot{\xi},\xi',\dot{\xi}')$ and $(\xi,p,\xi',-p')$, are called principal variables, and the original sets, $(x,\dot{x},z,\dot{z})$ and (x,p_z,z,p_x), are called central variables. The Euler-Lagrange equations from Eq. (3.17) can be shown to be Eqs. (3.2) twice, once for ξ and once for ξ'; while the canonical equations issuing from (3.5) are

$$\frac{dx^n}{dt} \approx F_2^n(x,p_z,t), \tag{3.20}$$

$$\frac{dp_{z^n}}{dt} \approx -F_{1n}\,(x,p_z,t). \tag{3.21}$$

The canonical equations that ensue from Eq. (3.18) for (ξ,p) are the same as these, as are those for $(\xi', -p')$, with the obvious substitutions of ξ's for x's, and p's for p_z's, etc. Thus, with principal variables we find duplicate realizations of the equations of motion.

It can be that a "dynamical" gauge transformation

$$f_{2n}(x,\dot{x},t) \rightarrow f_{2n} + A_n(x,t), \tag{3.22a}$$

$$f_{1n}(x,\dot{x},t) \rightarrow f_{1n} + \frac{dA_n}{dt} \tag{3.22b}$$

exists which sends f_{1n}, f_{2n} to the form of a gradient on the (usual) velocity phase space, i.e. of a function $L(x,\dot{x},t)$. This is the traditional realm of mechanics, and when it is the case, it is always possible to arrange $\delta\mathcal{L}\approx 0$ and $\delta\mathcal{H}\approx 0$ in Eqs. (3.17) and (3.18). One has then, e.g. for Eqs. (3.20) and (3.21), using principal variable forms now,

$$\frac{d\xi^n}{dt} \approx \frac{\partial H}{\partial p^n}, \qquad \frac{dp_n}{dt} \approx -\frac{\partial H}{\partial \xi^n} \tag{3.23}$$

$$\frac{d\xi'^n}{dt} \approx \frac{\partial H'}{\partial(-p'_n)}, \qquad \frac{d(-p'_n)}{dt} \approx -\frac{\partial H'}{\partial \xi'^n}, \tag{3.24}$$

where in the last line I write a prime on H only to signify its dependence upon primed variables (cf. Eq. (3.18)). Formal quantization from principal variables leads here to the Schrödinger equation

$$i\hbar \frac{\partial\Phi}{\partial t} = [H(\xi,-i\hbar\partial/\partial\xi,t) - H(\xi',+i\hbar\partial/\partial\xi',t)]\Phi, \tag{3.25}$$

where $\Phi=\Phi(\xi,\xi',t)$ looks like a density function rather than a half-density (wave amplitude). Finally, Eqs. (3.15) and (3.16) show that duplicate realizations of the fundamental bracket relations will also result, a situation that recalls Wigner's theorem (see [23]). Equation (3.18) remains the general form for $\mathcal{H}$, however, where $\delta\mathcal{H}$ in general is irreducibly present, a circumstance reflecting the fact that not every (covariant! -- see below) vector (F_{1n}, F_2^n) is expressible as the gradient of a function H.

That is the basic picture. We have a universal constraint hypersurface of dimension 2N in a big space of 4N dimensions. There are two ways to represent the system dynamics, namely in central variables where a single system point moves in the universal hypersurface for all time, and in principal variables where the system is represented twice, by two points which execute constrained, identical, noninteracting, duplicate motions. And the two methods, central and principal variable representations, are rigorously equivalent. The next task is to streamline the notation, and describe the basic elements of the second level of generalization of Dirac's methods, yet to come.

4. COVARIANT FORMULATION

The constraint hypersurface of the big phase space is defined by

$$\mathbb{U}^{2N}: z^n \approx 0,\ p_{x^n} \approx 0,\ n = 1,\ldots,N. \tag{4.1}$$

Arranging indices, I write the constraint variables as

$$\delta^\mu \rightarrow (z^1,\ldots,z^N,p_{x^1},\ldots,p_{x^N}) \tag{4.2}$$

and, for the remaining canonical variables,

$$w^{\mu} \rightarrow (x^1,\ldots,x^N, p_{z^1},\ldots p_{z^N}), \tag{4.3}$$

so that the fundamental Poisson bracket relations read

$$\{w^{\mu},\delta^{\nu}\} \approx \varepsilon^{\mu\nu}, \tag{4.4}$$

where

$$\|\varepsilon^{\mu\nu}\| = \begin{pmatrix} 0 & 1 \\ -1 & 0 \end{pmatrix}, \tag{4.5}$$

with entries N×N unit and null matrices, and also

$$\{w^{\mu},w^{\nu}\} \approx \{\delta^{\mu},\delta^{\nu}\} \approx 0. \tag{4.6a,b}$$

In fact, Eqs. (4.4) and (4.6a,b) do not strictly follow since Eqs. (4.1) are only weak equations; but this will be fixed up in the transformation theory, most of the essential features of which appear below. See also [7].

Suppose points of the big space have been identified by a canonical central variable coordinate system, i.e., (w^{μ},δ^{μ}). Consider the class of transformations $(w^{\mu},\delta^{\mu}) \rightarrow (w'^{\mu},\delta'^{\mu})$,

$$w^{\mu} \rightarrow w'^{\mu}(w,0,t) = \tilde{w}^{\mu}(w,t), \tag{4.7a}$$

$$\delta^{\mu} \rightarrow \delta'^{\mu}(w,0,t) = 0, \tag{4.7b}$$

with

$$\det \|\partial\tilde{w}^{\mu}/\partial w^{\nu}\| \neq 0. \tag{4.8}$$

Extending Dirac's methods again as promised, I introduce a representation of these transformations by a set of strong equations,

$$w^{\mu} \rightarrow w'^{\mu} \simeq \tilde{w}^{\mu}(w,t) + \frac{\partial\tilde{w}^{\mu}}{\partial w^{\nu}}\,\alpha^{\nu}_{\lambda}(w,t)\delta^{\lambda}, \tag{4.9a}$$

$$\delta^{\mu} \rightarrow \delta'^{\mu} \simeq \frac{\partial\tilde{w}^{\mu}}{\partial w^{\nu}}\,\beta^{\nu}_{\lambda}(w,t)\,\delta^{\lambda}. \tag{4.9b}$$

These transformations leave U^{2N} invariant, and fall into equivalence classes specified by the choice of functions, $\tilde{w}^{\mu}(w,t)$, $\alpha_{\lambda}^{\nu}(w,t)$ and $\beta_{\lambda}^{\nu}(w,t)$. Starting from here, I have introduced a subclass, called $\mathcal{U}$-canonical transformations[7], for which

$$\alpha_{\lambda}^{\nu} = 0, \tag{4.10a}$$

$$\det\|\beta_{\lambda}^{\nu}(w,t)\| \neq 0, \tag{4.10b}$$

and for which also the transformations are required to be canonical, obeying Eqs. (4.4) and (4.6a,b), in the sense of strong specification. In this way, I have derived in Ref. [7] a general coordinate transformation structure intrinsic to U^{2N}. In particular, the $\tilde{w}$ receive interpretation as coordinates for points in U^{2N}, and the quantities

$$\delta_{\mu} \equiv \varepsilon_{\mu\nu}\delta^{\nu}, \quad \|\varepsilon_{\mu\nu}\| \equiv \|\varepsilon^{\mu\nu}\|^{-1} \tag{4.11}$$

have been shown to transform covariantly, like components of a 1-form; viz., under

$$w^{\mu} \to \tilde{w}^{\mu}(w,t) \tag{4.12}$$

one has

$$\delta_{\mu} \to \frac{\partial w^{\lambda}}{\partial \tilde{w}^{\mu}}\, \delta_{\lambda}. \tag{4.13}$$

Also, the $\varepsilon^{\mu\nu}$ receive the interpretation the notation suggests, as components in canonical coordinates on U^{2N} of a symplectic (Poisson) tensor. I denote its components in general coordinates by $e^{\mu\nu}(w,t)$ and introduce

$$\overset{\circ}{w}{}^{\mu} \equiv e^{\mu\nu}(w,t)\delta_{\nu}, \tag{4.14}$$

which is a directional field attached to points in U^{2N}. Furthermore, it was shown that if principal variables are defined by

$$\eta^{\mu} \simeq w^{\mu} + \frac{1}{2}\,\overset{\circ}{w}{}^{\mu}, \tag{4.15a}$$

$$\eta'^{\mu} \simeq w^{\mu} - \frac{1}{2}\,\overset{\circ}{w}{}^{\mu}, \tag{4.15b}$$

instead of Eqs. (3.15) and (3.16), then the transformation (4.12) induces transformations of η^{μ} and η'^{μ} given by

$$\eta^{\mu} \to \tilde{w}^{\mu}(\eta,t) \text{ and } \eta'^{\mu} \to \tilde{w}^{\mu}(\eta',t), \tag{4.16}$$

so that the central and principal variable structure is preserved. Moreover, the bracket relations among the η and η' are

$$\begin{gathered}\{\eta^{\mu},\eta^{\nu}\} \approx e^{\mu\nu}(\eta,t), \ \{\eta'^{\mu},\eta'^{\nu}\} \approx -e^{\mu\nu}(\eta',t), \\ \{\eta^{\mu},\eta'^{\nu}\} \approx 0.\end{gathered} \tag{4.17}$$

The first two of these relations give as big-space Poisson brackets, separate forms for unprimed and primed variables alone which coincide with the natural bracket relations intrinsic to $\mathbb{U}^{2N}$. The last relation asserts the invariant bracket commutativity of η and η'. Corresponding to Eq. (3.18), the Hamiltonian in principal variables is

$$\mathcal{H} = \mathcal{H}(\eta,\eta',t) \simeq H(\eta,t) - H(\eta',t) + \delta\mathcal{H}(\eta,\eta',t). \tag{4.18}$$

Returning to central variables and writing

$$F_{\mu} \to (F_{11},\ldots,F_{1N},F_{2}^{1},\ldots,F_{2}^{N}), \tag{4.19}$$

we have

$$\mathcal{H} = \mathcal{H}(w,\mathring{w},t) \simeq F_{\mu}(w,t)\mathring{w}^{\mu} \tag{4.20}$$

and

$$\{w^{\mu},\mathring{w}^{\nu}\} \approx e^{\mu\nu}(w,t), \ \{w^{\mu},w^{\nu}\} \approx \{\mathring{w}^{\mu},\mathring{w}^{\nu}\} \approx 0. \tag{4.21}$$

The generalized canonical equations of motion from Eq. (4.20) give $d\mathring{w}^{\mu}/dt \approx 0$ and

$$\frac{dw^{\mu}}{dt} \approx e^{\mu\nu}(w,t)\, F_{\mu}(w,t) \equiv F^{\mu}(w,t), \tag{4.22}$$

as I promised in Section 1.

5. LINE ELEMENT GEOMETRY

I consider next a possible procedure suggested, although somewhat different, from the theory of Finsler spaces [24]. It involves a formal variation scheme drawn from the "Lagrangian"

$$\mathcal{L}[\sigma_1,\sigma_2] \equiv \int_{\sigma_1}^{\sigma_2} d\sigma \ell(y(t,\sigma),\overset{\circ}{y}(t,\sigma),t), \tag{5.1}$$

where y stands for the 2N quantities $(x,\dot{x})$ and $\overset{\circ}{y}$ for

$$\overset{\circ}{y} = \frac{\partial y(t,\sigma)}{\partial\sigma}. \tag{5.2}$$

It is supposed that a smooth curve $(\sigma_1,\sigma_2) \to V^{2N}$, where V^{2N} stands for the velocity phase space $(x,\dot{x})$ of the system, has been specified for each value of t on an interval (t_1,t_2) (and of course that the t-dependence is smooth). I intend that ℓ be the analogue of the fundamental function used in Finsler spaces to define arc-length, with $\mathcal{L}[\sigma_1,\sigma_2]$ rather than the action (!) as the analogue of the latter. In the most general case, ℓ is a homogeneous function of the first degree in the line element variables, $\overset{\circ}{y}$, which take the values specified in Eq. (5.2) for definition of $\mathcal{L}[\sigma_1,\sigma_2]$. We will recover the theory of the previous sections if we specialize to the case that ℓ is linear in the $\overset{\circ}{y}$ -- I have not explored the general homogeneous case. When ℓ is linear,

$$\frac{\partial\ell}{\partial\overset{\circ}{y}^\mu} \equiv \ell,_{\overset{\bullet}{\mu}} \equiv f_\mu(y,t), \tag{5.3}$$

and Eq. (5.1) gives

$$\mathcal{L}[\sigma_1,\sigma_2] = \int_{y(t,\sigma_1)}^{y(t,\sigma_2)} dy^\mu(t,\sigma)\ f_\mu(y(t,\sigma),t). \tag{5.4}$$

If $f_\mu(y,t)$ is the gradient of a function $L = L(y,t)$, then

$$\mathcal{L}[\sigma_1,\sigma_2] = L(y(t,\sigma_2),t) - L(y(t,\sigma_1),t). \tag{5.5}$$

Obviously, we are invited to make the identifications,

$$y(t,\sigma_2) = \zeta(t),\ y(t,\sigma_1) = \zeta'(t), \tag{5.6}$$

where $\zeta = (\xi,\dot{\xi})$ and $\zeta' = (\xi',\dot{\xi}')$, for now Eq. (5.5) reduces to Eq. (3.17) for the case $\delta\mathcal{L} = 0$.

But Eq. (3.17) would require $\mathcal{L}$ to be a function of only the end-point values ζ,ζ' in the case also that $f_\mu(y,t)$ is not integrable. It appears the only way to accomplish this result is to have something like $|y(t,\sigma_2) - y(t,\sigma_1)| = O(\varepsilon)$, where ε eventually is to tend to zero:

$$\mathcal{L}[\sigma_1,\sigma_2] \doteq (\zeta - \zeta')^\mu \, f_\mu(\tfrac{1}{2}\,(\zeta + \zeta'),t), \tag{5.7}$$

where $\doteq$ signifies omission of terms of order ε^2. By Taylor expansions of the right hand side one can show that Eq. (5.7) in fact gives

$$\mathcal{L}[\sigma_1,\sigma_2] \doteq \mathcal{L}(\zeta,\zeta',t), \tag{5.8}$$

so that Eq. (5.1) becomes

$$\mathcal{L}(\zeta,\zeta',t) \doteq \int_{\sigma_1}^{\sigma_2} d\sigma \ell(y,\mathring{y},t), \tag{5.9}$$

where again it is understood now that $|\sigma_1 - \sigma_2| \to 0$. Evidently, the directional element $\mathring{y}\cdot\Delta\sigma \doteq \zeta - \zeta'$ corresponds to the central variable constraints, $(z\ ,\dot{z}\ )$, and ℓ is the universal Lagrangian in central variable form. So, I can simply write

$$\mathcal{L}(\zeta,\zeta',t) \doteq \int_{\sigma_1}^{\sigma_2} d\sigma\, \mathcal{L}(y,\mathring{y},t) \tag{5.10}$$

since the linearity (homogeneity) of ℓ assures independence of σ_1 and σ_2.

The same kinds of considerations give a completely analogous result for the two forms for the universal Hamiltonian, viz.,

$$\mathcal{H}(\eta,\eta',t) \doteq \int_{\sigma_1}^{\sigma_2} d\sigma \mathcal{H}(w,\dot{w},t). \tag{5.11}$$

It is worth noting, in passing, that the most general fundamental function would be a central variable $\mathcal{H}$ not merely linear, but homogeneous of the first degree, in the components of the directional element; the same generalization applies for $\mathcal{L}$ in Eq. (5.8).

Thus, to construct an action condition based on Eq. (5.1) or Eq. (5.8) it appears necessary to implement a suitable limiting procedure, $\varepsilon \to 0$, if one can be found. Evidently, this should lead to the same constraint formalism as before.

6. HAMILTON-JACOBI THEORY

The $\mathcal{U}$-canonical transformations (see Section 4) possess a generating function formalism in which the effect on $\mathcal{H}$,

$$\mathcal{H} \to \mathcal{K} \simeq \mathcal{H} + \partial\mathcal{X}/\partial t, \tag{6.1}$$

where $\mathcal{X}$ is a universal generating function, induces transformations of the vector field F for the dynamics, e.g.,

$$F_\mu \to G_\mu = F_\mu - e_{\mu\nu} v^\nu, \tag{6.2}$$

where v^ν is the "velocity field of the transformation", viz.,

$$v^\nu(w,t) = \frac{\partial \chi^\nu(W,t)}{\partial t}(w,t) \tag{6.3}$$

in which W are the new general (!) coordinates for $\mathcal{W}$,

$$w^\mu = \chi^\mu(W,t). \tag{6.4}$$

The origin of Eqs. (6.2) - (6.4) is

$$G_\mu \approx \mathcal{K}_{,\dot{\mu}}, \quad F_\mu \approx \mathcal{H}_{,\dot{\mu}}, \quad \text{and } \mathcal{X} \simeq \chi^\mu(W,t)\,\delta_\mu, \tag{6.5}$$

together with Eq. (4.14).

Equation (6.4) shows that Eq. (6.3) is the Hamilton-Jacobi equation. Regarding W^μ as initial conditions for (6.3), the W^μ obey canonical equations for $\mathcal{K}$. Choosing $v^\mu \equiv F_0^\mu$ arbitrarily, we have

$$\frac{\partial \chi^\mu(W,t)}{\partial t} \equiv \frac{dw^\mu}{dt} = F_0^\mu(w,t) \tag{6.6}$$

and

$$\frac{dW^\mu}{dt} = \overline{G}^\mu(W,t) = \overline{F}^\mu(W,t) - \overline{F}_0^\mu(W,t), \tag{6.7}$$

where the over-bars are needed because the vectors must be referred to the new (W-) coordinate system.

Equation (6.6) specifies an arbitrarily moving coordinate system and Eq. (6.7) expresses the dynamical behavior from that coordinate system. It can be shown that

$$\mathring{E}_{\mu\nu}(w,t) = \dot{e}_{\mu\nu}(w,t) - dv_{\mu\nu}(w,t). \tag{6.8}$$

Here, $\mathring{E}$ and $\dot{e}$ refer to time derivatives of the symplectic tensor referred to "body" and "space" coordinate systems, respectively, where "body system" refers to new coordinates W, whose velocity field relative to the "space" system, which means old coordinates, is v^μ. The quantities $dv_{\mu\nu}$ are components of the exterior derivative of the object whose components are v_μ, i.e., $dv_{\mu\nu} = v_{\nu,\mu} - v_{\mu,\nu}$. In canonical coordinates, $e_{\mu\nu} = \varepsilon_{\mu\nu}$ = const, so I call the class of coordinate systems preserving $\dot{e}_{\mu\nu} = \dot{e}^C_{\mu\nu} \equiv 0$, "the canonical viewpoint". For these, $d(F-G) = dv = 0$, i.e., F and G differ by at most a gradient field. Another class, which I call "the Hamiltonian viewpoint", has $dG = 0$, or in other words, $\mathring{E} \equiv \dot{e}^H = -dv = d(G-F) = -dF$. This terminology is intended to parallel that of Heisenberg and Schrödinger pictures, where $v = 0$ and $v = dH$, respectively.

Owing to the t-dependence of e, if A^μ is a vector, $\partial A/\partial t$ is not. However, a covariant time derivative can be constructed; it is

$$\frac{\Delta A_\mu}{\Delta t} = \frac{\partial A_\mu}{\partial t} - \frac{1}{2}\,\Omega_{\mu\nu}[dv]A^\nu \,, \tag{6.9}$$

$$\frac{\Delta A^\mu}{\Delta t} = \frac{\partial A^\mu}{\partial t} + \frac{1}{2}\,\Omega^{\mu\nu}[dv]A_\nu \,, \tag{6.10}$$

where the connection is a tensor given by

$$\Omega_{\mu\nu}[dv] = -\,d(F-G)_{\mu\nu}, \tag{6.11}$$

and is a property of the "viewpoint". One has, in particular,

$$\frac{\Delta e}{\Delta t} = 0, \tag{6.12}$$

whose vanishing is analogous to that of the ordinary covariant derivative of the metric tensor in a Riemannian space.

Finally, I mention one further matter, regarding the generating function formalism. $\mathcal{K}(W,\delta,t)$ is a central variable form, in which the separation of old and new canonical variables coincides with a separation of general coordinates for $\mathcal{U}$, and constraint variables. As a result, Eq. (6.4) does not have the feature present in the generator R_0 in eq. (1.1) of mixed old and new variable forms. The mixed variable $\mathcal{U}$-canonical generators have this feature. An example is

$$\mathcal{R}(x,P_Z,z,P_X,t) = \rho_{1n}(x,P_Z,t)z^n + \rho_2^n(x,P_Z,t)\,P_{X^n}. \tag{6.13}$$

If a function $R(x,P_z t)$ exists such that

$$\rho_{1n} = \frac{\partial R}{\partial x^n} \text{ and } \rho_2^n = \frac{\partial R}{\partial P_z{}^n}, \tag{6.14}$$

then R can be written -- in principal variables now ! --

$$\mathcal{R} \simeq R(\xi,P,t) - R(\xi', - P't). \tag{6.15}$$

Correspondingly, the transformation is

$$p_{z^n} \simeq \rho_{1n} \simeq \frac{\partial R(x,P_Z,t)}{\partial x^n}, \tag{6.16a}$$

$$X^n \simeq \rho_2^n \simeq \frac{\partial R(x,P_Z,t)}{\partial p_{z^n}}, \tag{6.16b}$$

which is the usual one for a canonical transformation, and which has the form of a gradient! For a general discussion see [7]. Whether, e.g., for a general coordinate transformation one has $p_{z^n} \to \frac{\delta R}{\delta x^n}$, with $\delta/\delta x^n$ a covariant derivative of some sort, is not yet known. Together with things such as Eqs. (6.9) and (6.10), it would not be surprising to find a generalization of Eq. (1.1) in the Hamiltonian viewpoint, let us say, in which partial derivatives by t and q were replaced with covariant derivatives. Naturally, the interest in such generalizations is not primarily in the aid they might or might not provide in solving differential equations, but in basic geometrical relations they would express.

7. EXAMPLES

The linear oscillator

$$\frac{dx}{dt} = p , \tag{7.1}$$

$$\frac{dp}{dt} = - \omega_0^2 x - \gamma p \tag{7.2}$$

has a fixed point at the origin. The eigenvalues of the derivative map are solutions to

$$\begin{vmatrix} -\lambda & 1 \\ -\omega_0^2 & -\gamma \ -\lambda \end{vmatrix} = \lambda^2 + \gamma\lambda + \omega_0^2 = 0 , \tag{7.3}$$

$$\lambda = \lambda_{\pm} = -\frac{1}{2}\gamma \pm \frac{i}{2}\omega_1, \tag{7.4}$$

where $\omega_1 = (\omega_0^2 - (\gamma/2)^2)^{1/2}$, so the origin is a sink. If ω_1 is real, choose F_0^μ in Eqs. (6.6) as

$$F_0^\mu = (-\frac{1}{2}\gamma x, -\frac{1}{2}\gamma p), \tag{7.5}$$

which gives the Hamilton-Jacobi equations $dx/dt = -\frac{1}{2}\gamma x$ and $dp/dt = \frac{-1}{2}\gamma p$, and the transformation $X = x \exp \frac{1}{2}\gamma t$, $P = p \exp \frac{1}{2}\gamma t$. The new vector field is Hamiltonian,

$$G_\mu = (\omega_0^2 x + \frac{1}{2}\gamma p, p + \frac{1}{2}\gamma x) = H_{,\mu}, \tag{7.6}$$

with

$$H(x,p) = \frac{1}{2}p^2 + \frac{1}{2}\omega_0^2 x^2 + \frac{1}{2}\gamma p x. \tag{7.7}$$

From $H(x,p) = K(X,P)$, and $e^{\mu\nu} = e^{\gamma t}\varepsilon^{\mu\nu}$ from the transformation, we have

$$K = e^{-\gamma t}[\frac{1}{2}P^2 + \frac{1}{2}\omega_0^2 X^2 + \frac{1}{2}\gamma P] \tag{7.8}$$

and the generalized canonical equations

$$\dot{X} = [X,K] = P + \frac{1}{2}\gamma X, \tag{7.9a}$$

$$\dot{P} = [P,K] = -\omega_0^2 X - \frac{1}{2}\gamma P, \tag{7.9b}$$

which are true Hamiltonian equations, for an <u>un</u>damped resonance-frequency shifted oscillator. The expansion of the phase volume due to the $\exp \frac{1}{2}\gamma t$ factors in the two directions identified in Eq. (7.4) compensates the contraction of the flow to the attractor at the origin. See that the attractor has been exorcised.

Note also, from Eq. (7.8),

$$\frac{dK}{dt} = \frac{\partial K}{\partial t} = -\gamma K \rightarrow K(t) = K(0)\exp{-\gamma t}. \tag{7.10}$$

Consider instead the effect of the more or less arbitrary choice $F_0^\mu = (0, -\gamma p)$. Now $dx/dt = 0$, $dp/dt = -\gamma p$, so $X = x$, $P = p \exp \gamma t$, and again $e^{\mu\nu} = e^{\gamma t}\varepsilon^{\mu\nu}$. G_μ is Hamiltonian

$$G_\mu = (\omega_0^2 x, p) = H'_{,\mu}, \tag{7.11}$$

with

$$H' = \frac{1}{2}\,p^2 + \frac{1}{2}\,\omega_0^2 x^2, \tag{7.12}$$

which looks clean enough, but is not since

$$K' = H' = \frac{1}{2}\,e^{-2\gamma t}p^2 + \frac{1}{2}\,\omega_0^2\,X^2. \tag{7.13}$$

The canonical equations, owing to the exp γt factor from $e^{\mu\nu}$ are Hamiltonian with effective Hamilton function given by exp $\gamma t \cdot K'$, viz.,

$$H_{eff} = \frac{1}{2}\,e^{-\gamma t}\,P^2 + \frac{1}{2}\,e^{\gamma t}\,\omega_0^2\,X^2, \tag{7.14}$$

which is the Kanai-Havas Hamiltonian [25,26], much discussed in the physics literature. The equations obeyed by X and P are

$$\ddot{X} + \gamma\dot{X} + \omega_0^2\,X = 0, \tag{7.15a}$$

$$\ddot{P} - \gamma\dot{P} + \omega_0^2\,X = 0, \tag{7.15b}$$

neither one an undamped equation. Also

$$\frac{dK}{dt} = \frac{\partial K}{\partial t} = -\,\gamma e^{-2\gamma t}P^2, \tag{7.16}$$

which, unlike Eq. (7.10), cannot be integrated to give K(t) without knowledge of the solutions to the equations of motion.

A second example is the Duffing equation with (negative) stiffness -1 and damping $\delta \neq 0$,

$$\dot{x} = y \tag{7.17a}$$

$$\dot{y} = x - \delta y - x^3. \tag{7.17b}$$

There are three fixed points, two sinks on the x-axis, at $x = \pm\,1$, and a hyperbolic saddle at the origin. For the saddle the eigenvalues of the Jacobian matrix A for the associated linearized problem are $\lambda_{\pm} = -\frac{1}{2}\delta \pm [(\frac{1}{2}\delta)^2 + 1]^{1/2}$, one positive and one negative, corresponding to stable and unstable manifolds. Since A is symmetric, the eigenvectors are orthogonal, and a rotation,

$$x' = (1 + \lambda_+^2)^{-1/2}\,(x + \lambda_+ y), \tag{7.18a}$$

$$y' = (1 + \lambda_-^2)^{-1/2}(x + \lambda_- y) \tag{7.18b}$$

will produce a diagonalized form of the linearized problem:

$$\dot{x}' = \lambda_+ x', \tag{7.19a}$$

$$\dot{y}' = \lambda_- y'. \tag{7.19b}$$

If as the simplest strategy I choose the components of F_0^μ equal to the right-hand sides of Eqs. (7.19), the effects of contraction and expansion along the stable and unstable manifolds will be compensated, but only locally, i.e., at the origin, rather than everywhere, as was the case for the linear oscillator. Since the sum of the eigenvalues is $-\delta$, the same as the divergence of the vector field specifying Eqs. (7.17), the dynamics in the moving coordinate system,

$$X = x' \exp(-\lambda_+ t), \tag{7.20a}$$

$$Y = y' \exp(-\lambda_- t), \tag{7.20b}$$

found by solving Eqs. (7.19) as a Hamilton-Jacobi system, will be <u>area preserving</u>. Also, the transformation law for the symplectic tensor gives

$$E^{\mu\nu} = -e^{\delta t} \cdot \boldsymbol{\varepsilon}^{\mu\nu}, \tag{7.21}$$

where the minus sign is an (essentially irrelevant) accidental consequence of sign choices which caused the determinant of the rotation in Eqs. (7.18) to be -1 instead of +1. The result is that $G_\mu = F_\mu - F_{0\mu} = K_{,\mu}$, where the Hamiltonian is

$$K = -\frac{1}{4}(1 + \lambda_+^2)^{-2}(X\exp\lambda_+ t + \lambda_+ Y\exp\lambda_- t)^4, \tag{7.22}$$

and the equations for the new variables are

$$\dot{X} = -\exp\delta t \cdot \frac{\partial K}{\partial Y}, \tag{7.23a}$$

$$\dot{Y} = +\exp\delta t \cdot \frac{\partial K}{\partial X}, \tag{7.23b}$$

which are irredeemably nonautonomous. The basic reason, of course, is that the stable and unstable manifolds are not the x' - and y' - axes, but only tangent to these at the origin.

Comparing Eqs. (7.10) and (7.16), it is natural to pose this question: How can one identify and implement natural uniqueness criteria for definition of a "Heisenberg picture", when a system is not Hamiltonian? Equation (7.8) and the corresponding system appear very natural for the linear oscillator, and Eq. (7.10) shows K to have a natural interpretation as system energy. On the other hand, the picture set by the second choice of F^{μ}_{0} is unnatural. A second criterion is suggested by the contrast of these two cases, in Eqs. (7.10) and (7.16), viz., that K(t) have the property

$$\frac{\partial K}{\partial t} = f(K,t) \tag{7.24}$$

for some f, as this can allow direct determination of K(t) for all orbits, given only K(0). For the Duffing oscillator example, the Hamiltonian given in Eq. (7.22) does not seem to admit a relation like Eq. (7.24). I do not have a candidate "Heisenberg picture" for this problem.

8. GLOBAL FORMULATIONS OF HAMILTONIAN THEORY

The big (4N-dimensional) space $\mathcal{B}$ which I have introduced is not one of the usual symplectic arenas of Hamiltonian dynamics; nor is the manifold of a dynamical system normally taken to be that of a hypersurface $U \subset \mathcal{B}$. Hamiltonian mechanics is customarily formulated[27] in terms of a Hamiltonian vector field X on a symplectic manifold (P,ω), where $X \lrcorner\, \omega = -\, dH$, in which I use $\lrcorner$ to denote the left interior product: $X \lrcorner\, \omega = -\, i_X \omega$, with $H: P \to \mathbb{R}$ the Hamiltonian and ω a (nondegenerate) invariant symplectic 2-form.

There are three other modern formulations, the canonical, symplectic and deformed symplectic structure formulations, due to Lichnerowicz, Tulczyjew, and Sternberg, respectively. For a short review of the first two of these, see[28]; Sternberg's theory is in[29]. The first of the three, the canonical formulation, differs from the customary one described above in that its emphasis, instead of resting with the 2-form ω, lies with a (unique) 2-vector G essentially inverse to ω; Hamiltonian mechanics is then formulated on a Poisson manifold (P,G). Thus if (P,ω) is a symplectic manifold, G obeys $G \llcorner (u \lrcorner\, \omega) \equiv u$ for every vector u on P. The Hamiltonian vector field is then given explicitly, as

$X = - G \lrcorner dH$. The canonical formulation is equivalent to the customary one. One can think of ω as lowering indices via $\lrcorner$, and G as raising indices, like $\varepsilon_{\mu\nu}$ and $\varepsilon^{\mu\nu}$.

The other two approaches are somewhat different, however, and possess unique features analogous to different facets of the treatment I have given. Still, neither one is applicable to dissipative systems, and neither one in its local formulations reduces to the conservative case of the formalism in the present paper.

Sternberg's procedure[29] begins with the realization of the manifold P as a cotangent bundle, $P = T^*Q$, on which the usual natural symplectic structure $d\theta$ is identified from the canonical one-form θ, defined on T^*Q via the projection $\pi: T^*Q \to Q$. In local coordinates

$$\theta = p \cdot dq = p_i dq^i. \tag{8.1}$$

Let σ be a one-form on Q, to be considered also as a one-form on T^*Q via the pullback under π; in local coordinates

$$\sigma = A \cdot dq = A_i(q)\, dq^i. \tag{8.2}$$

σ is also a section of T^*M over M, so one can define a vertical translation $\phi_\sigma: T^*M \to T^*M$ by σ; in local coordinates

$$\phi_\sigma(p,q) = (p + A, q), \tag{8.3}$$

so that ϕ_σ amounts to a certain "change of variables". The pullback of $d\theta$ under ϕ_σ is

$$\phi_\sigma^*(d\theta) = d\theta + d\sigma. \tag{8.4}$$

Now suppose that X is the Hamiltonian vector field associated to $H \circ \phi_\sigma^{-1}$, where $H: T^*M \to \mathbb{R}$ is smooth; in local coordinates $H \Rightarrow H \circ \phi_\sigma^{-1}$ is the minimal substitution, $p \Rightarrow p - A$.

So, by definition,

$$X \lrcorner\, d\theta \equiv - d(H \circ \phi_\sigma^{-1}) \tag{8.5}$$

$$= - (\phi_\sigma^{-1})^* dH. \tag{8.6}$$

But by eqs. (8.4) and (8.6),

$$(\phi_\sigma^* X) \rfloor (d\theta + d\sigma) = (\phi_\sigma^* X) \rfloor (\phi_\sigma^*(d\theta))$$

$$= \phi_\sigma^*(X \rfloor d\theta)$$

$$= - dH, \qquad (8.7)$$

whence the pullback of X is the Hamiltonian vector field for the original Hamiltonian H, with no minimal substitution, but relative to the new (deformed) symplectic structure, $d\theta + d\sigma$. In local coordinates the vector field indices are lowered now (not with $d\theta$, but) with

$$d\theta + d\sigma = d(p_i + A_i) \wedge dq^i. \qquad (8.8)$$

An advantage of this, of course, is the elimination of the "potentials" (A_i) in favor of the "field strengths" (components of dA).

In the formalism I have constructed, $\mathcal{U}^{2N}$ does not arise as a cotangent bundle to a configuration manifold. Correspondingly, eq. (3.8), which for $f_2 \to \partial L/\partial \dot{q}$ would normally be the genesis of a Legendre transformation from coordinate variables $(q,\dot{q})$ for TQ to phase variables (q,p) for T^*Q, now specifies a change from coordinate variables $(x,\dot{x})$ for points in $\mathcal{W}^{2N}$: $z^n \approx 0, \dot{z}^n \approx 0$, to (x,p_z) for points in $\mathcal{U}^{2N}$: $z \approx 0$, $p_x \approx 0$.

Instead of having a "natural" symplectic structure ($d\theta$) defined via a cotangent bundle projection map, $\mathcal{U}^{2N}$ inherits a symplectic tensor from $\mathcal{B}$ via the "crossed-pair" associations, (x,p_z) and (z,p_x), between the constraints and their generators.

Still it is easy also to gain the effect described above using this paper's formalism, namely the absorption of the A_i into the symplectic tensor where it appears only as components of dA. One has only to perform a ("vertical translation") coordinate change, and the tensor transformation of ε gives the result, as it must[7].

In the customary treatment of Hamiltonian mechanics described at the outset, a dynamical system is a Hamiltonian vector field $X = X_H$ together with a symplectic manifold (P,ω). Tulczyjew's formulation is actually a certain generalization, depending on the notion of a Lagrangian submanifold defined from a natural symplectic structure for T^*P, which is independent of ω itself. (A Lagrangian submanifold N of a symplectic manifold (M,Ω) has $\dim N = \frac{1}{2} \dim M$ and $\Omega|_N = 0$.)

Thus, let (P,ω) be a symplectic manifold. The mapping

$$\beta:TP \to T^{*}P;\ X \to X \rfloor \omega,$$

whose definition depends upon ω, defines a vector bundle isomorphism. Let θ_p denote the canonical one form on $T^{*}P$, and let $\omega_p = d\theta_p$ denote the corresponding natural symplectic structure, which is independent of ω. The pullback of θ_p under β, viz. $\chi = \beta^{*}\theta_p$, and that of ω_p, $\rho = d\chi = \beta^{*}\theta_p$, depend upon both, and (TP,ρ) is a new symplectic manifold, "twice the size" of (P,ω). Tulczyjew now defines a dynamical system in (P,ω) as a Lagrangian submanifold $\mathring{D}$ of the symplectic manifold (TP,ρ). The idea is that as a submanifold of a tangent bundle, $\mathring{D} \subset TP$ can be interpreted as a system of differential equations. The Hamiltonian vector fields are included since the image of a Hamiltonian vector field $X_H:P \to TP$, $\mathrm{im}X \subset TP$, can be shown to be Lagrangian.

I know no examples of dissipative systems included in Tulczyjew's formulation, and it seems improbable that there are any. Thus, it can be shown that the Lagrangian submanifolds of $T^{*}M$ which project diffeomorphically onto M, where M is a differentiable manifold, are in one-to-one correspondence with the closed one-forms on M (see [27], page 410), and these will not correspond to nonconservative systems. In addition, while $\mathbb{U}^{2N}$ has half the dimension of $\mathcal{B}$, and the w^{μ} have mutually vanishing Poisson brackets, a condition that does give local expression to the vanishing of the restriction of a symplectic structure for $\mathcal{B}$ to $\mathbb{U}^{2N}$, the coordinate transformation theory essential to the structural features of my formalism is based on weak and strong equality procedures which have only local definition. It is not clear yet how a probably desirable goal of global realization of $\mathbb{U}^{2N}$ as some kind of Lagrangian subspace would be accomplished.

This brings me to the first of two final matters to be considered about globalization, mentioned at the end of Section 1, i.e. my extensions of Dirac's local methods of analysis. A starting point for a global formulation may be to understand $\mathcal{L}$ and $\mathcal{H}$ as local 1-jet-valued mappings. Thus if $\pi:\mathcal{B} \to \mathbb{U}$ denotes a suitable natural projection we would define $\mathcal{H}$ via mappings on the fibers, $\mathcal{H}_p:\pi^{-1}(p) \to J^1(\pi^{-1}(p), \mathbb{R})$, for every $p \in \mathbb{U}$; this suggests $\mathcal{H} \in J^1(\mathcal{B}, T\mathbb{U})$. In regard to the second matter, which is the generalized Hamilton-Jacobi theory of Section 6, what the formalism really seems to suggest is a radical extension of the notion of a dynamical system. The idea is that of a principal fiber bundle, $\mathcal{B}$, in which $\mathbb{U}^{2N}$ is the base manifold and the (additive) structure group (of velocity fields for coordinate transformations) is $V^{2N} \simeq \mathbb{R}^{2N}$, isomorphic to the translation group: $\mathcal{B}=\mathcal{B}(\mathbb{U}^{2N},V^{2N})$ with projection $\pi':\mathcal{B}\to\mathbb{U}^{2N}$. The "moving coordinate systems" would be bundle sections. We might then consider options that would surround the step of defining a dynamical system as a map, not of a manifold M, with $\phi:\mathbb{R}\times M\to M$, but as a map of a bundle, $\psi:\mathbb{R}\times\mathcal{B}\to\mathcal{B}$, with $\psi(0,p)=p$ and $\partial\psi(t,p)/\partial t = Z(\psi(t,p))$. There could be conflict between the jet and principal bundle formulations, however, and I have not tried to explore the possibilities. Nor have I tried to examine the role that could be played by the Finslerian approach described in Sectin 5. Finally, despite my remarks in Section 1 about time dependence, it is still possible (and perhaps necessary) to regard the treatment of nonautonomous systems as a separate issue. But, it would be nice if general non-autonomous systems could be accommodated within a framework of bundle mappings (flows) for purely autonomous systems.

9. CONCLUSION

In have described essential features of a local formalism of a generalized Hamiltonian dynamics applying to dissipative as well as conservative systems. The local methods make extensive use and generalization of techniques first introduced by Dirac; and I have reviewed Dirac's theory of constrained Hamiltonian dynamics, especially in regard to the problem of the secondary first class constraints. A suitable global realization of the generalized formalism of dynamics I have described does not yet exist. Global formulations have been discussed briefly at the end. The theory may be understood as a generalized symplectic geometry, in which the symplectic tensor plays a role more significantly paralleling that of the metric tensor in a Riemannian space than is customary. A time-dependent (general) coordinate transformation structure gives rise to a generalized Hamilton-Jacobi theory in which the vector field for the dynamics is decomposed arbitrarily into a sum of two vector fields specifying two generally non-autonomous systems, one describing the "motion of a coordinate system," and the other the "appearance of the dynamics" from the moving coordinate system. The symplectic tensor in general contains explicit time-dependence owing to that of the coordinates relative to those for which the system is originally defined. Correspondingly, the formalism is endowed with a novel covariant time-derivative, $\Delta/\Delta t$, which obeys

$\Delta e/\Delta t = 0$, where e is the symplectic tensor. I have described applications to simple systems such as the damped linear and Duffing oscillators. In the former, the attractor at the origin is easily "exorcised" by a suitable coordinate change, and a resonance-frequency-shifted undamped oscillator Hamiltonian results, with a reasonable interpretation as system energy. The corresponding trick fails for the nonlinear oscillator, essentially due to the nontrivial forms of stable and unstable manifolds (for the hyperbolic saddle at the origin). The treatment of the time parallels that customary in Dirac's unified picture for quantum theory, one of a number of features that make the scheme interesting.

REFERENCES

[1] C. P. Lanczos, Variational Principles of Mechanics (Univ. of Toronto, 1966).
[2] C. L. Siegel and J. K. Moser, Lectures on Celestial Mechanics (Springer, Berlin, 1971).
[3] G.-I. Hori, Publ. Astron. Soc. Japan 18, 287 (1966) and Andre Deprit, Cel. Mech. 1, 12 (1969).
[4] A. A. Kamel, Cel. Mechs. 3, 90 (1970) and G.-I. Hori, Publ. Astron. Soc. Japan 23, 567 (1971).
[5] There are variations of the concept of a flow, e.g., where R is replaced by a topological group T, or a suitable subset S R, open in R x M. Also, I am considering only continuous systems here, not discrete ones. For something of the flavor of global dynamics, see Jacob Palis, Jr., and Wellington de Melo, Geometric Theory of Dynamical Systems (Springer, New York, 1982).
[6] R. K. Miller and G. R. Sell, in Dynamical Systems, Vol. 1, L. Cesari, J. K. Hale and J. P. LaSalle, eds., Academic, 1976.
[7] R. Cawley, Phys. Rev. A20, 2370 (1979); J. Math. Phys. 21, 2350 (1980); and Phys. Rev. D22, 859 (1980).
[8] P.A.M. Dirac, Canad. J. Math. 2, 147 (1950).
[9] E.C.G. Sudarshan and N. Mukunda, Classical Dynamics: A Modern Perspective (Wiley, New York, 1974).
[10] P. G. Bergmann and I. Goldberg, Phys. Rev. 98, 531 (1955).
[11] P.A.M. Dirac, Proc. Roy. Soc. A246, 326 (1958).
[12] P.A.M. Dirac, Lectures on Quantum Mechanics (Yeshiva Univ. Press, New York, 1964).
[13] G. R. Allcock, Phil. Trans. Roy. Soc. (London) A279, 33 (1975).
[14] Robert Cawley, Phys. Rev. Letters 42, 413 (1979).
[15] A. Frenkel, Phys. Rev. D21, 2986 (1980).
[16] R. Cawley, Phys. Rev. D21, 2988 (1980).
[17] G. R. Allcock, Kinam 2, 335 (1980).
[18] R. diStefano, Phys. Rev. D27 1752 (1983).
[19] J. Sniaticki, Ann. Inst. H. Poincare A20, 365 (1974).
[20] M. J. Gotay, J. M. Nester, and G. Hinds, J. Math. Phys. 19, 2388
[21] R. Sugano and H. Kamo, Prog. Theor. Phys. 67, 1966 (1982). See also "Pathological Dynamical Systems with Constraints," R. Sugano and T. Kimura, Horoshima University preprint 82-14, November, 1982.
[22] R. Skinner and R. Rusk, J. Math. Phys. 24. 2589 and 2595 (1983).
[23] R. Hagedorn, Nuovo Cim. Supp. 12, 73 (1959).
[24] H. Rund, The Differential Geometry of Finsler Spaces (Springer, Berlin, 1959).

[25] E. Kanai, Prog. Theor. Phys. 3, 448 (1948).
[26] P. Havas, Nuovo Cim. Supp. 5, 363 (1957).
[27] R. Abraham and J. Marsden, Foundations of Mechanics, Benjamin/Cummings, Reading, Massachusetts, second edition, 1978.
[28] W. M. Tulczyjew, "Hamiltonian, Canonical and Symplectic Formulations of Dynamics," in Bifurcation Theory, Mechanics and Physics, C. P. Bruter, A. Aragnol, and A. Lichnerowicz, eds., D. Reidel, Dordrecht, 1983 (p. 1)
[29] S. Sternberg, Proc. Nat. Acad. Sci. 74, 5253 (1977).

This work was supported by the NSWC Independent Research Program and by ONR.

PARTICLE CHANNELING IN CRYSTALS AND THE METHOD OF AVERAGING

H.S. Dumas and J.A. Ellison*
Department of Mathematics
University of New Mexico
Albuquerque, NM 87131

1. Introduction

The purpose of this paper is twofold: first, to introduce readers to the field of energetic particle channeling, and second, to bring attention to the Krylov-Bogoliubov method of averaging. Particle channeling is an important aspect of particle-solid interactions leading to its own interesting mathematical problems in Hamiltonian systems of 1, 2, and 3 degrees of freedom, perturbation theory, probability and stochastic processes, and ergodic theory. The method of averaging is a powerful procedure which, unlike most other perturbation techniques, allows for an estimate of the difference between solutions to approximate and exact problems.

After a brief overview of channeling in Section 2, we discuss the mathematical framework for particle channeling and introduce the perfect crystal model. In Section 3, we introduce the continuum model as an approximation to the perfect crystal model. A scaling is introduced so that the two models can be compared and it turns out that this scaling leads to equations in a standard form for the method of averaging. The first order theory for averaging is discussed in detail in Section 4 in general terms, independent of the channeling context. The second-order theory, a straightforward generalization of the first-order, is applied to the channeling problem in Section 5. Finally, in Section 6, we give a rigorous treatment of a model problem that has some relevance to channeling, providing the reader a chance to see details of the comparison between exact and approximate problems in a simplified context.

2. Channeling: Crystal Lattice Influence on Particle Motion

When a particle beam is incident on a crystaline target in a random direction, the target can be treated, to a good approximation, as if it were amorphous. However, if a well-collimated beam of energetic positive particles enters a crystal (for example MeV protons or GeV positrons entering silicon), within a small angle of a densely packed (low index) crystal axis or plane, the crystaline structure strongly influences the motion of these particles. This arises from the fact that sequential collisions of a particle with the lattice are strongly correlated, giving rise to a gentle steering of the beam along

* This work was supported by the National Science Foundation under Grant No. DMR-8214301.

crystal rows, whereas this does not occur along random directions. For example, in going from a random to an aligned incident direction, the number of particles backscattered per unit time can decrease by nearly a factor of 100, the depth at which particles in the beam are stopped in the target is significantly increased, and the number of incident particles transmitted through a thin crystal is substantially increased.

The influence of the crystal lattice on the motion of energetic charged particles is called channeling, and channeling effects become important when a particle is aligned near crystal axes at angles less than [1]

$$\psi_a = (4z_1z_2e^2/dpv)^{1/2} \tag{2.1}$$

and near crystal planes at angles less than

$$\psi_p = (4z_1z_2e^2Nd_p\,Ca/pv)^{1/2} \,. \tag{2.2}$$

Here z_1 and z_2 are the atomic numbers of the incident particles and the target, respectively, d is the distance between atoms in an atomic string, d_p is the interplanar distance, p and v are the incident particle momentum and velocity, respectively, N the atomic density, C is a constant of order $\sqrt{3}$, and $a \simeq a_0z_2^{-1/3}$ is a screening length in the particle-atom potential. If d is in Angstroms and pv is in eV then the square of the electronic charge is $e^2 = 14.4$ eV –Å . For 1 MeV protons along the $\{110\}$ axis of Si $\psi_a = 0.59\,^\circ$ and for 1 GeV/c positrons $\psi_a = 460\mu$rad.

Channeling can be readily observed because nuclear physicists have developed the art of producing well-collimated energetic charged particle beams and solid state physicists have become adept at producing nearly-perfect single crystal solids. A qualitative explanation of the channeling effect for alignment near a crystal axis (axial channeling) starts by viewing the crystal as a collection of strings of lattice atoms parallel to the crystal axis. If an energetic positive particle enters the crystal surface near one of these strings with a large, nearly parallel velocity, it will be gently pushed away by a series of small angle collisions with the atoms in the string as shown in Fig. 1a. It then moves relatively freely through the lattice until it comes close to another string, at which time the gentle steering process repeats itself. This is illustrated in Fig. 1b by the particle path labeled A. The motion has been projected onto a plane perpendicular to the atomic strings and the string positions in the transverse plane are shown by the large circles. Each collision with a string is the result of small angle collisions with many atoms. The gentle steering keeps the particles away from the strings and thus decreases the probability of a close encounter with a lattice atom. This decrease in the close-encounter probability explains the examples mentioned at the beginning of this section. This phenomenon was referred to as the "string effect" in the work of Lindhard [2] . The current term "channeling" is somewhat of a

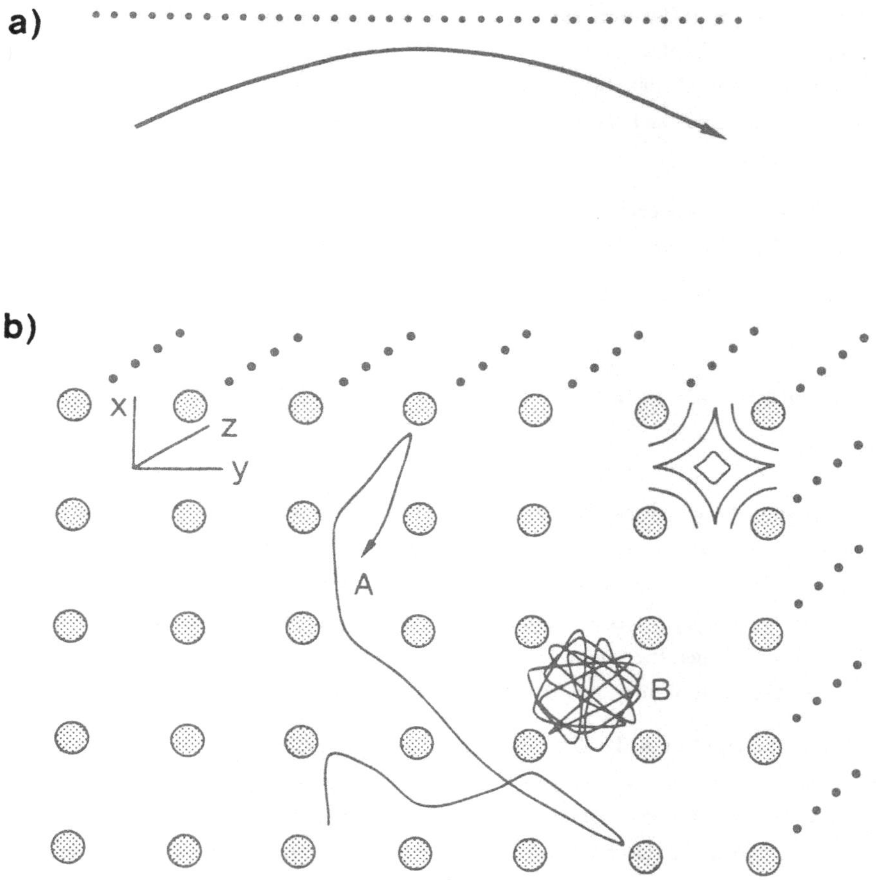

Fig. 1. (a) Channeled particle gently steered by an atomic string. (b) Crystal atomic strings for the {100} axis of diamond shown schematically by the large circles in the (x, y) plane. The small black dots represent the atomic strings going off in the z direction. Path A represents a channeled trajectory projected on the transverse (x, y) plane as it moves from the influence of one string to another. Path B shows the projection of a proper channeled particle in its motion confined to one channel. The potential contours in the upper right corner are sketched in for the continuum model and are shown in more detail in Fig. 3.

misnomer in this context, as particles do not typically stay in "channels" bounded by strings. However, the terminology has become standard. A particle whose path stays in one channel is said to be proper channeled and its path, labeled B, is illustrated in Fig. 1b. A similar explanation for planes leads to the concept of planar channeling; although weaker than its axial counterpart, the effect is stronger than proper channeling and is an important physical aspect of beam-crystal interactions. Channeling effects with negative particles have also been studied extensively.

Channeling was discovered in the early sixties, and a review article [3] and book [4] give general discussions, applications, and an extensive bibliography of the literature up to 1974. Since then many original papers have been published and several review articles, conference proceedings and books [5-6] have appeared. Reference [5] contains a well organized bibliography of the literature up to 1982.

The channeling effect provides a significant test of our understanding of the theory of particle interaction with condensed matter, as in energy loss and range distribution studies for example. Also, the channeling motion is sensitive to atomic potentials as they combine to create a crystal potential [7]. Channeled electrons and positrons emit electromagnetic radiation which is very sensitive to the crystal structure. The study of this "channeling radiation" has been extensive in the last few years [8-10]. Enhanced pair production is predicted [11] when a high energy photon beam is aligned with a crystal axis and experiments are underway at CERN to verify this prediction [12-13].

The channeling effect has led to important applications in physics and technology. It has become an important material analysis tool for studying crystal defects and other lattice disorder, for studying surfaces and interfaces, and for measuring the lattice location of impurity ions. These are discussed in detail in Ref. [5]. It has been used to measure nuclear lifetimes [14], and to bend high energy particle beams at Dubna [15], CERN [16], and Fermilab [17]. More recently [18], at Fermilab, a bent crystal has been used to replace a magnet in a secondary charged particle beam. The maximum momentum of particles that could be transmitted to an experimental location was raised from the magnetic septum limit of 225 GeV/c to the full primary beam momentum of 400 GeV/c. Channeling radiation is being proposed as a solid state probe [9] and recent work at CERN [19] shows that planar channeling radiation from GeV/c positrons gives rise to a monoenergetic gamma source of MeV photons. Most recently, the channeling effect has been used to study the strain in strained-layer-superlattices [20], a new material which promises some interesting applications in electronic and optical devices.

In addition to the fundamental physical insights and the above applications, the study of the channeling effect has led to many interesting and varied mathematical problems, as mentioned in the introduction. In order to discuss some of the problems relevant to these proceedings, the remainder of this section is used to formulate a mathematical framework for channeling phenomena.

A complete understanding of channeling and its applications requires detailed knowledge of the channeled particle's motion as it moves through the lattice. The starting point for theoretical discussions of particle channeling is the perfect crystal model [21]. In this model the motion of a nonrelativistic particle is governed by a three-degree of freedom Hamiltonian

$$H = \frac{1}{2m}(p_x^2 + p_y^2 + p_z^2) + U(x,y,z) . \tag{2.3}$$

Here U is a periodic potential reflecting the crystal symmetry and is usually taken to be the sum of screened Coulombic atomic potentials with the lattice atoms fixed at the perfect crystal lattice sites. For relativistic particles the Hamiltonian is given by $H = c(p_x^2+p_y^2+p_z^2+m^2c^2)^{1/2} + U(x,y,z)$ where c is the speed of light.

This model contains the effect of electrons only to the extent that they screen the nucleus of the lattice atoms; it does not contain the scattering of a channeled particle by individual electrons. Furthermore, this model does not contain the effect of the thermal motion of the lattice atoms. This does not mean these two effects are to be ignored; it does mean that it is H which is primarily responsible for the motion and that electron multiple scattering and thermal vibrations are to be treated as perturbations to the motion. In this sense the perfect crystal model relates to the study of particle channeling as the ideal gas relates to the study of real gases, or as the ideal crystal does to solid state studies. A number of approaches have been used to incorporate these effects into the perfect crystal model. However, it is not clear which approach is best. What is clear is that there are interesting mathematical problems in stochastic processes, diffusion processes, stochastic differential equations, and integro-differential equations. For example, a careful study of channeling Hamiltonian systems excited by white noise would be valuable here.

Within the perfect crystal model, the first problem concerns the use of classical mechanics. It has been argued that channeling becomes more classical as the beam energy increases [2,22]. This is apparently related to the correlated collisions discussed earlier and is in contrast to single scattering in which quantum effects become more important with increasing energy. More mathematical work is needed to understand the classical limit in this context. The next problem concerns the characterization of the motion of channeled particles. Classically, this raises the usual questions concerning Hamiltonian systems, namely: periodic motions, stable and random motions, ergodicity and related statistical questions, and basic perturbation procedures to obtain accurate approximations to the motion over long time intervals. The remainder of this paper will be concerned primarily with the classical case.

3. Perfect Crystal and Continuum Models and the Method of Averaging

When a well-collimated beam is aligned with a crystal axis the particles enter the crystal with a distribution that is roughly uniform in space and Gaussian, with small variance, in angle. Thus it is important to understand the motion generated by H for a fairly large ensemble of initial conditions. Even with extensive numerical calculations this is difficult, and thus in most channeling investigations one or two further assumptions have been made. The first assumption is that the atomic strings can be replaced by continuum strings in which the charge is smeared out along each string; more precisely, the Hamiltonian H of Eq. (2.3) is replaced by

$$H_A = \frac{1}{2m}(p_x^2+p_y^2+p_z^2) + \overline{U}(x,y) , \tag{3.1}$$

where $\overline{U}(x,y) = [\int_0^d U(x,y,z)dz]/d$ is the z average of U and d is defined in Eq. (2.1). Since the z momentum is now conserved, the motion can be studied in the transverse phase space using the two degree of freedom Hamiltonian

$$H_\perp = \frac{1}{2m}(p_x^2+p_y^2) + \overline{U}(x,y) \tag{3.2}$$

with its associated conservation of transverse energy

$$E_\perp = H_\perp(p_x,p_y,x,y) . \tag{3.3}$$

For most potentials $\overline{U}$ of interest in channeling, $H_\perp$ is not integrable.

We shall discuss in detail, in Sections 4, 5 and 6, how the Krylov-Bogoliubov method of averaging can be used to make precise the relation between the classical motions defined by H and H_A . Furthermore, this method gives a systematic perturbation scheme which approximates the solutions of H using the solutions of H_A and replaces, in an approximate way, the effect of the short term periodicity which has been averaged out. Thus we are able to characterize a certain class of solutions of H in terms of the solutions of $H_\perp$. There are qualitative differences between solutions of H and H_A ; for example, trajectories of H can penetrate atomic strings but trajectories of H_A cannot. This is because $\overline{U}(x,y)$ is infinite at string positions. In this light, the most general goal of the method of averaging is to determine the conditions under which the trajectories of the perfect crystal model can be described in terms of the trajectories of the continuum model. Work is in progress to extend this method to the quantum case and some success has been achieved on a model problem [23].

Because of the relation between solutions of H and $H_\perp$ it is important to understand the solutions of $H_\perp$; some progress has been made but more work needs to be done. For example, surface of section calculations have been done [24] in the low transverse energy regime, that is, for those values of $E_\perp$ that trap the particles in a single channel as shown in Fig. 1b. These are the so-called proper channeled particles mentioned earlier. The surface of section portraits show the usual transition from orderly to chaotic behavior as found , for example, in the well known Hénon-Hiles system [25]. This transition is shown for the $\{100\}$ axis of a diamond crystal in Fig. 2. In Fig. 2a through 2d, $E_\perp$ increases from $E_{\perp 1}$ through $E_{\perp 4}$; the location of these energy contours is shown in Fig. 3. For small $E_\perp$ (Fig. 2a), there is the appearance of a second integral of the motion and the trajectories move on two-dimensional invariant tori. In Fig. 2b, the region related to the two hyperbolic points has been replaced by some local chaotic behavior and some new structures have appeared. In both cases there are periodic motions which play a major role in the global behavior, but in general the motion appears to be quasiperiodic. For larger $E_\perp$ (Figs. 2c and 2d) regions of chaotic behavior are evident, however, regions of quasiperiodic motion still exist. The scattered points in 2c and 2d result from one and two trajectories respectively. Calculations have also been done in the $\{110\}$ case [24] and these two cases will be discussed in detail elsewhere. The basic mathematical problems here involve the existence of invariant tori, understanding of periodic solutions, and limit theorems for the phase space density. However, in channeling experiments most channeled particles are not proper channeled, and thus it is important to characterize the motions that wander from channel to channel. One expects these motions to have a strong random character and work is in progress to understand this in a way which will be useful in channeling investigations.

Even the solutions of the two degree of freedom Hamiltonian $H_\perp$ are difficult to characterize, and a second assumption is sometimes made. If $\rho(x,y,p_x,p_y,t)$ denotes the channeled particle phase space density defined by the initial ensemble density $\rho_0 = \rho(x,y,p_x,p_y,0)$ and $H_\perp$, then this density evolves according to the Liouville equation. The assumption is that ρ can be replaced by a time independent density, ρ_{SE} , calculated using the ergodic hypothesis. That is, particles become uniformly distributed on thin energy shells, thus

$$\rho_{SE}(x,y,p_x,p_y)\Delta V(E_\perp) \cong \int_{\Delta V(E_\perp)} \rho_{SE}\, d\mu = \int_{\Delta E_\perp} f(E_\perp)dE_\perp \cong f(E_\perp)\Delta E_\perp \ ,$$

where $V(E_\perp) =$ volume of $\{x,y,p_x,p_y)|H_\perp \leq E_\perp\}$ and f is the transverse energy density. Dividing by $\Delta V(E_\perp)$ and taking the limit as $\Delta E_\perp \to 0$ gives

$$\rho_{SE}(x,y,p_x,p_y) = \frac{f(H_\perp(x,y,p_x,p_y))}{V'\,(H_\perp(x,y,p_x,p_y))}\ . \tag{3.4}$$

The transverse energy density f , which does not change with time, is given by

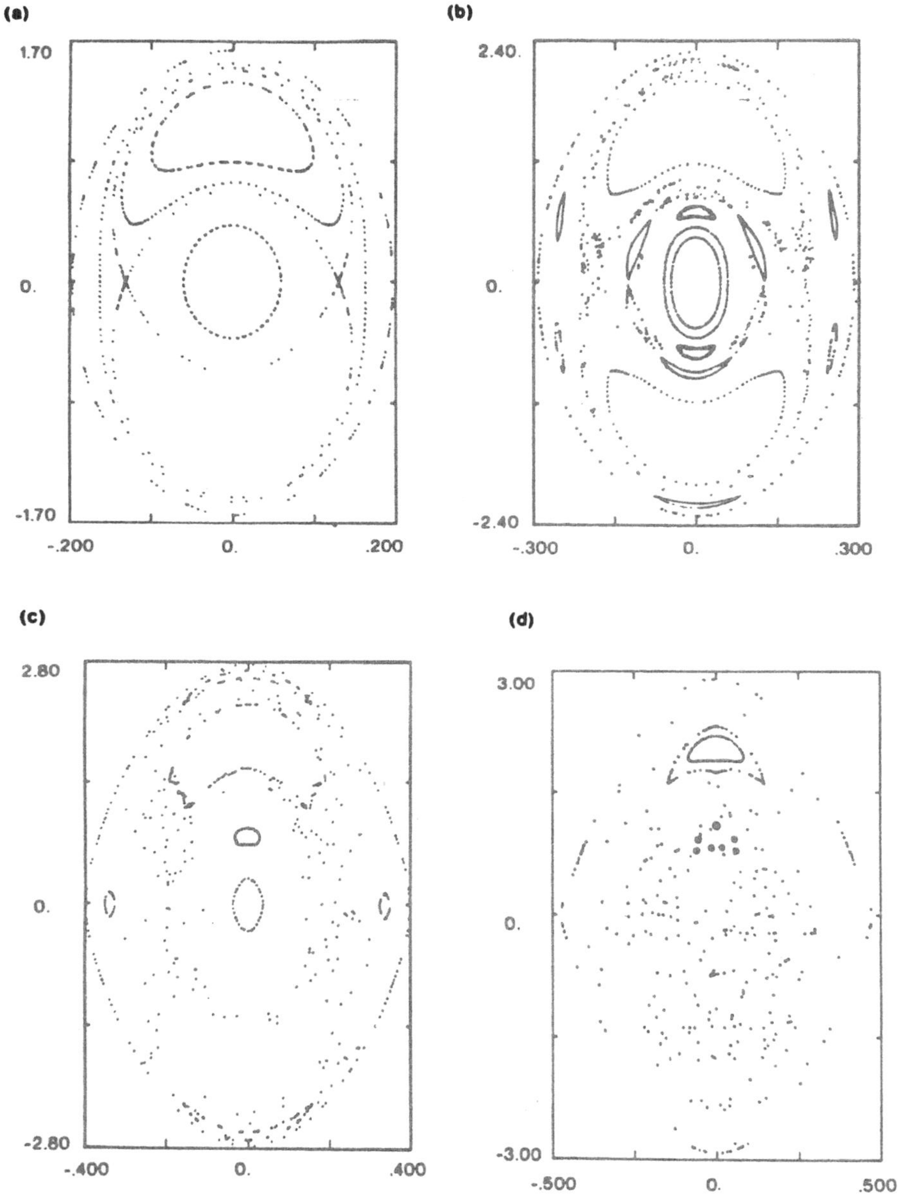

Fig. 2. Surface of section calculations for $\{100\}$ diamond with the Hamiltonian given in (3.2) and the lattice as in Fig. 1. The coordinates are (y, p_y) with $p_z > 0$ and a, b, c, and d correspond to transverse energies $E_{\perp 1}$, $E_{\perp 2}$, $E_{\perp 3}$, and $E_{\perp 4}$ respectively as shown in Fig. 3. The figures are actually symmetric in both y and p_y, but some details have been left off. The reader not familiar with this type of calculation is referred to Ref. [25] or recent books on nonlinear dynamics.

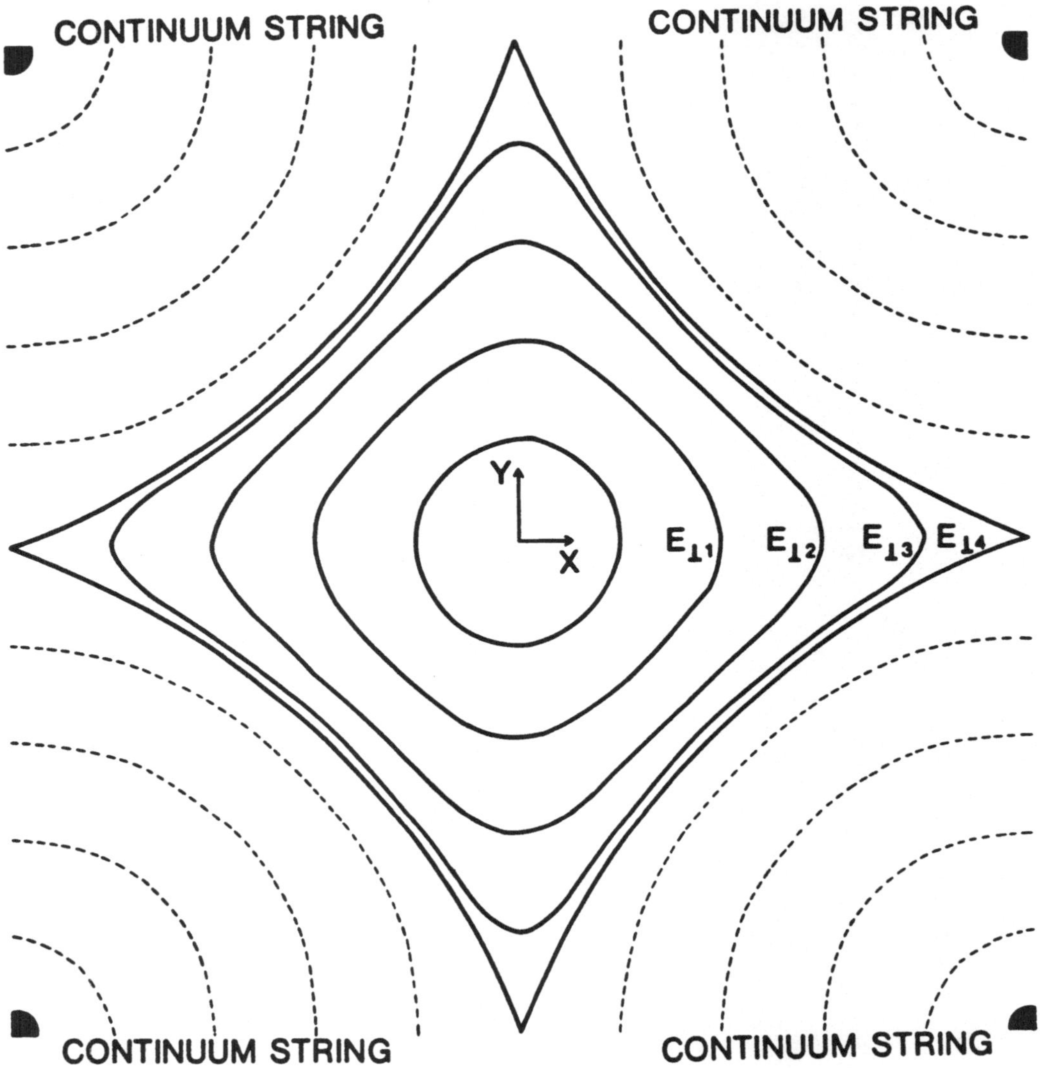

Fig. 3. Equipotential contours for the {100} axis of a diamond crystal for the potential of Eq. (3.2). The continuum strings are shown in the four corners and the contours are also shown in Fig. 1. The four energy contours labled $E_{\perp 1}$, $E_{\perp 2}$, $E_{\perp 3}$, and $E_{\perp 4}$ are the ones referred to in Fig. 2.

$$f(E_\perp) = \lim[\int_{\Delta V(E_\perp)} \rho_0 d\mu]/\Delta E_\perp \text{ as } \Delta E_\perp \rightarrow 0 .$$

Because of the assumptions made up to this point, the relation between ρ_{SE} and the actual channeled particle density is not clear. However, ρ_{SE} has been used with success in lattice location studies [26].

Although the limit of ρ as $t \rightarrow \infty$ does not exist, the time average, $\lim_{T\rightarrow\infty} T^{-1}\int_0^T \rho dt$ does exist and in the ergodic case the limit is ρ_{SE} . However, this is not particularly interesting from a channeling point of view and furthermore, the surface of section calculations indicate that $H_\perp$ is not ergodic, at least for small $E_\perp$. What is interesting in channeling are coarse-grained limits. These are being studied and a special case is described in Ref. [27].

In order to apply the method of averaging to H , the equations of motion need to be put in a standard form for the method. The scaling is motivated by the continuum model; it's ultimate justification is in the error bounds obtained. Let K be the value of the continuum potential at approximately the thermal vibration amplitude away from a string. For $z_1 = 1$ this value is typically about 100 eV. The scaling is then given by

$$\frac{p_x}{\sqrt{2mK}} , \frac{p_y}{\sqrt{2mK}} , \frac{p_z}{\sqrt{2mE}} , \frac{U}{K} , \frac{x}{d} , \frac{y}{d} , \frac{z}{d} , \frac{t}{t_0} , \tag{3.5}$$

where t_0 is the time required to go one lattice spacing in the potential free case and the x , y , z coordinates are shown in Fig. 1b. Letting $KW = U$ but not changing notation for the other variables gives

$$\begin{aligned} \frac{dx}{dt} &= \epsilon\, p_x , \quad \frac{dp_x}{dt} = -\frac{1}{2}\epsilon\, W_x(x,y,z) , \\ \frac{dy}{dt} &= \epsilon p_y , \quad \frac{dp_y}{dt} = -\frac{1}{2}\epsilon\, W_y(x,y,z) , \\ \frac{dz}{dt} &= p_z , \quad \frac{dp_z}{dt} = -\frac{1}{2}\epsilon^2 W_z(x,y,z) , \end{aligned} \tag{3.6}$$

where $\epsilon = (K/E)^{1/2}$ and the conservation law becomes $1 = \epsilon^2 p_x^2 + \epsilon^2 p_y^2 + p_z^2 + \epsilon^2 W(x,y,z)$. For 1 MeV protons, $\epsilon \simeq 0.01$. Eqs. (3.6) are now in the standard form for the method of averaging. For $\epsilon << 1$, x , p_x , y , p_y and p_z are slowly varying functions of time and therefore intuition suggests that over a long interval in the scaled time, the z average of the right hand side of (3.6) should determine most of the long term change in the dependent variables, while the effect of the small oscillations about the average should be less important. The averaged equations for (3.6) are just the equations of motion for H_A scaled according to Eq. (3.5). Notice that Eqs. (3.6) are conservative but not Hamiltonian. A more detailed discussion of the above scaling is given in Refs. [28-29].

In the next section the first order theory of averaging is discussed in general and the procedure for obtaining error bounds is outlined. In Sec. 5 the method of averaging will be applied to Eqs. (3.6).

4. The Method of Averaging

The method of averaging which is used in this channeling study is based on a more general theory developed by N.N. Bogoliubov and Y.A. Mitropolski [30]. An interesting discussion and simplification of some of their proofs is contained in [31] and a treatment by Arnold is given in [32]. Here we follow Kyner's presentation in his Lecture Notes on Nonlinear Resonance [33], which is restricted to the first-order theory. The second-order theory will be applied to the channeling problem in the next section. Understanding the first-order theory should make this generalization straightforward. The notation in this section is independent of the other sections.

We consider a system of ordinary differential equations

$$\frac{dx}{dt} = \epsilon\, X(x,y,\epsilon)\ ,\ x = (x_1, \ldots, x_M)\ ,$$

$$\frac{dy}{dt} = \omega(x) + \epsilon\, Y(x,y,\epsilon)\ ,\ y \text{ a scalar}\ , \tag{4.1}$$

with initial conditions $x(0) = a$, $y(0) = b$. The functions, $X(x,y,\epsilon)$, $Y(x,y,\epsilon)$, are assumed to be smooth and to have period 1 in y . Notice that Eq. (3.6) is of this form and that from the previous discussion the averaged equations associated with Eq. (4.1) would be obtained by averaging $X(x,y,0)$ and $Y(x,y,0)$ over the y variables. The x_m are called slow variables and y is called the fast variable, since if $\epsilon = 0$,

$$\begin{aligned} x_m &= a_m\ ,\ 1 \le m \le M\ , \\ y &= \omega(a)t + b\ . \end{aligned} \tag{4.2}$$

Both Kyner and Arnold discuss the more general case where y is a vector. We treat the scalar case because the vector case is complicated by the possibility of resonance and for the channeling problem presented here there is only one fast variable.

The goal of the method of averaging is to construct a near identity transformation,

$$x = u + \epsilon P(u,v) \ ,$$
$$y = v + \epsilon Q(u,v) \ , \tag{4.3}$$

so that Eqs. (4.1) become

$$\frac{du}{dt} = \epsilon\, U(u) + \epsilon^2 G_1(u,v,\epsilon) \ ,$$
$$\frac{dv}{dt} = \omega(u) + \epsilon\, V(u) + \epsilon^2 G_2(u,v,\epsilon) \ . \tag{4.4}$$

Here u is an M-vector, v a scalar and P and Q have period 1 in v. The initial conditions are $u(0) = a'$, $v(0) = b'$ where a' and b' are defined implicitly by (4.3) with $x(0) = a$ and $y(0) = b$. Notice that the fast variable has been eliminated to $O(\epsilon^2)$ in going from (4.1) to (4.4). As we shall now show, this elimination is an averaging procedure; in fact, we can take

$$U(u) = \int_0^1 X(u,y,0)dy$$
$$V(u) = \int_0^1 Y(u,y,0)dy \ . \tag{4.5}$$

Approximate solutions to (4.1) can be constructed by solving the first order averaged equations,

$$\frac{du^*}{dt} = \epsilon\, U(u^*) \ , \quad u^*(0) = a' \ ,$$
$$\frac{dv^*}{dt} = \omega(u^*) + \epsilon\, V(u^*) \ , \ v^*(0) = b' \ , \tag{4.6}$$

and substituting the solution into (4.3). Equations. (4.6) are obtained from (4.4) by ignoring the $O(\epsilon^2)$ terms.

If we differentiate (4.3) and make use of (4.1) and (4.4), then, after a little algebra, we find the following linear equations for G_1 and G_2 :

$$(I_m + \epsilon P_u)G_1 + \epsilon P_v G_2 = \frac{1}{\epsilon}[X(u+\epsilon P, v+\epsilon Q,\epsilon) - U(u) - P_v\omega(u)] - [P_u U(u) + P_v V(u)] \ ,$$
$$\epsilon Q_u G_1 + (1+\epsilon Q_v)G_2 = \frac{1}{\epsilon^2}[\omega(u+\epsilon P) - \omega(u)] \tag{4.7}$$
$$+ \frac{1}{\epsilon}[Y(u+\epsilon P, v+\epsilon Q,\epsilon) - V(u) - Q_v\omega(u)] - [Q_u U(u) + Q_v V(u)] \ .$$

In order for G_1 and G_2 to be $O(1)$ we must require

$$X(u,v,0) - U(u) = P_v \omega(u) \ ,$$
$$\omega_u(u)P + Y(u,v,0) - V(u) = Q_v \omega(u) \ . \tag{4.8}$$

These equations will determine P and Q, and then (4.7) defines G_1 and G_2.

If P is to be periodic in v, the left side of (4.8a) must have zero mean. From this requirement, we have the first Eq. (4.5). If, in addition, we choose $P(u,v)$ to have zero mean in v, then in order to solve for $Q(u,v)$, we must have the second Eq. (4.5). If we assume the scalar function $\omega(u)$ is not zero and choose $Q(u,v)$ to have zero mean also, then

$$P(u,v) = \frac{1}{\omega(u)} \int_0^1 v' \, [X(u,v+v',0) - U(u)]dv' \ ,$$
$$Q(u,v) = \frac{1}{\omega(u)} \int_0^1 v' \, [\omega_u(u)P(u,v+v') + Y(u,v+v',0) - V(u)]dv' \ , \tag{4.9}$$

as is easily verified by noting that the right hand sides have zero mean and that differentiation with respect to v followed by an integration by parts leads to (4.8). Another representation is obtained by expanding $X(u,v,0)$ and $Y(u,v,0)$ in a Fourier series in v. It is not always convenient to require that $P(u,v)$ and $Q(u,v)$ have zero mean. Morrison [34] has developed a generalized method of averaging in which the arbitrary additive functions play an essential role.

An approximate solution of (4.1) is defined by

$$x^* = u^* + \epsilon \, P(u^*,v^*) \ ,$$
$$y^* = v^* + \epsilon \, Q(u^*,v^*) \ , \tag{4.10}$$

where $u^*(t)$ and $v^*(t)$ satisfy the averaged Eq. (4.6). If, in the domain of interest, P, Q and their derivatives can be bounded by the same constant C, then subtracting (4.10) from (4.3) yields

$$|x-x^*| \leq (1+\epsilon C)|u-u^*| + \epsilon \, C|v-v^*| \ ,$$
$$|y-y^*| \leq \epsilon \, C|u-u^*| + (1+\epsilon C)|v-v^*| \ . \tag{4.11}$$

Thus the comparison of (x,y) and (x^*,y^*) is in terms of $u - u^*$ and $v - v^*$. Let $r = u - u^*$ and $s = v - v^*$. Then subtracting (4.6) from (4.4) yields

$$\frac{dr}{dt} = \epsilon[U(u)-U(u^*)] + \epsilon^2 G_1(u,v,\epsilon)$$
$$\frac{ds}{dt} = \omega(u) - \omega(u^*) + \epsilon(V(u)-V(u^*)) + \epsilon^2 G_2(u,v,\epsilon) \ . \tag{4.12}$$

Assuming that in the domain of interest the derivatives of U, V and ω and the functions G_1 and G_2 are all bounded by the same constant C as before gives the inequalities

$$|r(t)| \leq \epsilon C \int_0^t |r(t')|dt' + \epsilon^2 Ct$$
$$\tag{4.13}$$
$$|s(t)| \leq (1+\epsilon)C \int_0^t |r(t')|dt' + \epsilon^2 Ct .$$

By the generalized Gronwall's inequality [35]

$$|r(t)| \leq \epsilon e^{\epsilon Ct}(1-e^{-\epsilon Ct}) \leq \epsilon^2 Ct e^{\epsilon Ct} . \tag{4.14a}$$

Putting this in the inequality for $|s(t)|$ yields

$$|s(t)| \leq (1+\epsilon)e^{\epsilon Ct}(1-e^{-\epsilon Ct}) \leq (1+\epsilon)\epsilon Ct e^{\epsilon Ct} . \tag{4.14b}$$

Combining (4.14) with (4.11) yields the following bounds on an interval $0 \leq t \leq T/\epsilon$, T fixed,

$$|x(t) - x^*(t)| \leq \epsilon^2 tC^* ,$$
$$\tag{4.15}$$
$$|y(t) - y^*(t)| \leq \epsilon tC^* ,$$

where the constant C^* depends on the bounds C and T. Notice that the errors in x are $O(\epsilon^2)$ for $t \in [0,T]$ and $O(\epsilon)$ for $t \in [0,T/\epsilon]$, whereas the error in the fast variable y is $O(\epsilon)$ on $[0,T]$ and $O(1)$ on $[0,T/\epsilon]$.

In general, the estimates (4.15) are the best possible. However, by taking advantage of the structure of the specific problem improved bounds can sometimes be obtained [37]. Approximation on an infinite interval can be constructed only under the most exceptional circumstances.

5. Second-Order Averaging for Axial Channeling: Formal Discussion

First-order averaging for channeling is discussed in some detail in [28] and [29]. Error bounds in the axial case, under the assumption that perfect crystal trajectories stay away from the strings, are obtained in [29]. Here we discuss second-order averaging in a formal way and in the next section we give a rigorous discussion in a special case. Because the y dependence adds no essential complication in the axial case, we omit the y dependence in Eqs. (3.6). Thus the IVP for Eq. (3.6) becomes

$$\frac{dx}{dt} = \epsilon p_x \ , \qquad x(0) = \xi \ ,$$

$$\frac{dp_x}{dt} = -\frac{1}{2}\epsilon W_x(x,z) \ , \qquad p_x(0) = \eta \ ,$$

$$\frac{dz}{dt} = p_z \ , \qquad z(0) = 0 \ , \tag{5.1}$$

$$\frac{dp_z}{dt} = -\frac{1}{2}\epsilon^2 W_z(x,z) \ , \qquad p_z(0) = [1 - \epsilon^2(\eta^2 + W(\xi,0)]^{1/2} \ ,$$

with conservation law $1 = \epsilon^2 p_x^2 + p_z^2 + \epsilon^2 W(x,z)$. It is natural in the channeling problem to be given $p_x(0)$ and $p_z(0)$, to consider ξ,η to be distributed uniformly over a transverse spatial cell, and to let these define the total energy E . However, from a mathematical point of view, it is more natural to analyze all the trajectories of H for a given energy E , and that is the approach adopted here. Equations. (5.1) can be written in the form

$$\frac{d}{dt}\begin{pmatrix} x \\ p_x \\ p_z \end{pmatrix} = \epsilon \begin{pmatrix} p_x \\ -\frac{1}{2}W_x(x,z) \\ 0 \end{pmatrix} + \epsilon^2 \begin{pmatrix} 0 \\ 0 \\ -\frac{1}{2}W_z(x,z) \end{pmatrix} \ ,$$

$$\frac{d}{dt} z = p_z \ , \tag{5.2}$$

where W_x and W_z have period 1 in z and W_z has zero mean in z . It is convenient to write $W(x,z) = \overline{W}(x) + \phi(x,z)$ where $\overline{W}$ is the z average of W and ϕ has zero mean.

The idea of second order averaging is to look for a transformation of the form

$$\begin{pmatrix} x \\ p_x \\ p_z \end{pmatrix} = \begin{pmatrix} u_1 \\ u_2 \\ u_3 \end{pmatrix} + \epsilon P(u,v) + \epsilon^2 R(u,v) \ ,$$

$$z = v + \epsilon\, Q(u,v) + \epsilon^2 S(u,v) \ , \tag{5.3}$$

with P , Q , R and S periodic in v such that (5.2) becomes its average plus $O(\epsilon^3)$ terms:

$$\frac{d}{dt}\begin{pmatrix}u_1\\u_2\\u_3\end{pmatrix} = \epsilon\begin{pmatrix}u_2\\ -\frac{1}{2}\overline{W}'(u_1)\\ 0\end{pmatrix} + \epsilon^3 G(u,v,\epsilon) \ , \tag{5.4}$$

$$\frac{dv}{dt} = u_3 + \epsilon^3 H(u,v,\epsilon) \ .$$

Differentiating (5.3) and using (5.2) and (5.4) yields

$$\epsilon^3 A(\epsilon)\binom{G}{H} = \begin{pmatrix}\epsilon p_x\\ -\frac{1}{2}\epsilon W_x(x,z)\\ -\frac{1}{2}\epsilon^2 W_z(x,z)\\ p_z\end{pmatrix} - A(\epsilon)\begin{pmatrix}\epsilon u_2\\ -\frac{1}{2}\epsilon\overline{W}'(u_1)\\ 0\\ u_3\end{pmatrix} \ , \tag{5.5}$$

where

$$A(\epsilon) = \begin{bmatrix} I_3 + \epsilon P_u + \epsilon^2 R_u & \epsilon P_v + \epsilon^2 R_v \\ \epsilon Q_u + \epsilon^2 S_u & 1 + \epsilon Q_v + \epsilon^2 S_v \end{bmatrix}$$

and where x , z , p_x , and p_z are given in terms of u and v by Eqs. (5.3). This is analogous to Eq. (4.7). Since G and H are to be $O(1)$ and A is $O(1)$, the right side of (5.5) must vanish through $O(\epsilon^2)$. The $O(1)$ term is identically zero and the $O(\epsilon)$ term gives

$$\begin{aligned} u_3 P_{1v} &= 0 \ , \\ u_3 P_{2v} &= \frac{1}{2}(\overline{W}' - W_x) = -\frac{1}{2}\phi_{u_1}(u_1,v) \ , \\ u_3 P_{3v} &= 0 \ , \\ u_3 Q_v &= P_3 \ . \end{aligned} \tag{5.6}$$

If we require that P and Q have zero mean in v , then $P_1 = P_3 = Q = 0$ and

$$P_2(u_1,u_3,v) = \frac{-1}{2u_3} \int_0^1 v' \, \phi_{u_1}(u_1,v+v')dv' \equiv -\frac{1}{2u_3}\frac{\partial}{\partial u_1}\mathbf{I}_v\phi(u_1,v) \, . \tag{5.7}$$

Here $\mathbf{I}_v$ is the operator which takes a zero mean periodic function of v into its zero mean integral. The results in (5.6) and (5.7) are identical to that obtained in first order averaging. The $O(\epsilon^2)$ term, taking into account the above results, gives

$$\begin{aligned}
u_3R_{1v} &= P_2 \, , \\
u_3R_{2v} &= -u_2\frac{\partial}{\partial u_1}P_2 \, , \\
u_3R_{3v} &= -\frac{1}{2}W_v(u_1,v) = -\frac{1}{2}\phi_v(u_1,v) \, , \\
u_3S_v &= R_3 \, .
\end{aligned} \tag{5.8}$$

If we require R and S to have zero mean then

$$\begin{aligned}
R_1(u_1,u_3,v) &= -\frac{1}{2u_3^2}\frac{\partial}{\partial u_1}\mathbf{I}_v^2\phi \, , \\
R_2(u_1,u_2,u_3,v) &= \frac{u_2}{2u_3^2}\frac{\partial^2}{\partial u_1^2}\mathbf{I}_v^2\phi \, , \\
R_3(u_1,u_3,v) &= -\frac{1}{2u_3}\phi \, , \\
S(u_1,u_3,v) &= -\frac{1}{2u_3^2}\mathbf{I}_v\phi \, ,
\end{aligned} \tag{5.9}$$

where the arguments u_1 and v of ϕ have been suppressed.

In summary, the transformation (5.3) is now defined and the solutions of (5.1) are given approximately by

$$
\begin{aligned}
x \cong x^* &= u_1^* && +\epsilon^2 R_1(u_1^*,u_3^*,v^*)\ , \\
p_x \cong p_x^* &= u_2^* + \epsilon P_2(u_1,^* u_3^*,v^*) && + \epsilon^2 R_2(u_1^*,u_2^*,u_3^*,v^*)\ , \\
p_z \cong p_z^* &= u_3^* && + \epsilon^2 R_3(u_1^*,u_3^* v^*)\ , \\
z \cong z^* &= v^* && + \epsilon^2 S(u_1^*,u_3^*,v^*)\ ,
\end{aligned}
\tag{5.10}
$$

where u^* and v^* are defined by

$$
\begin{aligned}
\frac{du_1^*}{dt} &= \epsilon u_2^*\ , && u_1^*(0) = u_1(0)\ , \\
\frac{du_2^*}{dt} &= -\frac{1}{2}\epsilon \overline{W}'\ (u_1^*)\ , && u_2^*(0) = u_2(0)\ , \\
\frac{du_3^*}{dt} &= 0\ , && u_3^*(0) = u_3(0)\ , \\
\frac{dv^*}{dt} &= u_3^*\ , && v^*(0) = v(0)\ ,
\end{aligned}
\tag{5.11}
$$

and where $u_1(0)$, $u_2(0)$, $u_3(0)$ and $v(0)$ are determined from (5.3). Notice that the continuum model equations are still the basic equations through second order. Error bounds for first order averaging have been discussed in [29] and in the next section we give a detailed error analysis for second order averaging in the case of a simplified potential.

6. Detailed Analysis of a Simple Problem

A. Statement of the Problem

The problem we wish to consider here consists of Eqs. (5.1) where the potential has the specific form

$$
W(x,z) = (A+\frac{1}{2}Bx^2) + (A+\frac{1}{2}Cx^2)\cos 2\pi z\ . \tag{6.1}
$$

Without loss of generality, we require A , C and $B - C$ to be positive, so that $W(x,z) \geq 0$ for all x and z , and $W_{xx}(0,z) = B + C\cos 2\pi z > 0$ for all z . With this potential, Eqs. (5.1) become

$$
\begin{aligned}
&\frac{dx}{dt} = \epsilon p_x \,, && x(0) = \xi \,, \\
&\frac{dp_x}{dt} = -\frac{1}{2}\epsilon(B+C\cos 2\pi z)x \,, && p_x(0) = \eta \,, \\
&\frac{dp_z}{dt} = \pi\epsilon^2(A+\frac{1}{2}Cx^2)\sin 2\pi z \,, && p_z(0) = [1-\epsilon^2(\eta^2+W(\xi,0))]^{1/2} \,, \\
&\frac{dz}{dt} = p_z \,, && z(0) = 0 \,,
\end{aligned}
\tag{6.2}
$$

with conservation law $1 = \epsilon^2 p_x^2 + p_z^2 + \epsilon^2 W(x,z)$.

This problem was chosen to model axial channeling in a simplified way so that the reader's view of the averaging method would not be obscured by the ensuing calculations. Despite its simplicity, the problem retains many of the features of a more complex version to be treated elsewhere by the authors [38]. The origins of this model can be found in Refs. [39-41]. In [39], a theory of resonance dechanneling is developed. In [40], the effect of the z-periodicity in Eq. (3.6) on channeling radiation is investigated. Since the averaging technique provides a systematic way to incorporate this periodicity into the continuum model in a rigorous way, it is an alternate approach to the study in [40]. This will be discussed in detail in Ref. [38]. In Ref. [41] the stability of special channeling trajectories in a two-dimensional lattice with a very general potential is proved using Poincaré maps and the Moser twist theorem. Work is in progress to continue this study by using the averaging procedure to approximate the Poincaré maps [42].

To see the simplification present Eq. (6.1), note that in general, the potential of a two-dimensional lattice which is periodic and symmetric in z has the form $W(x,z) = \sum_{n=0}^{\infty} W_n(x)\cos 2\pi nz$. If this series is truncated at the second term, and if the remaining coefficients are replaced by their Taylor expansions through second order, then we obtain the potential of (6.1).

B. Qualitative Results

Before proceeding to the approximate solution obtained by averaging it is useful to gather together what is known about the exact solution; in fact, the bounds on the solution given at the end of this subsection will be essential to what follows.

(1) Existence and Uniqueness

From the conservation law $1 = p_z^2 + \epsilon^2[p_x^2 + W(x,z)]$ and the form of W, it follows that

$$|p_x| \le \frac{1}{\epsilon}, \quad |p_z| \le 1, \quad |x| \le \frac{1}{\epsilon}\left(\frac{2}{B-C}\right)^{1/2}.$$

Then, using $\frac{dz}{dt} = p_z$, $z(0) = 0$ gives $|z| \le t$. Since the right-hand side of (6.2) is smooth, it is Lipschitz on any compact subset of its domain; we thus have local existence and uniqueness for any set of admissible initial conditions. Using the solution bounds just obtained, this local existence and uniqueness can be made global via standard continuation arguments [43]. We thus have existence and uniqueness of solutions to (6.2) for all times t.

(2) The Equilibrium Solution

We note that the special initial conditions $\xi = \eta = 0$ reduce (6.2) to $x(t) = p_x(t) = 0$ and

$$\frac{dz}{dt} = p_z \quad , \quad z(0) = 0 \, ,$$

$$\frac{dp_z}{dt} = \pi\epsilon^2 A \sin 2\pi z \quad , \quad p_z(0) = [1-\epsilon^2 W(0,0)]^{1/2} \, ,$$

which is a pendulum equation in z. For ϵ small the initial conditions are real and the pendulum is in the rotation regime. The approximate solution obtained in later sections will therefore contain a perturbation expansion of the solution to this problem as a special case. It is the stability of this type of solution that Sáenz [41] has treated rigorously for a class of nonharmonic potentials. Unfortunately, the problem here does not have enough nonlinearity to apply his results.

(3) Mathieu's Equation

Another special case of (6.2) is perhaps of interest. If $O(\epsilon^2)$ terms are ignored, we obtain $p_z = 1$, $z = t$ and

$$\frac{dx}{dt} = \epsilon p_x \ , \ x(0) = \xi \ ,$$
$$\frac{dp_x}{dt} = -\frac{1}{2}\epsilon(B+C\cos 2\pi t)x \ , \ p_x(0) = \eta \ .$$

This is a Mathieu equation and it has unbounded soltuions in certain parameter regions. However, it is easy to show that for $B > C > 0$ and ϵ small the solutions are bounded and this is consistent with the fact that the x and p_x components of (6.2) are bounded.

(4) Refined Solution Bounds

The error estimates for comparing the approximate and exact solutions will require sharper bounds on the exact solution than those given by energy considerations at the outset of this subsection. For this purpose we consider the equivalent integral equation for x and p_x in (6.2)

$$\begin{pmatrix} x(t) \\ p_x(t) \end{pmatrix} = \Phi(t)\begin{pmatrix} \xi \\ \eta \end{pmatrix} + \int_0^t \Phi(t-s)\begin{pmatrix} 0 \\ -\frac{1}{2}\epsilon C[\cos 2\pi z(s)]x(s) \end{pmatrix} ds \ , \qquad (6.3)$$

where $\Phi(t) = \begin{bmatrix} \cos\epsilon wt & 1/w\sin\epsilon wt \\ -w\sin\epsilon wt & \cos\epsilon wt \end{bmatrix}$ and $w = (B/2)^{1/2}$. Taking absolute values on the x-component of (6.3) and applying a version of Gronwall's inequality [35] gives

$$|x(t)| \le (\xi^2+\frac{\eta^2}{w^2})^{1/2}\exp(\epsilon\frac{C}{2w}t) \equiv M_1(\epsilon t) \ . \qquad (6.4)$$

Taking absolute values on the p_x-component and using (6.4) gives

$$|p_x(t)| \le (\eta^2+w^2\xi^2)^{1/2}\exp(\epsilon\frac{C}{2w}t) \equiv M_2(\epsilon t) \ . \qquad (6.5)$$

Integrating the last two equations in (6.2) from 0 to t and using (6.4) gives

$$|p_z(t) - p_z(0)| \le \epsilon M_3(\epsilon t) \qquad (6.6)$$

where $M_3(\epsilon t) = A\pi\epsilon t + \frac{1}{2}\pi w(\xi^2+\frac{\eta^2}{w^2})(exp(\epsilon\frac{C}{w}t)-1)$, and $|z(t) - p_z(0)t| = O(\epsilon^2 t^2)$. It is important to note that for T fixed M_1 , M_2 and M_3 are $O(1)$ on the t-interval $[0, T/\epsilon)$ and on this half open interval $|x(t)| < M_1(T)$, $|p_x(t)| < M_2(T)$ and $|p_z(t) - p_z(0)| < \epsilon M_3(T)$.

C. Averaging Transformation and Approximate Solution

Using the results of Sec. 5 and the potential in (6.1), the averaging transformation becomes

$$
\begin{aligned}
x &= u_1 && && + \epsilon^2 \frac{Cu_1}{8\pi^2 u_3^2}\cos 2\pi v \ , \\
p_x &= u_2 && - \epsilon \frac{Cu_1}{4\pi u_3}\sin 2\pi v && - \epsilon^2 \frac{Cu_2}{8\pi^2 u_3^2}\cos 2\pi v \ , \\
p_z &= u_3 && && - \epsilon^2 \frac{A + \frac{1}{2}Cu_1^2}{2u_3}\cos 2\pi v \ , \\
z &= v && && - \epsilon^2 \frac{A + \frac{1}{2}Cu_1^2}{4\pi u_3^2}\sin 2\pi v \ .
\end{aligned}
\tag{6.7}
$$

This transforms Eq. (6.2) into its average plus $O(\epsilon^3)$ terms:

$$
\begin{aligned}
\frac{du_1}{dt} &= \epsilon u_2 && + \epsilon^3 G_1(u, v, \epsilon) \ , \\
\frac{du_2}{dt} &= -\frac{1}{2}B\epsilon u_1 && + \epsilon^3 G_2(u, v, \epsilon) \ , \\
\frac{du_3}{dt} &= && + \epsilon^3 G_3(u, v, \epsilon) \ , \\
\frac{dv}{dt} &= u_3 && + \epsilon^3 H(u, v, \epsilon) \ ,
\end{aligned}
\tag{6.8a}
$$

with initial conditions

$$u_1(0) = \xi(1-\epsilon^2\frac{C}{8\pi^2}) + O(\epsilon^3) \equiv a + O(\epsilon^3) ,$$

$$u_2(0) = \eta(1+\epsilon^2\frac{C}{8\pi^2}) + O(\epsilon^3) \equiv b + O(\epsilon^3) ,$$

$$u_3(0) = 1 - \frac{1}{2}\epsilon^2[\eta^2+A+\frac{1}{2}B\xi^2] + O(\epsilon^3) \equiv c + O(\epsilon^3) , \tag{6.8b}$$

$$v(0) = O(\epsilon^3) .$$

Dropping the $O(\epsilon^3)$ terms gives the averaged IVP

$$\frac{du_1^*}{dt} = \epsilon u_2^* , \qquad u_1^*(0) = a ,$$

$$\frac{du_2^*}{dt} = \frac{1}{2}\epsilon B u_1^* , \qquad u_2^*(0) = b ,$$

$$\frac{du_3^*}{dt} = 0 , \qquad u_3^*(0) = c , \tag{6.9}$$

$$\frac{dv^*}{dt} = u_3^* , \qquad v^*(0) = 0 ,$$

with solution

$$\begin{pmatrix} u_1^*(t) \\ u_2^*(t) \end{pmatrix} = \Phi(t)\begin{pmatrix} a \\ b \end{pmatrix} ,$$

$$u_3^*(t) = c , \tag{6.10}$$

$$v^*(t) = ct ,$$

where $\Phi(t)$ is defined in (6.3). Finally, as seen in (5.10) and (5.11), the approximate solution of (6.2) is obtained by inserting (6.10) into the transformation (6.7).

In the special case where $\xi = \eta = 0$, we have $u_1^* = u_2^* = 0$, $u_3^*(t) = 1$ and $v^*(t) = t$. Thus the approximate solution of (6.2) becomes

$$\begin{aligned} x^* &= 0 , \\ p_x^* &= 0 , \\ p_z^*(t) &= 1 - \epsilon^2 \frac{A}{2} \cos 2\pi t , \\ z^*(t) &= t - \epsilon^2 \frac{A}{4\pi} \sin 2\pi t , \end{aligned} \tag{6.11}$$

which gives the previously mentioned approximate solution of the pendulum equation.

D. Error Bounds

The ultimate test of any approximation scheme lies in how close the approximate solution is to the actual solution. In this final subsection we give error estimates which prove that for the slow variables the averaging procedure is good to order ϵ^3 on time intervals of order one, or order ϵ^2 on intervals $[0, T/\epsilon)$ for fixed T.

The exact and approximate solutions are related by

$$\begin{aligned} x - x^* &= u_1 - u_1^* && + \epsilon^2(R_1(u,v) - R_1(u^*,v^*)) , \\ p_x - p_x^* &= u_2 - u_2^* + \epsilon(P_2(u,v) - P_2(u^*,v^*)) && + \epsilon^2(R_2(u,v) - R_2(u^*,v^*)) , \\ p_z - p_z^* &= u_3 - u_3^* && + \epsilon^2(R_3(u,v) - R_3(u^*,v^*)) , \\ z - z^* &= v - v^* && + \epsilon^2(S(u,v) - S(u^*,v^*)) , \end{aligned} \tag{6.12}$$

where P_2 , R_1 , R_2 , R_3 and S are given in Eq. (6.7). To proceed with the error analysis, we need to find a subset $\Omega_u \subset \mathbb{R}^4$ such that $(u(t),v(t))$ and $(u^*(t),v^*(t))$ are in Ω_u for $0 \le t < T/\epsilon$, where these are solutions of (6.8) and (6.9) respectively, and such that P , R and S and their derivatives are bounded by some constant C on Ω_u . Assuming this for the moment (6.12) yields

$$
\begin{aligned}
|x-x^*| &\le (1+\epsilon^2 C)|u_1-u_1^*| + \epsilon^2 C|u_3-u_3^*| + \epsilon^2 C|v-v^*| , \\
|p_x-p_x^*| &\le \epsilon C|u_1-u_1^*| + (1+\epsilon^2 C)|u_2-u_2^*| + (\epsilon C+\epsilon^2 C)|u_3-u_3^*| + (\epsilon C+\epsilon^2 C)|v-v^*| , \\
|p_z-p_z^*| &\le \epsilon^2 C|u_1-u_1^*| + (1+\epsilon^2 C)|u_3-u_3^*| + \epsilon^2 C|v-v^*| , \\
|z-z^*| &\le \epsilon^2 C|u_1-u_1^*| + \epsilon^2 C|u_3-u_3^*| + (1+\epsilon^2 C)|v-v^*| .
\end{aligned}
\tag{6.13}
$$

Because $(u(t),v(t))$ and $(u^*(t),v^*(t))$ satisfy (6.8) and (6.9) respectively, it follows that

$$
\begin{aligned}
\begin{pmatrix} u_1(t)-u_1^*(t) \\ u_2(t)-u_2^*(t) \end{pmatrix} &= \begin{pmatrix} u_1(0)-u_1^*(0) \\ u_2(0)-u_2^*(0) \end{pmatrix} + \epsilon^3 \int_0^t \Phi(t-s) \begin{pmatrix} G_1(s) \\ G_2(s) \end{pmatrix} ds , \\
u_3(t)-u_3^*(t) &= u_3(0)-u_3^*(0) + \epsilon^3 \int_0^t G_3(s)\,ds , \\
v(t)-v^*(t) &= v(0)-v^*(0) + \int_0^t (u_3(s)-u_3^*(s))\,ds + \epsilon^3 \int_0^t H(s)\,ds ,
\end{aligned}
\tag{6.14}
$$

where G and H are functions of s through their dependence on u and v as indicated in (6.8a). From Eq. (6.8b) the difference in initial data is $O(\epsilon^3)$, from Eq. (6.3) $\Phi(t)$ is bounded and if we also assume that $G(u,v,\epsilon)$ and $H(u,v,\epsilon)$ are bounded for $(u(t),v(t))$ in Ω_u and $0 \le t < T/\epsilon$, then

$$
\begin{aligned}
|u_1(t)-u_1^*(t)| &= O(\epsilon^3) + O(\epsilon^3 t)\,, \\
|u_2(t)-u_2^*(t)| &= O(\epsilon^3) + O(\epsilon^3 t)\,, \\
|u_3(t)-u_3^*(t)| &= O(\epsilon^3) + O(\epsilon^3 t)\,, \\
|v(t)-v^*(t)| &= O(\epsilon^3) + O(\epsilon^3 t^2) + O(\epsilon^3 t)\,.
\end{aligned}
\tag{6.15}
$$

for $0 \le t < T/\epsilon$. These bounds are better than the exponential type bounds obtained in the general case in Sec. IV because of the representation (6.14) using $\Phi(t)$. This improvement is the same as obtained by Kirchgraber [37] in a more general setting.

Combining (6.13) and (6.15) gives

$$
\begin{aligned}
|x-x^*|\,,\ |p_x-p_x^*|\,,\ |p_z-p_z^*| &= O(\epsilon^3 t)\,, \\
|z-z^*| &= O(\epsilon^3 t^2)\,,
\end{aligned}
\tag{6.16}
$$

for $0 \le t < T/\epsilon$, which is the desired result. That is, x^* , p_x^* , p_z^* are $O(\epsilon^3)$ approximations on order one intervals and $O(\epsilon^2)$ approximations on order $\frac{1}{\epsilon}$ intervals. z^* is an $O(\epsilon^3)$ approximation on order one intervals but only $O(\epsilon)$ on $\frac{1}{\epsilon}$ intervals. Order constants can be determined but this suffices to illustrate the procedure.

We now proceed to find the domain Ω_u . From (6.4) through (6.6) it follows that the solutions of (6.2) stay in the domain

$$
\Omega_x = \{(x,p_x,p_z,z) : |x| < M_1(T)\ ,\ |p_x| < M_2(T)\ ,\ |p_z-\delta| < \epsilon M_3(T)\ ,\ z \in \mathbb{R}\}
$$

for $0 \le t < T/\epsilon$ where $\delta = p_z(0)$. Since (6.7) is a near identity transformation for ϵ small, the solutions of (6.8) are expected to remain in a slightly larger domain. With this in mimd, we roughly double the size of Ω_x and let

$$
\Omega_u = \{(u_1,u_2,u_3,v) : |u_1| < 2M_1(T)\ ,\ |u_2| < 2M_2(T)\ ,\ |u_3-\delta| < \frac{1}{2}\delta\ ,\ v \in \mathbb{R}\}\,.
$$

The requirement $|u_3-\delta| < \frac{1}{2}\delta$ keeps u_3 away from zero so that (6.7) is well defined. Furthermore, it is easy to check that the solutions of (6.9) given in (6.10) remain in Ω_u , that P , R , S and their derivatives are bounded on Ω_u and that G and H are bounded on Ω_u . It remains to show that $(u(t),v(t))$ remains in Ω_u for $0 \leq t < T/\epsilon$. To show this it suffices to show that for ϵ sufficiently small the mapping $T : \Omega_u \rightarrow \mathbb{R}^4$ defined by (6.7) is such that $T^{-1}\Omega_x \subset \Omega_u$ and is a bijection from $T^{-1}\Omega_x$ to Ω_x . More precisely, let X denote (x,p_x,p_z,z) and U denote (u_1,u_2,u_3,v) , then given $X \in \Omega_x$, there is a unique $U \in \Omega_u$ such that $X = \mathbf{T}U$. To prove this, fix $X \in \Omega_x$ and define the mapping $Q_X : \Omega_u \rightarrow \mathbb{R}^4$ by $Q_X(U) = X + U - \mathbf{T}U$. Since $\mathbf{T}$ is a near-identity transformation for small ϵ , one easily checks the existence of an ϵ_1 , independent of $X \in \Omega_x$, such that Q_X is a contraction map [44] for $0 \leq \epsilon \leq \epsilon_1$. Thus for each $X \in \Omega_x$ there exists a unique $U \in \Omega_u$ such that $Q_X(U) = U$, but this implies there is a unique U such that $X = \mathbf{T}U$. Since the solution of (6.2) stays in Ω_x , and for each $X \in \Omega_x$, $U = \mathbf{T}^{-1}X \in \Omega_u$, it follows that the solution of (6.8) stays in Ω_u . This completes the error analysis.

Acknowledgments

Discussions with A. Ben-Lemlih, W.T. Kyner, S.T. Picraux, A.W. Sáenz, and E. Uggerhøj are gratefully acknowledged.

References

[1] The axial critical angle was derived by Lindhard and the planar critical angle was derived using Lindhard's standard potential [2]. For nonrelativistic particles $pv = mv^2$ and pv is usually replaced in the formulas by $2E$ where E denotes the beam kinetic energy.

[2] J. Lindhard, Kgl. Danske Videnskab. Selskab., Mat.-fys. Medd **34**, No. 14 (1965).

[3] D.S. Gemmell, Rev. Mod. Phys. **46**, 129 (1974).

[4] *Channeling,* edited by D.V. Morgan (Wiley, New York, 1973).

[5] L.C. Feldman, J.W. Mayer and S.T. Picraux, *Materials Analysis By Ion Channeling: Submicron Crystallography* (Academic Press, New York, 1982).

[6] Y.H. Ohtsuki, *Charged Beam Interaction with Solids* (Taylor and Francis, New York, 1983).

[7] See, e.g., W.M. Gibson and J.A. Golovchenko, Phys. Rev. Lett. **28**, 1301 (1972).

[8] V.V. Beloshitsky and F.F. Komarov, Phys. Rep. **93**, 117 (1982).

[9] J.U. Andersen, E. Bonderup and R.H. Pantell, Ann. Rev. Nucl. Part. Sci. **33**, 453 (1983).

[10] *Coherent Radiation Sources,* edited by A.W. Sáenz and H. Überall, *Topics in Current Physics* , 38 (Springer, Berlin, Heidelberg 1985).

[11] J.C. Kimbell, N. Cue, L.M. Roth, and B.B. Marsh, Phys. Rev. Lett. **50**, 950 (1983). See also the two articles by Cue and Kimball in Nucl. Instrum. Meth. **B2**, 25-34 (1984).

[12] N. Cue and J.C. Kimble, private communication.

[13] E. Uggerhøj, private communication.

[14] J.U. Andersen, A.S Jensen, K. Jorgensen, E. Laegsgaard, K.O. Nielsen, J.S. Forster, I.V. Mitchell, D. Ward, W.M. Gibson, and J.J. Cuomo, Kgl. Danske Videnskab. Selskab, Mat.-fys. Medd., **40**, No. 7 (1980).

[15] A.F. Elishev et al., Phys. Lett. **88B**, 387 (1979).

[16] J.F. Bak, P.R. Jensen, H. Madsbøll, S.P. Møller, H.E. Schiott, E. Uggerhøj, J.J. Grob and P. Siffert, Nucl. Phys. **B242**, 1 (1984). See also J. Bak et al., Phys. Lett. **93B**, 387 (1980).

[17] R.A. Carrigan, Jr., W.M. Gibson, C.R. Sun and E.N. Tsyganov, Nucl. Instr. and Meth. **194**, 205 (1982); J.A. Ellison, S.I. Baker, R.A. Carrigan, Jr., J.S. Forster, I.V. Mitchell, W.M. Gibson, I.J. Kim, M. Pisharody, S. Salman, C.R. Sun, and R. Wijayawardana, Nucl. Instrum. Meth. **B2**, 9, (1984). See also Chapter 10 of [10] and references therein.

[18] S.I. Baker, R.A. Carrigan, Jr., R. Schailey, T.E. Toohig, W.M. Gibson, I.J. Kim, F. Sun, C.R. Sun, V. Tanikella, R. Wijayawardana, J.S. Forster, H. Hatton, I.V. Mitchell, and J.A. Ellison, Nucl. Instrum. Meth, (to be published).

[19] J. Bak, J.A. Ellison, B. Marsh, F.E. Meyer, O. Pedersen, J.B.B. Petersen, E. Uggerhøj, K. Østergard, S.P. Møller, A.W. Sørensen, and M. Suffert, Nucl. Phys. **B** (to be published).

[20] W.K. Chu, J.A. Ellison, S.T. Picraux, R.M. Biefeld, and G.C. Osbourn, Phys. Rev. Lett. **52**, 125 (1984); S.T. Picraux, L.R. Dawson, G.C. Osbourn, and R.M. Biefeld, Appl. Phys. Lett. **43**. 1020 (1983), and references therein.

[21] See p. 9-11 of [2].

[22] P. Lervig, J. Lindhard, and V. Nielsen, Nucl. Phys. **A96**, 481 (1967); L.T. Chadderton, J. Appl. Cryst. **3**, 429 (1970); M.V. Berry, J. Phys. **C4**, 697 (1971).

[23] A. Ben-Lemlih and J.A. Ellison, private communication.

[24] J. Su and C. Seal, master's degree work at the University of New Mexico (unpublished).

[25] M. Hénon and C. Heiles, Astron. J. **69**, 73 (1964).

[26] See Chapter 3 of [5] and in particular Secs. 3.7 and 3.8.

[27] J.A. Ellison and T. Guinn, Phys. Rev. **B18**, 5963 (1978).

[28] J.A. Ellison, University of New Mexico Technical Report No. 300, 1974 (unpublished).

[29] T.J. Burns and J.A. Ellison, Phys. Rev. **B29**, 2790 (1984).

[30] N.N. Bogoliubov and Y.A. Mitropolsky, *Asymptotic Methods in the Theory of Nonlinear Oscillations* (Gordon and Breach, New York, 1961), Chaps. 5 and 6.

[31] P. Swinnerton-Dyer, Proc. London Math. Soc. (3) **34**, 385 (1977).

[32] V.I. Arnold, *Geometrical Methods in the Theory of Ordinary Differential Equations* (Springer, New York, 1983).

[33] W.T. Kyner, *Third Compilation of Papers On Trajectory Analysis and Guidance Theory,* NASA Report, PM-81, August 1969. See also SIAM J. Appl. Math. **13**, 136 (1965).

[34] J.A. Morrison, SIAM J. **13**, 96 (1965).

[35] See, e.g., [36], Lemma 6.2, p. 36.

[36] J.K. Hale, *Ordinary Differential Equations* (Wiley-Interscience, New York, 1969).

[37] See , e.g., U. Kirchgraber, Celestial Mechanics, **14**, 351 (1976).

[38] H.S. Dumas and J.A. Ellison, (in preparation).

[39] M.A. Kumakhov and R. Wedell, Phys. Stat. Sol. (b) **76**, 119 (1976); K. Lenkeit and R. Wedell, Phys. Stat. Sol. (b) **98**, 235 (1980).

[40] G. Kurizki and J.K McIver, Phys. Lett. **89A**, 43 (1982).

[41] A.W. Sáenz, Phys. Lett. **93A**, 271 (1983) and J. Math. Phys. (to be published).

[42] J.A. Ellison and A.W. Sáenz, private communication.

[43] See ,e.g., [36], Theorem 2.1, p.17.

[44] W. Rudin, *Principles of Mathematical Analysis* (McGraw-Hill, New York, 1976), p. 220.

RIGOROUS STABILITY RESULTS ON CRYSTAL CHANNELING VIA CANONICAL MAPS

A. W. Sáenz
Naval Research Laboratory
Washington, D.C. 20375-5000
and
Physics Department
Catholic University
Washington, DC 20064

1. INTRODUCTION

Suppose that a fast positively charged particle is injected into a plane lattice of atoms. Under reasonable assumptions on the repulsive interactions between the particle and the atoms of the array, is it possible for the particle to travel stably, in some sense, between two lines of atoms? Numerical studies suggest strongly an affirmative answer in the framework of classical mechanics [1,2]. The present investigation was motivated by the desire to answer in a mathematically rigorous way questions of this type in the context of well-defined classical Hamiltonian models.

Motions of fast charged particles in crystals which, e.g., remain confined between lines or planes of atoms, at least over distances which are large compared to the pertinent atomic spacings, are called channeling motions, or simply channeling. This is an old subject in physics to which an enormous number of experimental and theoretical investigations have been devoted (see, e.g., [1,2]). Nevertheless, channeling phenomena are for the most part not well understood mathematically at the present time.

Channeling motions in crystals have been studied rigorously in [3-5]. The first rigorous theorems on channeling stability were obtained in [3,4] by a fixed-point method. Ellison and his collaborators (see, e.g., [5] and the bibliography therein) have derived important rigorous results on channeling motions by averaging-theory arguments. The theory of averaging is of very general applicability in a channeling context, but save in exceptional circumstances, its conclusions are only known to hold over finite, although typically long time intervals, as is the case with the results obtained in [5]. On the other hand, the above fixed-point approach leads to stability theorems holding over infinite time intervals, but up to now it has only been possible to apply it to rather special channeling motions. It seems that much could be gained in our theoretical understanding of channeling phenomena by adroitly combining these two methods.

It is well known that in the present state of mathematical knowledge strong stability theorems for Hamiltonian systems generally can only be proved for systems with $\leqslant 2$ degrees of freedom. Instability results for such conservative systems are typically much easier to prove, no limitation on the numbers of degrees of freedom being necessary in general.

A prime objective of the present work has been to prove the orbital stability of certain distinguished

Remark: Not surprisingly from an intuitive viewpoint, one can show that if conditions (I), (II), and $A_2 < 0$ obtain, the distinguished solution defined in Theorem 1 is orbitally unstable for each $\rho \in \mathbb{R}$ if E is sufficiently large. By the Hartman-Grobman theorem for maps, this statement follows by proving that under these conditions the point $\zeta_H(\rho,E)$, considered as a fixed point of the map $\mathcal{P}(\rho,E)$ (see Section 2A) for $H = H_{NR}$ and of an analogous map for $H = H_R$, is hyperbolic at each such ρ,E. On the other hand, the proof of the orbital stability property asserted in the present theorem for the case when $A_2 > 0$ is more involved, since stability properties of solutions of nonlinear Hamiltonian systems cannot generally be proved on the basis of linearized treatments.

B. Strategy of Proof

Idea of the Proof of Theorem 1:

We will only sketch the proof of this theorem for the NR-model, since that for the R-model is similar.

For large enough E, there is a well defined canonical map $\mathcal{P}(\rho,E):(\xi,\eta) \rightarrow (\xi',\eta')$ for (ξ,η) in a small enough neighborhood of the origin in $\mathbb{R}^2$. Here $(\rho,\xi,p_1 > 0,\eta)$ is the point on the solution curve $(x_1^H(t;z,E),x_2^H(t;z,E),p_1^H(t;z,E),\ p_2^H(t;z,E))$ at which this curve intersects the hyperplane $x_1 = \rho$ at $t = 0$ and $(\rho + 1,\xi',p_1 > 0,\eta')$ the point on this curve at which it intersects the hyperplane $x_1 = \rho + 1$.

In the present case $H = H_{NR}$, we will denote $x_i^H(t;z,E)$, $p_i^H(t;z,E)$, t_0^H, and U_H by $x_i(t;z,E)$, $p_i(t;z,E)$, t_0 and U, dropping the subscripts and superscripts H. If E_1 is a sufficiently large constant, then for each $z \in \mathbb{R} \times U$, $E > E_1$ the function

$$u = x_1(t;z,E)$$

is a diffeomorphism which maps $(-t_0,t_0)$ bijectively onto an open interval containing $[\rho,\rho + 1]$. The existence of this diffeomorphism allows us to define the functions

$$X(u;z,\epsilon) = x_2(t;z,E),$$

$$P(u;z,\epsilon) = p_2(t;z,E),$$

which satisfy the differential equations

$$\frac{dX}{du} = \epsilon\,\frac{P}{\sigma(u,X,P,\epsilon)}, \tag{8a}$$

$$\frac{dP}{du} = \epsilon\,\frac{f_2(u,X)}{\sigma(u,X,P,\epsilon)} \tag{8b}$$

channeling motions described by the Hamiltonian models of two degrees of freedom defined below. Models of this type are capable of describing a rich variety of two- and three-dimensional channeling phenomena, and hence are of physical as well as mathematical interest. Furthermore, the canonical map methods used in this paper can be applied to realistic three-dimensional channeling models, leading to interesting conclusions on channeling instability, as will be discussed in a later publication.

One of the channeling models considered here is nonrelativistic (NR-model) and the other is relativistic (R-model). They are described by the respective Hamiltonians H_{NR} and H_R, which are smooth real-valued functions defined at each $(x_1,x_2,p_1,p_2) \in \mathbb{R}^4$ such that (x_1,x_2) is in a neighborhood of the line $x_2 = 0$ in $\mathbb{R}^2$:

$$H_{NR}\,(x_1,x_2,p_1,p_2) = \frac{1}{2}\,(p_1^2 + p_2^2) + V(x_1,x_2),$$

$$H_R\,(x_1,x_2,p_1,p_2) = \sqrt{p_1^2 + p_2^2 + 1} + V\,(x_1,x_2).$$

We will assume throughout that the potential $V(x_1,x_2)$ is a real-valued analytic function of x_1,x_2 jointly* in such a neighborhood, has period unity in x_1, is such that the x_2-component of the force on the particle of interest vanishes *on the average* along the line $x_2 = 0$, and that the average of the x_2-derivative of this force component along this line is nonzero. The third of these assumptions is substantially weaker than the corresponding condition imposed in previous work [3,4] that this force component vanish *pointwise* along the latter line.

Theorem 1 of Section 2A asserts that at high enough particle energy E there exist for all time certain distinguished solutions of the equations of motion for H_{NR} and H_R whose corresponding phase-space orbits are periodic in x_1 and are such that their x_2 and p_2 coordinates tend to zero as $E \to \infty$. Because we do not require that the above condition on the pointwise vanishing of the x_2-component of the force hold, we can no longer assert trivially, as in [3,4], the existence of such distinguished solutions at sufficiently large E. Their existence can be proved by using well known theorems ([6], Theorem 3.2, p. 194 or [7], Theorem 3).** The latter theorem is particularly convenient for our present purposes and is the main ingredient used to prove our Theorem 1.

Under the assumptions of Theorem 1 plus certain natural linear and nonlinear stability conditions, the orbital stability of the distinguished solutions defined in that theorem is asserted by Theorem 2 of Section 2A, which is our main result. In its proof, an essential role is played by an appropriate version of Moser's twist theorem in [8], p. 228.

*In this paper, analyticity of the relvant functions will always be understood as joint analyticity in the pertinent arguments.

**A theorem encompassing these two fundamental theorems has been proved recently by the author. The method of proof is simpler than in the original treatments in [6] and [7]. It is based on elementary inequalities and a simple contraction-mapping argument; no appeal to results of the theory of averaging is made in the proof.

2. STATEMENT OF PRINCIPAL RESULTS AND STRATEGY OF PROOF

A. Principal Results

In this subsection, we will state our assumptions on V more precisely. Throughout the paper, we will always suppose that V satisfies the following two conditions:

(I) V is a real-valued function which is analytic in the real variables x_1,x_2 in a strip

$$S \in \{(x_1,x_2) \in \mathbb{R}^2 : |x_2| < \alpha\}, \tag{1}$$

for some constant $\alpha > 0$.

(II) In the strip (1), V is periodic in x_1 with period unity and

$$A_1 = 0, \tag{2}$$

where

$$A_j = \int_0^1 \partial^j V(x_1,0)/\partial x_2^j \; dx_1.$$

In the proof of our main theorem we will assume, in addition, that

(III) The following inequalities hold:

$$A_2 > 0, \tag{3a}$$

$$A_2 A_4 - \frac{5}{3} A_3^2 \neq 0. \tag{3b}$$

Conditions (I) and (III) were also imposed in [3,4], where their motivation was discussed. In particular, Eqs. (3a) and (3b) are linear and nonlinear stability conditions, respectively. Condition (2) evidently states that the x_2-component of the force on the particle vanishes on the average along the line $x_2 = 0$.

In what follows, H will stand for H_{NR} or H_R.

For each quadruple ρ,ξ,η,E of real numbers with $(\rho,\eta) \in S$ for which they exist, we denote by $x_i^H(t;z,E)$, $p_i^H(t;z,E)$ $(i = 1,2)$ functions having the following properties at each t in a maximal open interval of $\mathbb{R}$ containing the origin: they are differentiable in t, are such that $(x_1^H(t;z,E),\ x_2^H(t;z,E)) \in S$, satisfy the Hamiltonian equations of motion

$$x_i = \frac{\partial H}{\partial p_i},\ p_i = -\frac{\partial H}{\partial x_i},\ i = 1,2, \tag{4}$$

and the initial conditions

$$x_1^H(0;z,E) = \rho,\ x_2^H(0;z,E) = \xi,$$

$$p_1^H(0;z,E) > 0,\ p_2^H\ (0;z,E) = \eta, \tag{5}$$

and lie on the energy surface, $H = E$. Here $z = (\rho,\zeta)$, where $\zeta = (\xi,\eta)$.

One can show that system (4) has exactly one solution $x_i^H(t;z,E), p_i^H(t;z,E)$ $(i = 1,2)$ for $|t| < t_0^H$, $z \in \mathbb{R} \times U_H$, $E > E_0^H$, which is analytic in t,ρ,ξ,η,E at these values of its arguments. Here $U_H \subset \mathbb{R}^2$ is a neighborhood of the origin and E_0^H a sufficiently large constant, t_0^H and U_H being independent of ρ,E. At each such z, E, there is a unique time $\tau_H(z,E)$ at which this solution of (4), emerging from the hyperplane $x_1 = \rho$ at $t = 0$, cuts the hyperplane $x_1 = \rho + 1$, i.e.,

$$x_1^H(\tau_H(z,E);z,E) = \rho + 1.$$

The motivation for considering these two hyperplanes (surfaces of section) is that $V(x_1,x_2)$ is 1-periodic in x_1.

As preparation for the statement of our main theorem, we state

Theorem 1: Let V obey conditions (I), (II), and

$$A_2 \neq 0. \tag{6}$$

Then there exist a positive constant E_H and a neighborhood D_H of the origin in $\mathbb{R}^2$, such that at each $\rho \in \mathbb{R}, E > E_H$ there is exactly one point $\zeta_H(\rho,E) \in D_H$ with the properties that for $\zeta = \zeta_H(\rho,E)$ the solution $x_i^H(t;z,E)$, $p_i^H(t;z,E)$ $(i = 1,2)$ exists for all $t \in \mathbb{R}$ and that $x_2^H(t;z,E)$ (and hence $p_2^H(t;z,E)$) is periodic in t with period $\tau_H((\rho, \zeta_H(\rho,E)),E)$. Moreover, writing $\zeta_H(\rho,E) = (\xi_E(\rho,E),\eta_H(\rho,E))$, ξ_H and η_H are analytic functions of ρ,E for $\rho \in \mathbb{R},E > E_H$ having the asymptotic behavior

$$\xi_H(\rho,E) = O(E^{-1}),\ H = H_{\mathrm{NR}},H_{\mathrm{R}}, \tag{7a}$$

$$\eta_H(\rho,E) = \begin{cases} (2E)^{-1/2}h(\rho) + O(E^{-3/2}),\ H = H_{\mathrm{NR}}, \\ h(\rho) + O(E^{-1}),\ \ H = H_{\mathrm{R}}, \end{cases} \tag{7b}$$

for $E \to \infty$, uniformly with respect to ρ on $\mathbb{R}$, where

$$h(\rho) = \int_\rho^{\rho+1} u \frac{\partial V(u,0)}{\partial x_2}\ du.$$

Denote the distinguished solution in Theorem 1 by $\tilde{x}_i^H(t;\rho,E)$, $\tilde{p}_i^H(t;\rho,E) (i = 1,2)$. Let $O_+(z,E)$ be the phase-space orbit, if it exists, swept out during the time interval $t \geqslant 0$ by a particle of energy E subject to the initial conditions (5), i.e., $O_+(z,E) = \{(x_1,x_2,p_1,p_2):x_i = x_i^H(t;z,E),\ p_i = p_i^H(t;z,E)$ for all $i = 1,2$ and some $0 \leqslant t < \infty\}$. We then have

Theorem 2: Let V satisfy conditions (I)-(III). Then there exists a constant $E'_H \geqslant E_H$ such that the solution $\tilde{x}_i^H(t;\rho_0,E_0)$, $\tilde{p}_i^H(t;\rho_0,E_0)$ $(i = 1,2)$ is (future) orbitally stable for $\rho_0 \in \mathbb{R},E_0 > E'_H$. That is, given $\epsilon > 0$, $\rho_0 \in \mathbb{R},E_0 > E'_H$, there exists $\delta = \delta(\epsilon,\rho_0 E_0)$ such that the orbit $O_+(z,E)$ exists and each of its points p satisfies $d(p,O_+(z_0,E_0)) < \epsilon$ if $d((z,E),(z_0,E_0)) < \delta$, where d is the usual $\mathbb{R}^4$ distance function and $z_0 = (\rho_0,\zeta_H\ (\rho_0,E_0))$.

of Hamiltonian type, where $\epsilon = (2E)^{-1/2}$, $f_2 = -\partial V/\partial x_2$, and

$$\sigma(u,X,P,\epsilon) = \sqrt{1-\epsilon^2[2V(u,X)+P^2]}. \tag{9}$$

This is a simple example of isoenergetic reduction (see, e.g., [9], Secs. 180-182).

The integral equations

$$X(u;z,\epsilon) = \xi + \epsilon \int_\rho^u \frac{P(u',X,P\epsilon)}{\sigma(u',X,P,\epsilon)} du', \tag{10a}$$

$$P(u;z,\epsilon) = \eta + \epsilon \int_\rho^u \frac{f_2(u',X)}{\sigma(u';X,P,\epsilon)} du', \tag{10b}$$

in which we have omitted the arguments of the functions X,P on the right sides and which are equivalent to (8) plus the relevant initial conditions (5), play an important role in the proof of Theorems 1 and 2 for $H = H_{NR}$, as do the analogous integral equations for the case $H = H_R$ in the proof of these theorems for that case.

By (2), (6), and (9), in particular, we can apply Theorem 3 of [7] to Eqs. (8), thus concluding that for $\epsilon > 0$ sufficiently small these equations have a solution $\tilde{X}(u,\epsilon)$, $\tilde{P}(u,\epsilon)$ which is 1-periodic in u, is such that $\tilde{X}(u,\epsilon)$, $\tilde{P}(u,\epsilon) = O(\epsilon^2)$, uniformly with respect to u in $\mathbb{R}$, and is the only 1-periodic solution of (8) in a small enough ϵ-independent neighborhood of the origin. This solution corresponds to a unique fixed point $\zeta_1(\rho,E) = (\tilde{X}(\rho,\epsilon),\tilde{P}(\rho,\epsilon))$ of $\mathcal{P}(\rho,E)$, where $\zeta_1(\rho,E)$ coincides with $\zeta_H(\rho,E)$ (see Theorem 1) for $H = H_{NR}$. Iterating the fixed-point equation $\mathcal{P}(\rho,E)(\zeta) = \zeta$ and using $||\zeta_1(\rho,E)|| = O(\epsilon^2)$ and asymptotic formulas for $E \to \infty$ (obtained by differentiating (10) and applying Gronwall-type inequalities) for the first derivatives of $\xi' = X(\rho+1;z,E)$, $\eta' = P(\rho+1;z,E)$ with respect to ξ,η evaluated at $z = (\rho,\zeta_1(\rho,E))$, the desired asymptotic formulas (7) follow for the case $H = H_{NR}$. The remaining assertions of Theorem 1 follow straightforwardly.■

Idea of the Proof of Theorem 2:

Given the existence and other properties of the fixed point $\zeta_H(\rho,E) = (\xi_H(\rho,E),\eta_H(\rho,E))$ asserted in Theorem 1 of Section 2A, one can prove Theorem 2 of that section by arguments analogous to those used to prove the less general Theorem 1 of [1]. We will restrict the proof to the case $H = H_{NR}$, since no new ideas are involved in proving Theorem 2 for $H = H_R$. An essential role in the proof is played by suitable asymptotic formulas in the limit $E \to \infty$ for the derivatives of ξ',η' with respect to ξ,η of orders $\leqslant 3$ evaluated at $z = (\rho,\zeta_1(\rho,E))$. For the derivatives of second and third orders, these formulas can be derived by the same approach mentioned for the corresponding first order derivatives in the preceding paragraph. As in [1], the use of Eqs. (10) to obtain these asymptotic results enormously simplifies the task.

Using these asymptotic formulas one proceeds to calculate the Floquet multipliers $\lambda(\rho,E)$, $\overline{\lambda(\rho,E)}$ of the map $\mathcal{G}(\rho,E)$ at sufficiently large E, as well as the first twist coefficient $\mu(\rho,E)$ appearing in the Birkhoff normal form of this map at such E values:

$$z' = \lambda(\rho,E)z[1 + \mu(\rho,E)|z|^2] + O(|z|^4), z \to 0. \tag{11}$$

Here $z,\bar{z}$ (respectively, $z',\bar{z}'$) are complex variables related to the real variables ξ,η (respectively, ξ',η') by an appropriate analytic transformation. Under the assumptions of Theorem 2, and provided that $\rho \in \mathbb{R}, E > E_2 \geqslant E_1$ where E_2 is a sufficiently large constant, $\lambda(\rho,E)$ is nonreal, lies on the unit circle, and satisfies certain low-order nonresonance conditions imposed in the Moser twist theorem. Under the same assumptions, $\mu(\rho,E) = KE^{-1/2} + O(E^{-1})$ for $E \to \infty$, uniformly in ρ on $\mathbb{R}$, where K is nonvanishing and independent of ρ,E. These properties of the map (11) and suitable analyticity properties thereof allow us to apply an appropriate version of the twist theorem in [8], p. 228 to this map, and thus to complete the proof of Theorem 2.■

REFERENCES

[1] C. Lehmann, *Interaction of Radiation with Solids and Elementary Defect Production* (North-Holland, Amsterdam, 1977).

[2] D.S. Gemell, Rev. Mod. Phys. **56**, 129 (1974).

[3] A.W. Sáenz, Phys. Lett. **93**A, 337 (1983).

[4] A.W. Sáenz, "Canonical Map Approach to Channeling Stability in Crystals", J. Math. Phys. **26**, 1925 (1985).

[5] H.S. Dumas and J.A. Ellison, this volume.

[6] J. K. Hale, *Ordinary Differential Equations* (Krieger, Huntington, New York, 1980), 2nd ed.

[7] P. Swinnerton-Dyer, Proc. London Math. Soc. (3) **34**, 385 (1977).

[8] C. L. Siegel and J.K. Moser, *Lectures on Celestial Mechanics* (Springer, Berlin, 1971).

[9] A. Wintner, *The Analytical Foundations of Celestial Mechanics* (Princeton University Press, 1941).

SOME CONSIDERATIONS FOR A THEORY OF APPROXIMATE INVARIANTS*

Laurence S. Hall
Magnetic Fusion Energy Program
Lawrence Livermore National Laboratory
P.O. Box 5511, Livermore, CA 94550

1. INTRODUCTION

Frequently a system will behave as if constrained by an unknown invariant, a constant of the motion in addition to those already known. For practical purposes, integrability is established experimentally. We ask for the form of the new true or approximate invariant and examine a variety of applicable considerations: the universality of the representation of an invariant as a polynomial in momenta, mild symmetries, competition among independent forms, temporary representations, and approximate generic integrability. Results of computational experiments are examined in connection with the theoretical descriptions.

Dynamical systems of interest to the physicist are almost never either truly orderly or truly chaotic. Even classically, we now realize the omnipresent microscopic fuzziness of certain critical regions of phase space; at times this disorderliness erupts to dominate macroscopic behavior. Yet at other times these selfsame systems exhibit a tenacious macroscopic orderliness. Nature always insists, certainly, upon the technical accuracy of Poincaré's celebrated theorem of the nonexistence of generic invariants [1]. Nevertheless, she will often admit the generic invariant for every practical purpose. The two truths are illustrated by Fig. 1; here we consider systems that are practically integrable.

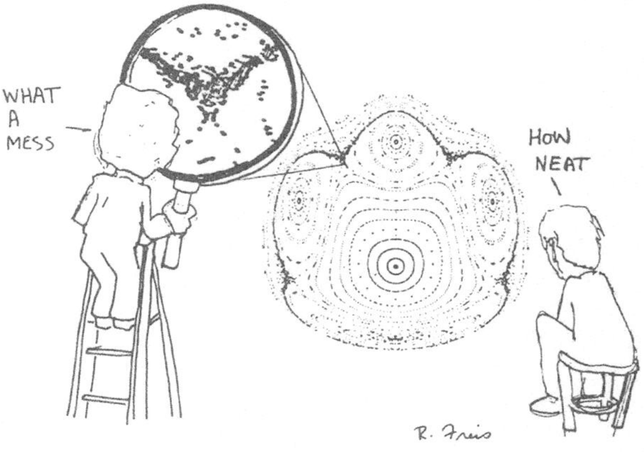

Figure 1. Two Truths

*Work performed under the auspices of the U.S. Department of Energy by the Lawrence Livermore National Laboratory under contract number W-7405-ENG-48.

We are concerned in particular with the "configurational invariants," defined macroscopically on the system as a whole and distinct from the microscopic adiabatic invariant that determines long-time behavior from short-time averages over rapid variations [2]. Our circumstances are also such that known symmetries are insufficient to yield integrable behavior immediately. Then, typically, the configurational invariant is found as a formal integral, for example, by a truncated perturbation series [3-5]. We will only quote results of perturbation calculations for our purposes here, however, since the theory is readily available elsewhere in this volume.

Configurational invariants are also computed by newer nonperturbative techniques [6,7]; we will look more closely at this since it is less familiar. The nonperturbative methods have an advantage that comes from their ability to recognize a new physical behavior at its point of entry, where perturbation theory does poorly. In addition, while perturbation theory is a powerful methodology, it is truly approximate. The nonperturbative methods on the other hand are capable of discovering exact invariants in closed form when they exist.

As a matter of nomenclature, it is useful to make a further distinction between the perturbative and nonperturbative methods. Characteristically, the perturbation theoretic approach expands the invariants of a $(2\nu+1)$-dimensional Hamiltonian system $H(p,q,t)$ as a power series in all components of p and q that span the 2ν-dimensional phase space. Terms are grouped together in equal <u>order</u>, i.e., by the sum of the powers of each component of p and q appearing in the given term. The presently successful nonperturbative approaches, on the other hand, develop the invariants as polynomials in the momenta alone. We call the sum of the powers of the components of momenta the <u>degree</u> of any term. Thus a term like $x^2yp_xp_y$ is of fifth order, but second degree.

When an invariant form is fully expanded in both coordinates and momenta, with terms grouped according to order, we shall say it is in "perturbation form." On the other hand, suppose an invariant is written

$$I_j = \sum_{n=0}^{N} \sum_{\{\Sigma n_k = n\}} \sum_{i_\nu \geq \cdots \geq i_1 \geq 1}^{\nu} A_{ji_1\cdots i_\nu} p_{i_1}^{n1} \cdots p_{i_\nu}^{n\nu} , \tag{1.1}$$

where each A-coefficient is a general function of position and time. Then we shall say that I_j is in "Whittaker's form," since Whittaker [3] considered Eq. (1.1) for $\nu = 2$ and $N \leq 2$.

In general, the perturbation form of an invariant truncated to any given order is inexact. Indeed, the convergence of the series itself is most often questionable, even if the system eventually turns out to be exactly integrable. It is of some importance, therefore, that the invariants of an exactly integrable system with Hamiltonian $H(p,q,t)$ may <u>always</u> be put in Whittaker's form.

2. THE UNIVERSALITY OF WHITTAKER'S POLYNOMIAL FORM OF THE INVARIANT

Suppose that a system with ν degrees of freedom is integrable; it admits ν general independent isolating integrals of the form $f_j(p,q,t) = c_j$. Then by the implicit function theorem, these equations may be solved (at least locally) for the ν components of p, i.e., $p_j = P_j(q,t;c)$. Of course, the process of inversion may be multi-valued, but until the orbit passes to a new sheet,

$$C_i^\mu\ (p,q,t;c) = \sum_{j=1}^{\nu} A_{ij}(q,t)\ [p_j - P_j(q,t;c)] = 0 \quad . \tag{2.1}$$

Here μ denotes the particular sheet in question, and at least temporarily the C_i^μ comprise ν separate, independent constants of the motion, linear in the momenta. Moreover, if the inversion is M-fold multivalued, $1 \leq \mu \leq M$, then a true invariant may be constructed as

$$J_i\ (p,q,t;c) = \prod_{\mu=1}^{M} C_i^\mu = 0 \quad . \tag{2.2}$$

We call M the "branching multiplicity." The J_i are ν separate, independent, analytic invariants of degree M in Whittaker's form.

It is clear that the search for the class of invariants of Whittaker's form, or indeed, the search for the temporary invariants linear in momenta which generate the J_i, is fundamental to the study of integrable systems [8]. It is likewise suggestive that the bigness or smallness of the branching multiplicity, which is also the degree of J_i, is closely related to observations of apparent integrability or apparent stochasticity in multiply-periodic systems.

Consider, for example, the autonomous two-dimensional system $H = \frac{1}{2} p_x^2 + \frac{1}{2} p_y^2 + V(x,y)$. Look at the orbit in configuration space. The speed, at any position, is determined by conservation of energy. Thus a second invariant serves merely to determine the direction of the orbit, i.e., the pitch-angle in velocity space. Examine the arrows along the orbits in Fig. 2.

The multiplicity of the inversion of the invariant to obtain p determines the variety of directions along which a given orbit can approach a given spatial position. If the system is ergodic, eventually every direction of approach is traversed. Otherwise, there is a constraint. Of course the useful invariants are those with small branching multiplicity. If M is large and there is no degeneracy, it will be hard to distinguish between allowed directions and those directions that are not allowed. To the physicist, the system then might as well be ergodic.

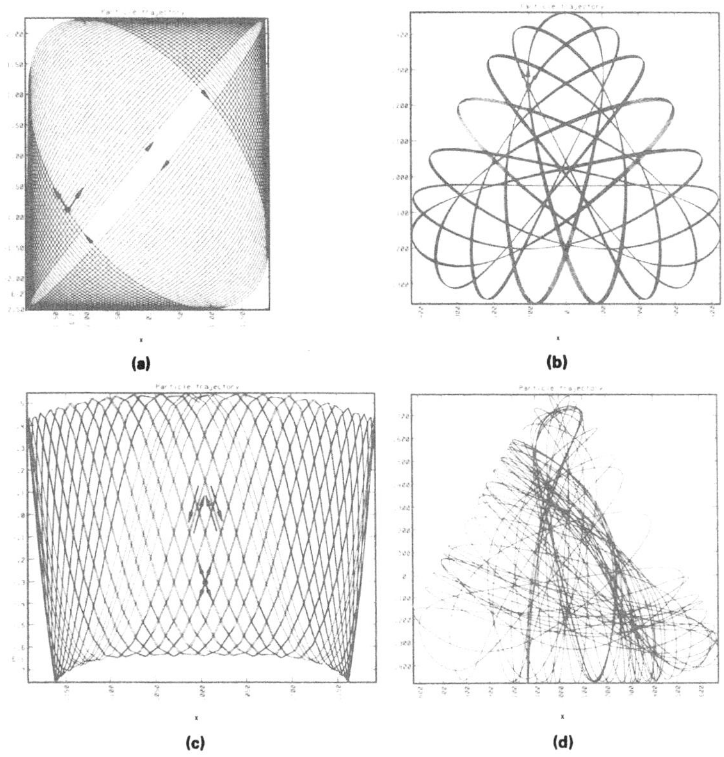

Fig. 2. Configuration space trajectories for systems with exact or approximate (a) second-degree invariant, (b) third-degree invariant, (c) fourth-degree invariant, and (d) stochastic orbit. [7]

3. PRACTICALLY INTEGRABLE, AUTONOMOUS, TWO-DIMENSIONAL SYSTEMS

Whittaker's form for the time-independent two-dimensional invariant is

$$I(p_x,p_y,x,y) = \sum_{j=0}^{N} \sum_{k=0}^{j} I_{jk}(x,y)\, p_x^{j-k}\, p_y^{k} \quad , \tag{3.1}$$

which we apply to the system

$$H(p_x,p_y,x,y) = \frac{1}{2}(p_x - A_x)^2 + \frac{1}{2}(p_y - A_y)^2 + V \quad , \tag{3.2}$$

where A_x, A_y and V each depend upon position. Already, H = E is a constant of the motion in Whittaker's form and we may use it to eliminate $\dot{x}^2 + \dot{y}^2 = (p_x - A_x)^2 + (p_y - A_y)^2$ from Whittaker's form of the second invariant. Then

$$I(p_x,p_y,x,y) = \mathrm{Re}\left[\frac{1}{2} F_0(x,y) + \sum_{j=1}^{N} F_j(x,y)\,(\dot{x} - i\dot{y})^j\right] \quad . \tag{3.3}$$

We call Eq. (3.3) the "standard invariant" form in two dimensions. It is nothing more than the Fourier expansion of Eq. (3.1) in velocity pitch-angle: $\dot{x} - i\dot{y} = ue^{-i\gamma}$, $u^2 = 2[E - V(x,y)]$.

If $\dot{I} = 0$, then the system requires that 2N+3 first-order coupled partial differential equations in x and y be satisfied, involving the 2N+3 quantities: $\mathrm{Re}(F_j)$ for $0 \leq j \leq N$, $\mathrm{Im}(F_j)$ for $1 \leq j \leq N$, $\Omega = \partial A_y/\partial x - \partial A_x/\partial y$, and V. These equations, which are linear in the F's, have simple complex characteristics. Upon transformation to new coordinates $(x,y) \to (\zeta,\zeta^*)$, where $\zeta = -i(x + iy)$, and ζ^* is its complex conjugate, then the bulk of the requirements are that [7]

$$F_n(\zeta,\zeta^*) = f_n(\zeta) + \int_{\sigma(\zeta)}^{\zeta^*} [(n+1)i\Omega F_{n+1} + 2(E-V)\partial F_{n+2}/\partial\zeta - 2(n+2)F_{n+2}\,\partial V/\partial\zeta]d\zeta^*, \qquad 0 \leq n \leq N . \tag{3.4}$$

[We consider that $\sigma(\zeta)$, the lower limit of the integral, is specified to suit our convenience.] The $f_n(\zeta)$ are free functions -- the parameters of integration.

Since $F_n = 0$ for $n > N$, the system now can be integrated downwards to give each F_n, $0 \leq n \leq N$. For example, $F_N = f_N(\zeta)$ and $F_{N-1} = f_{N-1}(\zeta) + Ni\, f_N(\zeta) \int \Omega(\zeta,\zeta^*)d\zeta^*$. All F_n turn out to be linear in the f_m, $m \geq n$, and their derivatives, with coefficients that are explicit quadratures involving V and Ω.

Having satisfied Eqs. (3.4), one transforms back to real coordinates, finding that if I is given by Eq. (3.3), then

$$\dot{I} = \frac{1}{2}\left(\dot{x}\,\frac{\partial F_0^I}{\partial y} - \dot{y}\,\frac{\partial F_0^I}{\partial x}\right) + \frac{1}{2}\left[\frac{\partial}{\partial x}(u^2F_1^R) + \frac{\partial}{\partial y}(u^2F_1^I)\right] , \tag{3.5}$$

where superscripts R and I denote real and imaginary parts, respectively. Now, $\dot{I} = 0$ becomes a constraint equation, limiting the forms of V and Ω that lead to integrable systems. However, Eq. (3.5) is useful as it stands for the approximately integrable case. Nonvanishing $\dot{I}$ is a measure of the error, and we may choose our free parameters so as to minimize $\dot{I}^2$.

A. Velocity-independent Forces and Competing Invariant Forms

Mild symmetries, i.e., recognizable symmetries that are insufficient in themselves to fully determine an invariant, nevertheless can still play an important role in determining invariant structure. For example, if the forces are independent of velocity, then $\Omega = 0$ and the odd and even F_n coefficients decouple. Such systems admit two separate standard invariant forms:

$$I^e = \mathrm{Re}\left[\frac{1}{2}F_0 + \sum_{j=1}^{N^e/2} F_{2j}\ (\dot{x} - i\dot{y})^{2j}\right] \quad , \tag{3.6a}$$

$$I^o = \mathrm{Re}\left[\sum_{j=0}^{(N^o-1)/2} F_{2j+1}\ (\dot{x} - i\dot{y})^{2j+1}\right] \quad . \tag{3.6b}$$

Why two forms; are they independent? Yes, indeed, though they cannot be in involution, nor can we expect both to be isolating. But then, which do we use, or must we use some combination? The answer, of course, is that we must use the "best" combination, in some physically relevant sense. Since we have already developed the concept of $\dot{I}^2$ as error, it certainly makes sense to make use of all our options to minimize this error. The "best" combination will minimize $\dot{I}^2$.

B. Reflection Symmetry

Continuing on, reflection symmetry, $V(-x,y) = V(x,y)$, is of considerable importance when it occurs. Then (again we take $\Omega = 0$),

$$F_n(\zeta,\zeta^*) = f_n(\zeta) + 2\int_{\zeta}^{\zeta^*} [(E-V)\partial F_{n+2}/\partial\zeta - (n+2)F_{n+2}\ \partial V/\partial\zeta]d\zeta^* \quad . \tag{3.7}$$

At $x = 0$ the integral vanishes and $f_n(y)$ is a <u>real</u> function. As a result, I^e is symmetric upon time reversal or a change of sign of x, and I^o is antisymmetric to either operation. Moreover, if the potential is nonsingular then $\partial V/\partial x$ vanishes at the midplane so that any orbit for which $\dot{x} = x = 0$ initially preserves $\dot{x} = x = 0$ for all time. We set $I = 0$ for this integrable orbit, thereby eliminating from I any additive dependence upon H. Although this is no constraint in odd degree, for the even-degree form

$$f_0(y) = -2\sum_{j=1}^{N^e/2} (-1)^j f_{2j}(y)\ u_0^{2j} \quad , \quad u_0(y) = 2^{\frac{1}{2}}[E - V(0,y)]^{\frac{1}{2}} \quad . \tag{3.8}$$

C. A Sample Calculation of an Approximate Invariant

The time derivative of an approximate even-degree invariant is given by the terms in F_0^I of Eq. (3.5); the odd-degree derivative is determined by F_1. Since F_0^I is odd in x, by symmetry it is sufficient that $\partial F_0^I/\partial x = 0$ for $F_0^I = 0$ in I^e. The natural approximation is to expand $\partial F_0^I/\partial x$, setting the lowest order terms to zero. Let's apply this procedure to the Hamiltonian

$$H = \frac{1}{2}p_x^2 + \frac{1}{2}p_y^2 + \frac{1}{2}Ax^2 + \frac{1}{2}By^2 + C\ x^2y + \frac{1}{3}Dy^3 \quad . \tag{3.9}$$

Equation (3.9), corresponding to the symmetric-cubic potential, reflects the Taylor's series expansion about its minimum of a general potential with one symmetric coordinate, through lowest-order nonlinear terms in the force. If we desire, two constants can be taken into the scaling; only B/A and D/C are significant. (Both B and A in Eq. (3.9) are assumed to be positive.)

For a second-degree invariant, we write

$$f_2(y) = a_{20} + a_{21}y + a_{22}y^2 + \cdots ,$$

in which case

$$\partial F_0^I/\partial x = \alpha_{20}x^2 + \alpha_{21}x^2y + \alpha_{40}x^4 + \alpha_{22}x^2y^2 + \text{higher order terms} ,$$

where

$$\alpha_{20} = -8C\, a_{20} + 2(B - 4A)a_{21} - 24\, E\, a_{23} ,$$

$$\alpha_{21} = 4(D-6C)\, a_{21} + 16\,(B - A)\, a_{22} - 96\, E\, a_{24} , \quad \text{etc.}$$

For the simplest result of all, we put $a_{20} = 1/4$, $a_{21} = -C/(4A-B)$, and all other a-coefficients are set to zero. (An overall multiplicative factor is irrelevant.) Then $\alpha_{20} = 0$, $\alpha_{21} = 4C\,(6C-D)/(4A-B)$, and all other α-coefficients vanish. Computing the approximate invariant from Eqs. (3.6a), (3.7), and (3.8), we find

$$K_2 = \frac{1}{2}\dot{x}^2 + \frac{1}{2}Ax^2 + \frac{2AC}{4A-B}x^2y - \frac{2C}{4A-B}(\dot{x}y - x\dot{y})\dot{x}$$
$$+ \frac{C(9C-D)}{6(4A-B)}x^4 - \frac{C(4C-D)}{4A-B}x^2y^2 , \qquad (3.10)$$

$$\dot{K}_2 = \frac{2}{3}\left(\frac{6C-D}{4A-B}\right)(x^3\dot{x} - 3x^2y\dot{y}) . \qquad (3.11)$$

A third-degree form for this potential, which we will want to examine in combination with K_2, is computed by an alternative procedure in paragraph 6.3 of [7].

In summary of Section 3, we find that the combination of velocity-space symmetry ($\Omega = 0$), reflection symmetry, and the existence of a special integrable orbit permits us to display three <u>manifestly independent</u> invariant forms: H, I^e, and I^o. Both H and I^e are even upon reflection in x or upon time reversal, I^o is odd. Both I^e and I^o vanish identically at $p_x = x = 0$; H there is a nontrivial function of y and p_y. However, only H and some "best" combination of I^e and I^o can be expected to be "good" (isolating) invariants.

4. COMPUTATIONAL OBSERVATIONS FROM THE HÉNON-HEILES SYSTEM

The Hénon-Heiles system [9], a special case of Eq. (3.9), is a richly endowed example of a two-dimensional configuration possessing the symmetries just discussed. The well-studied Hamiltonian is

$$H = \frac{1}{2} p_x^2 + \frac{1}{2} p_y^2 + \frac{1}{2} x^2 + \frac{1}{2} y^2 + x^2 y - \frac{1}{3} y^3 \quad ; \tag{4.1}$$

there is a shallow triangular potential well at the origin with escape energy $E_{esc} = 1/6$. The orbits undergo a transition from apparently-ordered to apparently-chaotic behavior (on the energy shell $H = E$) at about $(3/4)\ E_{esc}$ (see Fig. 3).

A. The Gustavson Invariant

The authoritative perturbative evaluation of the formal invariant for the nonstochastic Hénon-Heiles regime has been given by Gustavson, who carried the calculation through eighth order [4]. He found that it was necessary to keep terms in the invariant through at least seventh order if the characteristic shape of the surface of section, Fig. 3a, is to be recovered. On the other hand, the lowest order terms of Gustavson's invariant are fourth-order. Can the results be simplified?

Gustavson presents his results in the form of numerical tables, rounded to six decimal places. Nevertheless, it is possible to work upward, order by order,

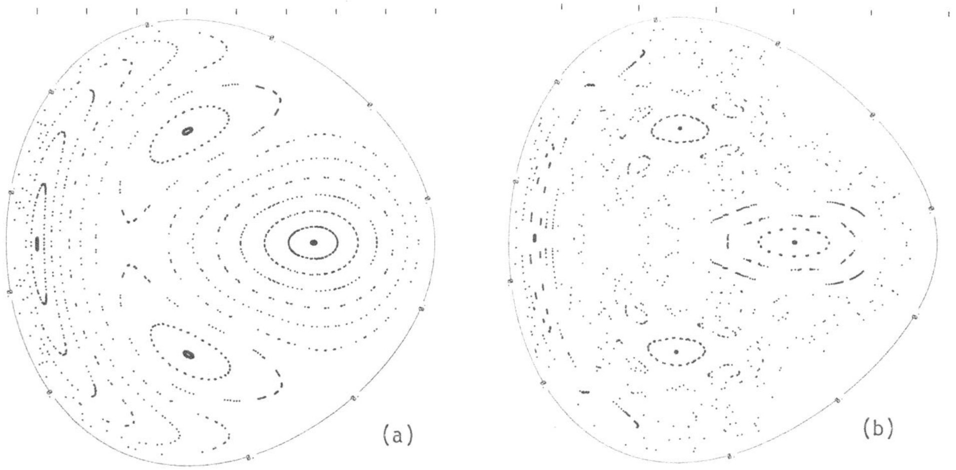

Fig. 3. Surface of section at x = 0 for the Hénon–Heiles system at (a) half excape-energy, and (b) three-fourths escape-energy.

recognizing the rational fractions represented decimally, to determine the analytic form of Gustavson's results. The calculations were performed using the MIT-MACSYMA algebraic manipulator [10]. We find that to within roundoff, through seventh order, Gustavson's invariant is

$$I_G = -\frac{5}{12}H^2\left[1 + \frac{77}{36}H\left(1 + \frac{221}{72}H\right)\right] + \frac{7}{12}\left[\left(1 + 2K - \frac{79}{24}H\right)J\right]^2 + \frac{7}{9}K(4K - 3H)^2 + \cdots \quad , \tag{4.2}$$

where

$$J = \boxed{(\dot{x}y - x\dot{y})\,[1 + 2(x^2 + y^2)] + (x^2 - y^2)\dot{x} - 2(xy + \dot{x}\dot{y})\dot{y} + \frac{2}{3}\dot{x}^3} + \frac{2}{15}\dot{x}[8(\dot{x}^2 - 3\dot{y}^2)\,(\dot{x}^2 + \dot{y}^2) + (5x^2 - 7y^2)\,(4\dot{x}^2 + 3x^2) - 3(x^2 + 5y^2)\,(4\dot{y}^2 + 3y^2)] + \frac{8}{5}y(4H - x^2 + y^2)\,(\dot{x}y - x\dot{y}) + \cdots \quad , \tag{4.3}$$

and

$$K = \boxed{\frac{1}{2}\dot{x}^2 + \frac{1}{2}x^2 + \frac{2}{3}x^2y - \frac{2}{3}\dot{x}\,(\dot{x}y - x\dot{y}) + \frac{5}{9}x^4 - \frac{5}{3}x^2y^2} + \cdots \quad . \tag{4.4}$$

At eighth order, roundoff was too severe to proceed further.

We notice first of all that I_G has an additive term in H alone that ought to be eliminated; it is diversionary and contributes nothing of significance. Its form is easily discovered by looking for the function of H that I_G becomes at $\dot{x} = x = 0$. Otherwise I_G depends on two forms, the antisymmetric J and the symmetric K. The boxes in Eqs. (4.3) and (4.4) contain all terms through fourth order: a third-degree form J_3 for J, and a second-degree form K_2 for K. In fact, J_3 is just the value computed in Ref. [7] for the third-degree invariant, when specialized to Hénon-Heiles. Likewise, K_2 is exactly the reduction of Eq. (3.10) for this case.

The form of I_G suggests the quantity

$$I = J^2 + (4/3)K\,(4K - 3H)^2 \tag{4.5}$$

should provide a good representation of the Hénon-Heiles system in the integrable regime. The surface of section of I for $J = J_3$ and $K = K_2$ (computed at half-escape-energy) is shown for comparison with I_G in Fig. 4, and for comparison with the orbital result in Fig. 3a.

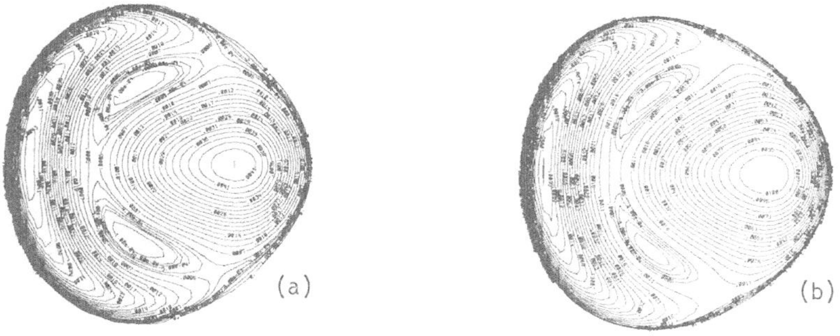

Fig. 4. Comparison of (a) the Gustavson invariant [4] and (b) $J_3^2 + (4/3)K_2(4K_2-3H)^2$ for the Hénon-Heiles system.

It is clear from these comparisons that an appropriate combination of the independent forms can give good results from much-lower-order calculations; thus the process is made much easier. It is also much easier to understand. However, the full program of making the best a priori combining choice is a subject of current research.

B. Experimentally-Temporary Invariants and Local Scattering Centers

W. P. Reinhardt and his collaborators [11-14] have carefully examined the Hénon-Heiles system in the "stochastic" regime through a series of numerical experiments. For example, Shirts and Reinhardt [12] have shown that the apparent stochasticity of Fig. 3b has an underlying order, see Fig. 5. Motion proceeds as if governed by one form of invariant, Fig. 5b, and then makes a transition to a different orbital behavior, Fig. 5d.

Jaffe and Reinhardt [13] conclude that transitions actually take place in specific spatial localities, Fig. 6. These scattering centers are points where the orbit approaches the limits of the energy shell, so that the motion is very slow. But these points are simultaneously where the curvature of the potential surface is negative along a line perpendicular to the flow [15]. The motion must be very sensitive to initial conditions there.

Helleman and Bountis [16] have made similar observations from a different point of view, also finding great sensitivity of the motion to the value of orbital parameters near the Jaffe-Reinhardt scattering centers. Indeed, our own notions of the time-derivative as a measure of error point up the same result. Contours of $\dot{I}^2$ averaged over velocity pitch angle are plotted alongside the surface of section of I in Fig. 7. Two cases are shown: (a) I is J_3^2 from the boxed terms of Eq. (4.3); and (b), I is a partially optimized (linear) combination of J_3^2 and a fourth degree invariant K_4 obtained by the procedures of Section 3C such that $\partial F_0^I/\partial x$ vanishes through sixth order. The same scattering localities found by Jaffe and Reinhardt are found here as well.

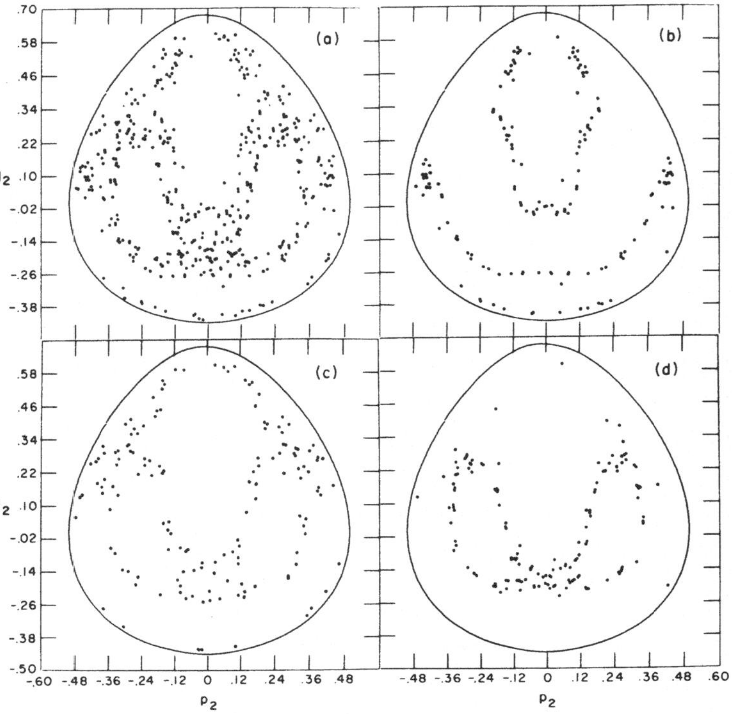

Fig. 5. Surface of section at x=0 for the Hénon-Heiles system for a single particle in the "stochastic" domain at three-fourths escape energy: (a) the entire long-time trajectory, (b) first-third of total time, (c) second-third of total time, (d) last-third of total time [12].

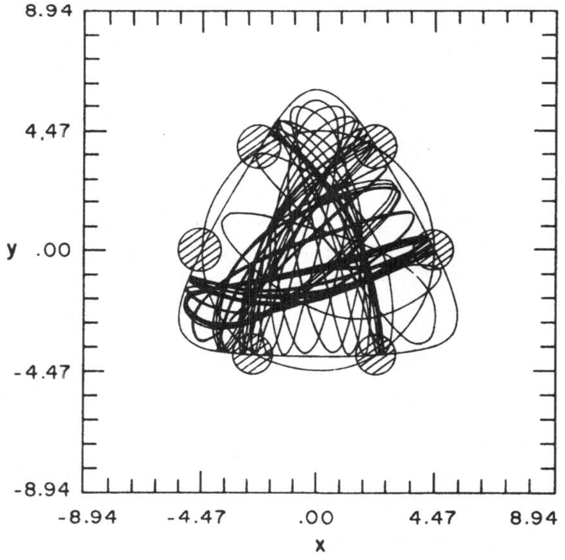

Fig. 6. Qualitative picture of coordinate space localization of chaos producing regions on the Hénon-Heiles surface [13].

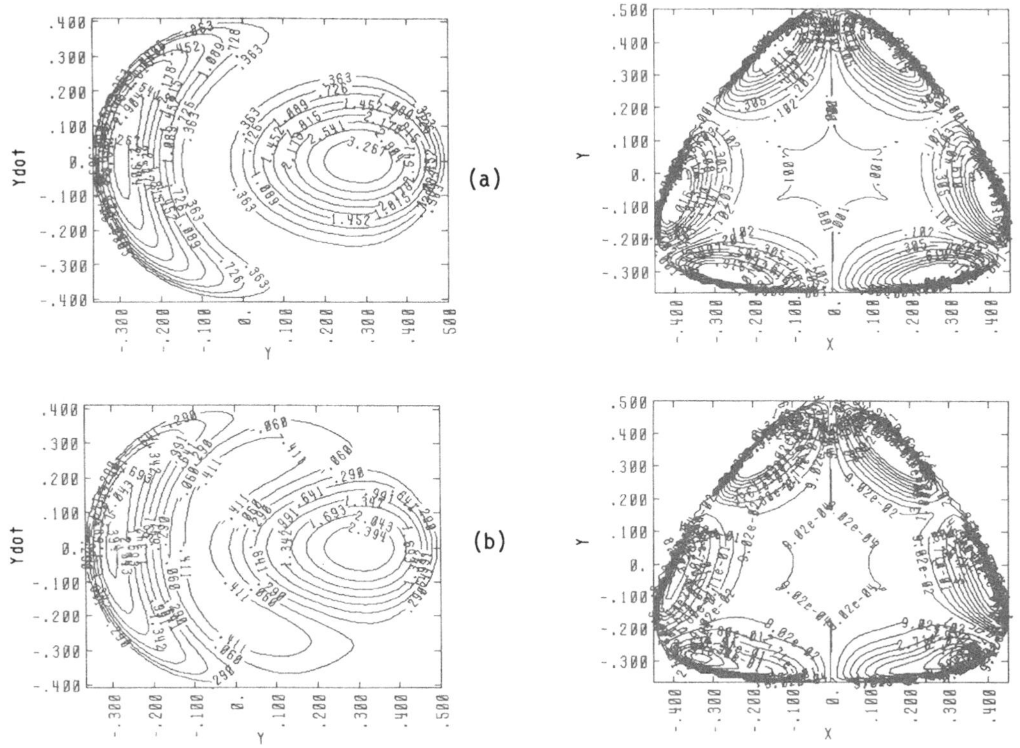

Fig. 7. Surface of section at x = 0 and the velocity-pitch-angle average of $\dot{I}^2$ for approximate invariant forms for the Hénon-Heiles system: (a) $I = J_3^2$, (b) $I = J_3^2 + \lambda K_4$ for near-optimal choices of K_4 and the constant λ.

5. GLOBAL FEATURES OF THE PARAMETER SPACE OF THE SYMMETRIC CUBIC POTENTIAL

While we have spent some time discussing the stochastic transition in the Hénon-Heiles system, in fact the Hénon-Heiles behavior is rather exceptional. Over most of the parameter plane of the symmetric cubic potential, i.e., Eq. (3.9), the behavior of particles trapped in the potential well is orderly for energies right up to the escape energy (and above).

A comparatively simple approximate fourth-degree invariant K_4 and its time derivative were computed in Ref. [7] for the system of Eq. (3.9). Suppose we define the dimensionless relative average error of this approximation: $\tau = A_s\langle\dot{K}_4^2\rangle/2E\langle K_4^2\rangle$, where A_s is the area of the energy shell in configuration space and the angular brackets signify the average over the energy shell. If we now plot τ versus parameters for a fixed ratio of the total energy to the escape energy, we have a picture of the relative accuracy of K_4 as an approximate invariant; R. C. White has made this computation. Figure 8 is White's contour plot of τ in the B/A vs.

D/C plane at 90% of the escape energy [17]. The error maximizes along a steep ridge at B/A = - D/C > 0. White notices that observable macroscopic indications of stochastic behavior are confined to the parameter region in the vicinity of the ridge, approximately within the contour indicated in Fig. 8. (Not all of the region inside the contour exhibits macroscopic stochastic behavior.)

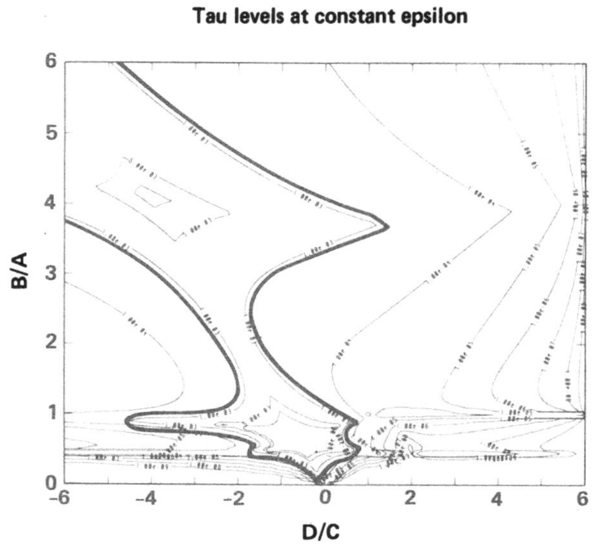

Fig. 8. Error diagram: Contour plot of τ vs. D/C and B/A for the symmetric cubic potential, Eq. (3.9). At 90% of the escape energy, stochastic behavior is macroscopically observable only within the heavy contour [17].

The ridge in the error diagram is widest and highest near B/A = 4, the 2:1 frequency resonance where perturbation theory finds the strongest resonance overlap. The ridge widens again around another peak at B/A = 1, Hénon-Heiles. But perhaps the most remarkable result is the large domain of orderly behavior. Away from the ridge, a fourth-degree invariant gives a good account of orbital behavior.

Figure 9a shows some typical plots of the orbit-computed surface of section versus parameters at 50% of the escape energy. The second column of Fig. 9a reflects points on the -D/C = B/A ridge. An improved fourth-degree invariant K_4 was also computed by the methods of Section 3C for an optimal selection of coefficients in the expansion of $f_2(y)$ and $f_4(y)$ subject to requiring $\partial F_0^I/\partial x = 0$ through sixth order. Figure 9b shows the comparison of this K_4 surface-of-section at half escape-energy. Away from the ridge, the agreement becomes quite good.

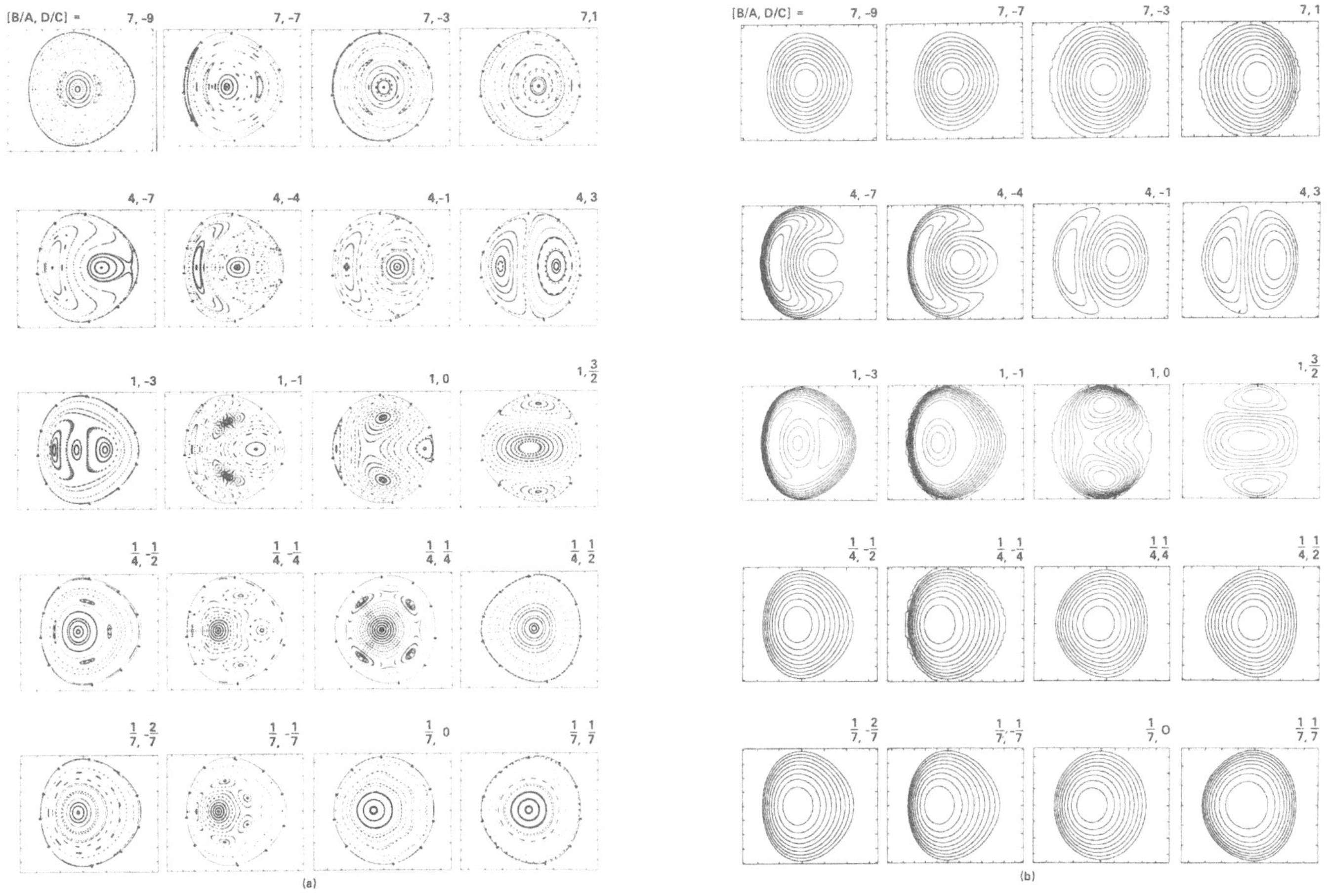

Fig. 9. Comparison of (a) the orbit-computed surfaces of section at x =0 with (b) surfaces of section of a fourth-degree invariant for which $\partial F_0^I/\partial x = 0$ through sixth order, for the symmetric cubic potential at half escape-energy. The second column, in each case, lies on the ridge $B/A = -D/C > 0$.

6. CONCLUSIONS

We have seen that despite Poincaré's pessimistic theorem [1], a dynamical system will frequently behave as if constrained by an unknown generic invariant. We have observed that when an exact invariant does exist, it can be written in Whittaker's form, and have then used this form to compute approximate invariants as well. We have also seen that a given dynamical system may exhibit behavior according to different approximations of low degree in different regions of phase space. Transitions between the characteristic behaviors may or may not occur as a function of time, depending upon whether the orbit does or does not approach certain critical points in phase space. The need for optimal selection among competing approximations has been illustrated. It is suggested that apparent stochastic transitions are the result of an increasing branching multiplicity. That is, chaotic behavior appears to result from the destruction of a degeneracy that permits a high-degree invariant to be represented by a low-degree approximate form.

ACKNOWLEDGMENTS

The author wishes to thank B. McNamara and R. C. White for their continuing interest and help, White for some of the numerical work, R. P. Freis for the illustration of Fig. 1, and W. P. Reinhardt for comments and the kind provision of original drawings for Figs. 5 and 6. He also thanks J. M. Wikkerink for her cheer and her rapid and accurate typing, formatting, and reformatting of the drafts of this paper.

REFERENCES

[1] H. Poincaré, Les méthodes nouvelles de la mécanique céleste (Gauthier-Villars, Paris, 1892), Vol. I, Chap. 5.

[2] N. N. Bogoliubov and Y. A. Mitropolski, Asymptotic Methods in the Theory of Non-Linear Oscillations (Gordon and Breach, New York, 1961); M. D. Kruskal, J. Math. Phys. 3, 806 (1962).

[3] E. T. Whittaker, A Treatise on the Analytical Dynamics of Particles and Rigid Bodies, 4th ed. (Cambridge Univ. Press, London, 1937).

[4] F. G. Gustavson, Astron. J. 71, 670 (1966).

[5] A. H. Nayfeh, Perturbation Methods (Wiley, New York, 1973). Also, for example, see the lectures by B. Abraham-Schrauner, A. Deprit, J. Finn, or A. Kamel in this volume.

[6] H. R. Lewis and P. G. L. Leach, J. Math. Phys. 23, 2371 (1982) and references therein. See also J. Goedert, this volume.

[7] L. S. Hall, Physica 8D, 90 (1983).

[8] L. S. Hall and J. Goedert (to be published).

[9] M. Hénon and C. Heiles, Astron. J. 69, 73 (1964).

[10] The MIT Mathlab Group, MACSYMA Reference Manual (Massachusetts Institute of Technology, Cambridge, Massachusetts, 1977).

[11] C. Cerjan and W. P. Reinhardt, J. Chem. Phys. 71, 1819 (1979).

[12] R. B. Shirts and W. P. Reinhardt, J. Chem. Phys. 77, 5204 (1982).

[13] C. Jaffe and W. P. Reinhardt, J. Chem. Phys. 77, 5191 (1982).

[14] W. P. Reinhardt, J. Phys. Chem. 86, 2158 (1982). See also his article in The Mathematical Analysis of Physical Systems, R. Mickens, Ed. (Van Nostrand-Reinhold, New York, 1984) (in press).

[15] R. Kosloff and S. A. Rice, J. Chem. Phys. 74, 1947 (1981).

[16] R. H. G. Helleman and T. Bountis, Stochastic Behavior in Classical and Quantum Mechanical Systems, G. Casati and J. Ford, Eds. (Springer, New York, 1979).

[17] R. C. White, B. McNamara, and L. S. Hall, paper 2U7, New Orleans Meeting, Bull. APS 27-11 (1982) 956 (unpublished).

EXACT INVARIANTS IN THE FORM OF MOMENTUM RESONANCES FOR PARTICLE MOTION IN ONE-DIMENSIONAL, TIME-DEPENDENT POTENTIALS*

João Goedert and H. Ralph Lewis
Los Alamos National Laboratory, MS-F642
Los Alamos, New Mexico 87545

1. INTRODUCTION

Exact invariants for Hamiltonian systems can be very useful for obtaining insight into the nature of solutions of the equations of motion and they can be helpful in computing solutions numerically. The search for invariants for specific systems has a long history. As an example, we mention the gravitational three-body problem, which is relevant to the motion of celestial bodies and has been studied extensively [1]. In 1887 Bruns showed that the ten so-called classical integrals of the three-body problem are the only invariants that exist that are algebraic functions of the coordinates, momenta, and time. In 1889 Poincaré showed for the restricted three-body problem that the only explicitly time-independent invariant that is periodic in the coordinates is an energy known as the Jacobian energy. These important results illustrate that an explicitly time-dependent invariant is likely to be outside the class of algebraic functions and that explicitly time-independent invariants are uncommon, even for autonomous systems.

In the search for exact invariants, attention traditionally has been concentrated on explicitly time-independent functions. Also, the systems considered have usually been autonomous; that is, the Hamiltonians usually have not depended on time explicitly. In this lecture, we consider the motion of a particle in a one-dimensional potential and allow the invariants as well as the potential to be <u>explicitly time-dependent</u>. For an autonomous one-dimensional Hamiltonian system, all invariants that are functionally independent of the energy are explicitly time-dependent. For any particular nonautonomous system, it may be that all invariants are explicitly

*Work performed under the auspices of the USDOE, UFRGS (Porto Alegre, Rio Grande do Sul, Brazil) and CNPq (Brazil).

time-dependent. In any event, any complete set of 2n invariants for an n-dimensional system must include at least one explicitly time-dependent invariant.

The equations of motion of any one-dimensional nonautonomous Hamiltonian system are equivalent to those of a corresponding autonomous two-dimensional Hamiltonian system. Therefore, it is always possible to treat one-dimensional nonautonomous systems by considering certain two-dimensional autonomous systems instead. The corresponding two-dimensional Hamiltonian depends linearly on a momentum variable that is associated with the energy of the one-dimensional system. In this regard, it should be noted that invariants derived by Holt [2], Hall [3], and others for particle motion in a two-dimensional, time-independent potential cannot be used to obtain invariants for one-dimensional, time-dependent potentials. The reason is that the Hamiltonian for the motion of a particle in a two-dimensional, time-independent potential depends quadratically on each momentum.

An important area in which explicitly time-dependent invariants for time-dependent potentials play a crucial role is the theory of collisionless plasmas [4]. When there is only one spatial dimension, the governing equations, which are the Vlasov-Poisson equations, describe a continuum of particles that move in the electric field generated by the particles themselves. The electric field is to be determined self-consistently along with the motion of the particles. The phase-space distribution function for the particles, which is a solution of the Vlasov equation, is a function of invariants of the motion of a single particle in the electric field. An exact or approximate invariant can be useful in connection with the Vlasov-Poisson equations if it applies to a class of electric fields that approximate the field associated with the exact solution.

Exact invariants for particle motion in classes of one-dimensional explicitly time-dependent potentials have been found for both linear and nonlinear equations of motion. For linear, arbitrarily time-dependent oscillators, invariants are known that are homogeneous quadratic forms in the coordinate and momentum [5]. For nonlinear equations of motion, an exact invariant is known that is quadratic in the momentum when the potential has a certain form that involves an arbitrary function of a time-dependent linear function of the spatial coordinate [6,7]. This invariant has been used [8] to find new exact solutions of the Vlasov-Poisson equations.

In this lecture we describe our use of a momentum-resonance ansatz of Lewis and Leach [9] to study exact invariants for time-dependent, one-dimensional potentials. This ansatz provides a framework for finding invariants admitted by a larger class of time-dependent potentials than was known previously. For a potential that admits an exact invariant in this resonance form, we have shown how to construct the invariant as a functional of the potential in terms of the solution of a definite linear algebraic system of equations. We have found a necessary and sufficient condition on the potential for the existence of an invariant with a given number of resonances. There exist more potentials that admit invariants with two resonances than were previously known, and we have found an example in parametric form of such a potential. We have also found examples of potentials that admit invariants with three resonances.

2. THE MOMENTUM-RESONANCE ANSATZ

We consider the Hamiltonian for a particle moving in a potential that depends on a coordinate q and time t,

$$H = \frac{1}{2} p^2 + V(q,t) . \tag{1}$$

The resonance ansatz of Lewis and Leach [9] postulates that the momentum dependence of an invariant $I(q,p,t)$ be expressed in terms of simple poles in the complex momentum plane,

$$I(q,p,t) = c(q,t) + \sum_{n=1}^{N} \frac{v_n(q,t)}{p - u_n(q,t)} . \tag{2}$$

This ansatz includes the case of invariants that are polynomials in p because any function of an invariant is also an invariant. Lewis and Leach showed that any function, $f(p)$, with a finite number of singularities located at points $p = p_i$, $1 \leq i \leq N$, can be expressed as some function of another function, $g(p)$, where the singularities of $g(p)$ are simple poles located at the points p_i. The functions of position and time that appear in the expression for the invariant in resonance form satisfy conditions that are decoupled to a remarkable degree. The condition that $I(q,p,t)$ be an invariant is

$$\frac{dI}{dt} \equiv \frac{\partial I}{\partial t} + p\,\frac{\partial I}{\partial q} - \frac{\partial V}{\partial q}\frac{\partial I}{\partial p} = 0 , \tag{3}$$

which implies that necessary and sufficient conditions on the functions c, v_n, and u_n such that $I(q,p,t)$ be an invariant are

$$\frac{\partial c}{\partial q} = 0 , \qquad \frac{\partial c}{\partial t} + \sum_{n=1}^{N} \frac{\partial v_n}{\partial q} = 0 ,$$

$$\frac{\partial v_n}{\partial t} + \frac{\partial}{\partial q}(u_n v_n) = 0 , \qquad \frac{\partial u_n}{\partial t} + u_n \frac{\partial u_n}{\partial q} = - \frac{\partial V}{\partial q} . \tag{4}$$

All potentials that admit an invariant with one pole ($N=1$) were determined by Lewis and Leach [9]. They also found a class of potentials that admit an invariant with two poles. In that case, the potential involves an arbitrary function of a time-dependent linear function of the spatial coordinate.

3. THE DISCRETE-MOMENT FORMULATION

There is a formulation of the momentum-resonance ansatz in terms of certain discrete moments that is very useful. Any invariant that can be written in resonance form can be constructed explicitly from these discrete moments as a functional of the potential. In addition, a necessary and sufficient condition for a potential to admit an invariant with N poles can be expressed in terms of the discrete moments. The kth moment $g_k(q,t)$ is defined by

$$g_k(q,t) = \sum_{n=1}^{N} u_n^k \, v_n . \tag{5}$$

If, for fixed q and t, we consider the quantities $v_n(q,t)$ to be the values of a function $v(q,p,t)$ that is defined at a discrete set of values of p given by $p = u_n(q,t)$ for $1 \leq n \leq N$, then $g_k(q,t)$ is the kth moment of $v(q,p,t)$ in that discrete space of values of p. By direct manipulation of (4), it can be shown that these moments satisfy the differential recursion relation

$$\frac{\partial g_k}{\partial q} = - \frac{\partial g_{k-1}}{\partial t} - (k-1) g_{k-2} \frac{\partial V}{\partial q} , \qquad k \geq 1, \tag{6a}$$

with the initial condition

$$g_0(q,t) = \alpha_1(t) q + \alpha_2(t) , \tag{6b}$$

where $\alpha_1(t) = - dc/dt$. Equation (6a) is precisely the recursion relation satisfied by the momentum moments of (3).* This equation defines the invariant $I(q,p,t)$ and, in the context of collisionless plasma physics, it is the Vlasov equation for the phase-space distribution function. Thus, it is natural to interpret $v(q,p,t)$ as the representation of an invariant or a phase-space distribution function in a discrete space of momentum values given by $p = u_n(q,t)$ for $1 \leq n \leq N$.

In addition to the recursion relation (6a), the moments also satisfy the algebraic recursion relation**

$$g_k = - \sum_{n=1}^{N} a_n g_{k-n} , \qquad k \geq N , \tag{7}$$

where the quantities a_n and u_n are related by

$$u_n^N + \sum_{k=1}^{N} a_k u_n^{N-k} = 0 . \tag{8}$$

This recursion relation is completely equivalent to the definition of

---

*We thank Daniel C. Barnes for mentioning this connection.
**We thank Gene H. Golub for pointing out that the definition of the discrete moments implies the existence of an algebraic recursion relation among them.

the moments given by (5), a fact that plays an important role in the numerical analysis of an ordinary differential equation with constant coefficients.

We now make two remarkable observations. The moments can be calculated as functionals of the potential by means of (6a); and then the results can be used in N of the recursion equations given by (7) to obtain a system of linear algebraic equations that determine the coefficients a_n. The second observation is that (2), which expresses the invariant in terms of the u_n and v_n, can be rewritten to express the invariant completely in terms of the moments and the coefficients a_n [13]:

$$I(q,p,t) = c(t) + \frac{\sum_{n=1}^{N} p^{N-n} \sum_{k=1}^{n} a_{k-1} g_{n-k}}{p^N + \sum_{n=1}^{N} a_n p^{N-n}}, \tag{9}$$

where we have defined

$$a_0 \equiv 1 . \tag{10}$$

The expressions for the invariant given by (2) and (9) are identically equal because of the algebraic recursion relation (7). These two observations prove the following linearization theorem:

Theorem: If $V(q,t)$ admits an invariant with N resonances, this invariant is given by equation (9) in terms of moments g_k and coefficients a_n. The g_k can be expressed as functionals of $V(q,t)$ by means of (6) and the a_n are determined from the g_k through the linear algebraic system of equations (7).

In addition to the linearization theorem, we have derived [13] a condition for a potential to admit an invariant in resonance form:

Necessary and Sufficient Condition on $V(q,t)$: The expression given by (9) in terms of the moments is an invariant ($dI/dt = 0$) if, and only if, the determinant

$$\Delta = \begin{vmatrix} g_0 & g_1 & \cdots & g_N \\ g_1 & g_2 & \cdots & g_{N+1} \\ \cdot & \cdot & \cdots & \cdot \\ \cdot & \cdot & \cdots & \cdot \\ g_N & \cdot & \cdots & g_{2N} \end{vmatrix} = 0 \qquad (11)$$

is satisfied. The matrix in this condition is a Hankel matrix [10-12]

4. SOME PRELIMINARY APPLICATIONS

A direct application of the condition (11) for the case $N = 1$ leads to a first-order linear differential equation for $V(q,t)$ that can be completely integrated to yield the previously known result for potentials that admit an invariant with only one resonance. For $N = 2$, Eq. (11) is a nonlinear integro-differential equation. In the special case for which $g_0 = 0$, the condition again becomes a linear partial differential equation for the potential that can be integrated. The result is the class of potentials found by Lewis and Leach [9]. However, condition (11) allows a wider class of potentials that admit an exact invariant with two resonances. In a future publication [13], we shall exhibit further solutions of (11) with $N = 2$ and $g_0 \neq 0$. They are obtained by recasting the necessary and sufficient condition with $N = 2$ into a form different than (11).

To illustrate the use of our condition (11), we present two potentials for which invariants with three resonances exist and we construct them by using (9). These examples are the result of a preliminary study of the $N = 3$ case.

We consider $N = 3$ and choose $g_0 = g_1 = g_3 = 0$ and $g_2 = 1$. This is consistent with (6). All other g_k can be calculated from (6a). The result is

$$g_4 = -3V(q,t) - \frac{3}{2} V_1(t) , \tag{12a}$$

$$g_5 = 3 \frac{\partial}{\partial t} \int V(q,t)dq + \frac{3}{2} \frac{dV_1}{dt} + V_2(t) , \tag{12b}$$

$$g_6 = - 3 \frac{\partial^2}{\partial t^2} \int^q dx \int^x dy V(y,t) - 3 \frac{d^2 V_1}{dt^2} q^2 - \frac{dV_2}{dt} q$$

$$- V_3(t) + \frac{15}{2} V_1 V(q,t) + \frac{15}{2} V^2(q,t) , \tag{12c}$$

where $V_1(t)$, $V_2(t)$ and $V_3(t)$ are unspecified functions of time. Condition (11) for this case is simply $g_6 - g_4^2 = 0$, which is the following integro-differential equation for the potential:

$$V^2(q,t) + V_1(t) V(q,t) + 2 \frac{\partial^2}{\partial t^2} \iint V(q,t)dq^2 = - \phi(q,t) , \tag{13}$$

where

$$\phi(q,t) = 2 \frac{d^2 V_1}{dt^2} q^2 + 2 \frac{dV_2}{dt} q + 2V_3 + \frac{3}{2} V_1^2 . \tag{14}$$

The search for solutions is simplified by starting with the equation

$$2 \frac{\partial^2 V}{\partial t^2} + \frac{\partial^2 V^2}{\partial q^2} + V_1(t) \frac{\partial^2 V}{\partial q^2} = - 4 \frac{d^2 V_1}{dt^2} , \tag{15}$$

which is obtained by taking the second spatial derivative of (13). Any solution of (15) can then be substituted into (13), which becomes an equation for the remaining unknowns. Particular solutions of (13)-(15) can be found easily. Two immediate solutions are:

$$V(q,t) = At\sqrt{q} , \quad V_1 = 0 , \quad V_2 = - \frac{1}{6}A^2 t^2 + B , \quad V_3 = 0 , \tag{16a}$$

and

$$V(q,t) = -\frac{A}{2} \pm \sqrt{(A^2 + 4C + 4Bq)} \quad ,$$

$$V_1 = A \; , \quad V_2 = -\frac{1}{2}Bt + D \; , \quad V_3 = -\frac{3}{4}A^2 - \frac{1}{2}C \; , \qquad (16b)$$

where A, B, C and D are arbitrary constants. Using those solutions in (12) to determine the g_k and, with the g_k calculating the a_n from (7), we obtain the corresponding invariants:

$$I = \frac{1}{p^3 + 3Apt\sqrt{q} - 2Aq\sqrt{q} + \frac{1}{2}A^2t^3 - 3B} \, , \qquad (17a)$$

and

$$I = \frac{1}{p^3 \pm 3p\sqrt{(A^2 + 4C + 4Bq)} + \frac{3}{2}Bt} \, , \qquad (17b)$$

respectively. It is interesting to notice that although both invariants are explicitly time-dependent, the first system is nonautonomous, but the second is autonomous. The latter possesses the energy invariant in addition to I and, therefore, we have a complete set of two invariants. The trajectories could be determined by, for example, solving the energy invariant for the momentum p and substituting this into I.

5. FINAL REMARKS

The momentum-resonance ansatz has been developed into an alternative form which, in an initial study, has exhibited additional advantages. We were able to show that an invariant in resonance form can be determined by only solving linear algebraic equations. We derived a closed condition that determines all potentials that admit invariants in resonance form. First applications of this condition reproduce previously known cases of exact invariants for one-dimensional potentials. In addition, we have found potentials that have three-resonance invariants.

The possibilities for applying our formalism are certainly not exhausted. A detailed study of the $N > 2$ cases is of immediate interest, including the meaning of the limit $N \to \infty$. Extensions of our analysis to include systems with more than one spatial dimension, systems with magnetic forces and quantum mechanical systems should be investigated.

REFERENCES

[1] E. T. Whittaker, A Treatise on the Analytical Dynamics of Particles and Rigid Bodies, 4th ed. (Cambridge Univ. Press, Cambridge, 1937).

[2] C. R. Holt, J. Math. Phys. 23, 1037 (1982).

[3] L. S. Hall, Physia 8D, 90 (1983).

[4] G. Schmidt, Physics of High Temperature Plasmas, 2nd. ed. (Academic, New York, 1979).

[5] H. R. Lewis, J. Math. Phys. 9, 1976 (1968).

[6] H. R. Lewis and P. G. L. Leach, J. Math. Phys. 23, 2371 (1982).

[7] W. Sarlet and J. R. Ray, J. Math. Phys., 22, 2504 (1981), Sec. 4.2.

[8] H. R. Lewis and K. R. Symon, Phys. Fluids 27, 192 (1984).

[9] H. R. Lewis and P. G. L. Leach, "A resonance formulation for invariants of particle motion in a one-dimensional time-dependent potential," Annals of Physics (to be published).

[10] U. Grenander and G. Szegö, Toeplitz Forms and Their Applications (University of California Press, Berkeley, 1958).

[11] G. Kowalewsky, Einführung in die Determinantentheorie (Walter de Gruyter, Berlin, 1954).

[12] A. S. Householder, The Theory of Matrices in Numerical Analysis, (Blaisdell, New York, 1964).

[13] In preparation.